Engineering Fluid Dynamics 2018

Engineering Fluid Dynamics 2018

Special Issue Editor

Bjørn H. Hjertager

MDPI • Basel • Beijing • Wuhan • Barcelona • Belgrade

MDPI

Special Issue Editor
Bjørn H. Hjertager
University of Stavanger
Norway

Editorial Office
MDPI
St. Alban-Anlage 66
4052 Basel, Switzerland

This is a reprint of articles from the Special Issue published online in the open access journal *Energies* (ISSN 1996-1073) from 2018 to 2019 (available at: https://www.mdpi.com/journal/energies/special_issues/eng_fluid_dyn2018).

For citation purposes, cite each article independently as indicated on the article page online and as indicated below:

LastName, A.A.; LastName, B.B.; LastName, C.C. Article Title. *Journal Name* **Year**, *Article Number, Page Range.*

ISBN 978-3-03928-112-1 (Pbk)
ISBN 978-3-03928-113-8 (PDF)

Contents

About the Special Issue Editor

Bjørn H. Hjertager, Professor, Dr. got his PhD from the University of Trondheim, Norway (NTH), in 1979, within the topic of Combustion, Heat Transfer, and Fluid Flow. Therafter, he stayed almost 10 years at Chr Michelsen Institute. Then, from the late 1980s and up to the late 1990s, he worked at the Telemark Institute of Technology (HIT-TF), and at the research institute Tel-Tek. He then stayed 11 years abroad, at Aalborg University in Denmark. Since 2008, he has been back to Norway as a Professor in Fluid dynamics at the University of Stavanger. He has published more than 180 papers in the field of fluid flow, heat transfer, combustion, gas explosions, and chemical reactors, and he has supervised 21 PhD candidates in Norway and Denmark.

Preface to "Engineering Fluid Dynamics 2018"

This book contains the successful submissions to a Special Issue of Energies on the subject area of "Engineering Fluid Dynamics 2018". The topic of engineering fluid dynamics includes both experimental as well as computational studies. Of special interest were submissions from the fields of mechanical, chemical, marine, safety, and energy engineering. We welcomed both original research articles as well as review articles. After one year, 28 papers were submitted, and 14 were accepted for publication. The average processing time was about 38 days. The authors had the following geographical distribution: China (9); Korea (3); Spain (1); and India (1).

Papers covered a wide range of topics, including analysis of fans, turbines, fires in tunnels, vortex generators, deep sea mining, as well as pumps.

I found the task of editing and selecting papers for this collection to be both stimulating and rewarding. I would also like to thank the staff and reviewers for their efforts and input.

Bjørn H. Hjertager
Special Issue Editor

energies

MDPI

Article

Source Term Modelling of Vane-Type Vortex Generators under Adverse Pressure Gradient in OpenFOAM

Iñigo Errasti *, Unai Fernández-Gamiz, Pablo Martínez-Filgueira and Jesús María Blanco

Nuclear Engineering and Fluid Mechanics Department, University of the Basque Country, E-01009 Vitoria-Gasteiz, Spain; unai.fernandez@ehu.eus (U.F.-G.); pablo.martinez@ehu.eus (P.M.-F.); jesusmaria.blanco@ehu.eus (J.M.B.)
* Correspondence: inigo.errasti@ehu.eus; Tel.: +34-945-013293

Received: 9 November 2018; Accepted: 6 February 2019; Published: 14 February 2019

Abstract: An analysis of the generation of vortices and their effects by vane-type vortex generators (VGs) positioned on a three-dimensional flat plate with a backward-facing ramp and adverse gradient pressure is carried out. The effects of a conventional vortex generator and a sub-boundary layer vortex generator are implemented by using a source term in the corresponding Navier-Stokes equations of momentum and energy according to the so-called jBAY Source Term Model. The influence of the vortex generator onto the computational domain flow is modelled through this source term in the Computational Fluid Dynamics (CFD) simulations using the open-source code OpenFOAM. The Source Term Model seems to simulate relatively well the streamwise pressure coefficient distributions all along the flat plate floor as well as certain parameters studied for vortex characterization such as vortex path, decay and size for the two vane-type vortex generators of different heights studied. Consequently, the implementation of the Source Term Model represents an advantage over a fully Mesh-Resolved Vortex Generator Model for certain applications as a result of a meaningful decrease in the cell number of the computational domain which implies saving computational time and resources.

Keywords: flow control; vortex generators; source term; Computational Fluid Dynamics (CFD); OpenFOAM; wind tunnel

---

1. Introduction

The need for bigger capacities and installed power of wind generation systems has allowed the implementation of larger rotor wind turbines. Nowadays, wind power generators of 5–7 MWatt may present rotor blades of 60 m long or even larger.

This meaningful increase in the rotor size and weight of wind turbines has made that the techniques used in the past years to control them have to be improved. Johnson et al. [1] summarized a big part of the most relevant flow control techniques applied in wind turbines to work in an optimal and safe way under diverse atmospheric conditions.

Several flow control devices have been developed in recent decades, Gad-el-Hak [2]. Most of them were firstly designed for aeronautical applications and consequently the initial starting point of research was aeronautics, Taylor [3]. According to Liu et al. [4], flow control devices are regularly also applied in turbo machinery systems. In recent years, researchers have been trying to optimize and install these devices in multi-megawatt wind turbines. Wood [5] and Johnson et al. [1] developed a four layer scheme to conceptually classify these flow control devices.

These control devices can be classified as active or passive ones depending on the control techniques applied as proposed by Aramendia et al. [6]. Passive flow control techniques enhance the turbine's efficiency and load reduction without external energy consumption. Active control techniques need a supplementary energy source to obtain the effect on the air flow and, unlike vortex generators (VGs) and other passive devices, active flow control requires elaborate algorithms to get the most favorable gain (Becker et al. [7]). Johnson et al. [1] carried out an investigation on around 15 control devices for wind turbines. Some of these flow control devices are still checked nowadays on full-scale wind turbines in working conditions.

Vortex generators (VGs) are passive flow control devices that modify the boundary layer fluid motion exchanging momentum from the outer into inner region. Due to this energy transport, the velocity of the inner region increases at the same time as the boundary layer thickness decreases and in turn the flow separation is delayed, Rao et al. [8]. Moreover, Lin et al. [9] proved the effect of drag reduction and lift increase induced by sub-boundary layer vortex generators (VGs).

According to Doerffer et al. [10] and Steijl et al. [11], the streamwise vortices generated by vane-type VGs produce an endless momentum to counter the momentum decrease of the natural boundary layer and the growth of its thickness induced by viscous friction and adverse gradients of pressure. These vane-type VGs are able to reduce or remove flow separation when there is a limited adverse pressure gradient as commented by Velte et al. [12,13]. If the flow separation even appears for cases of large adverse gradients of pressure, the vortices action of mixing will limit the reversed flow region in the shear layer and maintain, to some extent, pressure recovery all along the detached flow.

Fernández-Gamiz et al. [14] and Urkiola et al. [15] analysed the flow effects of a vane-type VG on a flat plate under conditions of negligible streamwise pressure gradient and the primary vortices generated to examine how the physics of the wake past the VGs can be simulated by means of Computational Fluid Dynamics (CFD) computations. In those studies, the CFD simulations were compared with experimental observations. For instance, Gao et al. [16] and Baldacchino et al. [17] focused on the increase of the maximum value of the lift coefficient of a 30% thick DU97-W-300 airfoil designed by the Delft University of Technology by implementing passive VG devices. When increasing the angle of attack, the lift and drag coefficients raised up to values higher than the ones obtained under steady-state conditions.

The so-called micro vortex generators were introduced by Keuthe [18]. These flow control devices were also named as sub-boundary layer vortex generators (SBVGs) according to Holmes et al. [19], submerged vortex generators in Lin et al. [20], low-profile vortex generators by Martinez-Filgueira et al. [21] and micro vortex generators in Lin et al. [22]. Kenning et al. [23] summarized in his study the potential applications of all these vortex generators, including the control of shock-induced separation, smooth surface separation or leading edge separation.

Vortex generators are fixed onto the blades of wind turbines in order to prevent or remove the flow separation as well as to decrease the blade sensitivity. VGs are usually mounted on the suction side of the blade in a spanwise array for instance when a wind turbine does not perform as expected. Accordingly, the implementation of VGs onto the blade could eventually be a valid way to enhance the efficiency of a wind turbine rotor, Schubauer et al. [24] and Bragg et al. [25].

In order to design and optimize the position of the vortex generator on a wind turbine blade, CFD tools can be used as for instance Fernández-Gamiz et al. [26] or Troldborg et al. [27] do. Nonetheless, modelling a full rotor by using a fully-meshed VG computation becomes excessively expensive. As a matter of fact, the size of the vortex generators is generally comparable to the local boundary layer thickness and this implies that a meaningful number of cells around the device in the computational domain is needed in order to accurately model the flow. An alternative way of modelling VGs in the CFD computations is to model the device effect on the boundary layer by using body forces.

Bender et al. [28] established a source term model, the so-called BAY model, based on the Joukowski Lift Theorem and the Thin Airfoil Theory which used body forces. This model was designed for simulating vane-type Vortex Generators in a finite volume Navier-Stokes code and ignores the condition to fully define the geometry in the mesh. In order to calibrate the model, Bender [28] created a test case to compare the results obtained for a source term modelled VG with those obtained for a fully-meshed VG. The test case was based on 24 VGs circumferentially mounted on a pipe in a co-rotating configuration and the results obtained were satisfying. For instance, Dudek [29] successfully evaluated the BAY model for different types of flows in a rectangular duct with a single vane-type VG.

Recently, a better version of the BAY model was designed by Jirasek [30], namely the jBAY model. The improved version was based on the Lift Force Theory based on Bender et al. [28] and brought a more suitable technique for simulating the flow when using VG arrays. Jirasek [30] applied a reduced method for determining the model control points to implement in a effortless way the model and obtain more correct results. The new version of the model was checked by reproducing a single vane-type VG on a flat plate in an S-duct air intake and a high-lift wing configuration. The results showed very good agreement between experimental data and the CFD simulations. Thus, Florentie et al. [31] studied the potential of the jBAY model to reproduce the effects of rectangular VGs positioned on a flat plate and proposed a model optimization.

The main goal of the current study is firstly the implementation of the so-called jBAY Source Term Model into OpenFOAM [32] and secondly to investigate how well the open source CFD simulations are able to mimic the physics of the flow behind modelled rectangular conventional and sub-boundary layer vortex generators (VGs) mounted on a three-dimensional flat plate with a backward-facing ramp under adverse pressure gradient conditions. The satisfactory implementation of the Source Term Model in OpenFOAM would be eventually able to bring new model applications of this open source CFD code. The theoretical background of this study is described in Section 2; Section 3 presents the experimental data; Section 4 describes in detail the computational configuration; the results and their discussion are covered in Section 5 and the conclusions are presented in Section 6.

2. Theoretical Background

2.1. Vortex Generator Setup

The model consists of a single rectangular VG positioned on a flat plate with a backward-facing ramp. The three-dimensional computational domain modelled in OpenFOAM represents the extended test section of the wind tunnel experiment performed by Lin [33] and is shown in the illustration of Figure 1. Note that the single VG is placed upstream the backward-facing ramp. The domain has been designed following a previous simulation study by Konig et al. [34].

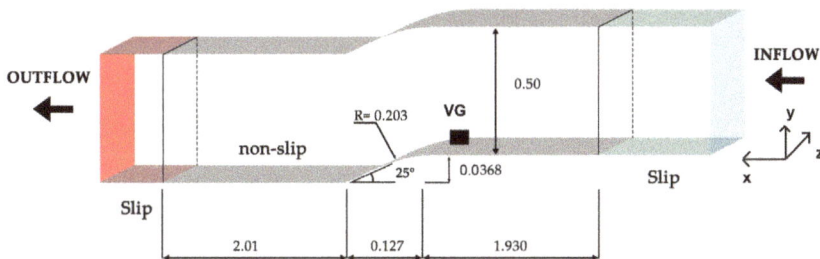

Figure 1. Description of the computational domain representing the extended wind tunnel test section. The domain dimensions are expressed in meters.

The geometry dimensions of the vane-type VG determined by a length L two times the VG height H are depicted in Figure 2a for two different heights $H_1 = 0.8\delta$ and $H_2 = 0.2\delta$ where δ is the local boundary layer thickness just upstream edge of the ramp. The two vortex generators, namely, conventional VG and sub-boundary layer VG of heights $H_1 = 0.8\delta$ and $H_2 = 0.2\delta$ respectively are positioned at distances of 5δ and 2.5δ from the upstream edge of the ramp. Figure 2b illustrates a classical configuration example of a pair of counter-rotating VGs. The rectangular vane has a constant thickness and no sharp edges. The parameter λ represents the spacing between VGs and lines A, B and C are measurement lines along the floor past the VG. The geometry and parameters of the two VG cases are summarized in the table of Section 4. Note that the computational domain has only one vane-type VG oriented to the main flow instead of two vanes in order to reduce the meshing and the computational resources needed. A boundary layer region develops over the flat plate induced by the viscous interaction between the wall and the air flow. The angle of incidence between the main flow direction and the VG vane is $\beta = 15°$ or $\beta = 25°$ depending on the VG case (see Figure 3) and the Shear Stress Transport (SST) turbulence model Menter [35] has been elected because of its capacity to solve whirling flows according to Liu et al. [4].

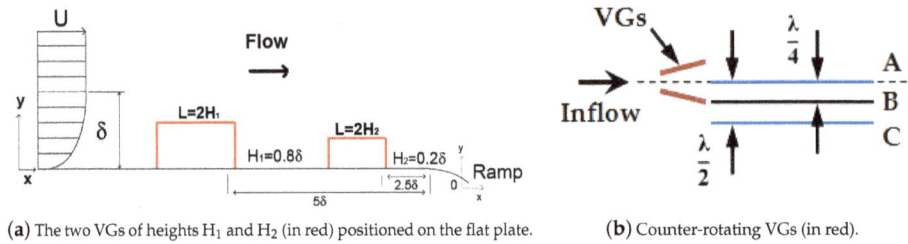

(a) The two VGs of heights H_1 and H_2 (in red) positioned on the flat plate.   (b) Counter-rotating VGs (in red).

Figure 2. (a) The two VGs studied of heights $H_1 = 0.8\delta$ and $H_2 = 0.2\delta$ on the flat plate with δ representing the local boundary layer thickness at the upstream edge of the ramp (not at scale); (b) Configuration example of a pair of similar counter-rotating VGs with three measurement lines A, B and C past the VG.

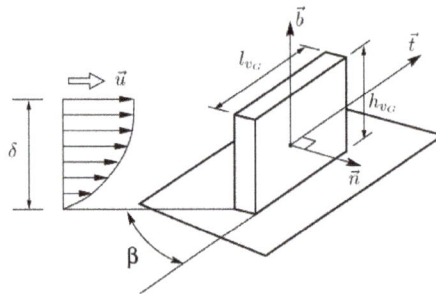

Figure 3. Unit vectors on a rectangular vortex generator (VG) according to the BAY Source Term Model.

The Reynolds number for the CFD simulations is $Re_\theta = 9100$ in order to mimic the experiments explained in the next section and is based on the local boundary local (BL) momentum thickness θ and computed by the following expression:

$$Re_\theta = \frac{U_\infty \cdot \theta}{\nu} \tag{1}$$

where $U_\infty = 40.23$ m·s^{-1} is the free stream velocity, $\nu = 1.9439 \times 10^{-5}$ m^2·s^{-1} the kinematic viscosity and the momentum thickness θ defined as:

$$\theta = \int_{-\infty}^{\infty} \frac{u}{U_\infty} \left(1 - \frac{u}{U_\infty}\right) dy \tag{2}$$

where u is the streamwise velocity component and y the vertical coordinate normal to the wall. Thus, the momentum thickness of the local boundary layer δ at the upstream edge of the ramp calculated by the previous equation is $\theta = 4.3967$ mm.

2.2. Source Term Model

Computational Fluid Dynamics (CFD) is a commonly method used to reproduce the air flow downstream the VGs and to estimate the blade efficiency. Furthermore, the combination of this technique of simulation and the corresponding experiments in wind tunnels can be considered as a reliable tool for parametric studies on VG layout. However, these computational methods are time-consuming when designing high-quality meshes and solving the corresponding flow equations.

In the current work, the so-called jBAY Source Term Model presented by Jirasek [30] and based on the widely used BAY Model developed by Bender et al. [28] is implemented. The BAY model was designed for simulating vane-type vortex generators into finite volume Navier-Stokes codes and allows substituting the VG geometry by a subdomain of similar size at the original VG location where a specific body force distribution is then applied. The model is integrated into the CFD code as a so-called source term in the Navier-Stokes momentum and energy equations.

In the model, a force normal is applied to the direction of the local flow and at the same time parallel to the surface. This normal force simulates the side force generated by a vane-type vortex generator. Bender et al. [28] designed this Source Term Model based on the Joukowski Lift Theorem and the Thin Airfoil Theory in order to model the VG influence. Taking into account a rectangular vortex generator (VG), the lift forces on the VG can be computed by the following expression:

$$\vec{L} = \rho(\vec{u} \times \vec{b}) \cdot \Gamma \cdot h_{VG} \tag{3}$$

where $\vec{L}$ is the lift force on the VG, ρ the local density, $\vec{u}$ the local velocity, $\vec{b}$ the unit vector defined as $\vec{b} = \vec{n} \times \vec{t}$ with $\vec{n}$ and $\vec{t}$ the unit vectors normal and tangential to the VG respectively (see Figure 3), h_{VG} the VG height and Γ the circulation defined as:

$$\Gamma = \beta \parallel \vec{u} \parallel l_{VG} \tag{4}$$

where β is the angle of incidence between the flow direction and the VG and l_{VG} the VG length. As expressed in Equation (3), the direction of the lift force $\vec{L}$ on the VG is obtained by means of the product of the local velocity $\vec{u}$ and the unit vector $\vec{b}$. According to the Joukowski Lift Theorem for 2D airfoils:

$$\vec{L} = \pi\rho(\vec{u} \times \vec{b})(\vec{u} \cdot \vec{n})S_{VG} \tag{5}$$

where S_{VG} is the vortex generator area $(l_{VG} \times h_{VG})$ and the lift force on a single cell $\vec{L}_{cell}$ is computed by the following expression:

$$\vec{L}_{cell} = \pi\rho(\vec{u} \times \vec{b})(\vec{u} \cdot \vec{n})S_{VG}\frac{V_{cell}}{V_s} \tag{6}$$

where V_{cell} is the volume corresponding to a single cell and V_s the total volume of the grid cells where the model is applied. According to Bender et al. [28] a new term $(\frac{\vec{u} \cdot \vec{t}}{\parallel \vec{u} \parallel})$ is introduced in the previous equation in order to obtain the final lift force on a single cell:

$$\vec{L}_{cell} = C_{VG} \cdot \rho(\vec{u} \times \vec{b})(\vec{u} \cdot \vec{n})\frac{\vec{u} \cdot \vec{t}}{\parallel \vec{u} \parallel}S_{VG}\frac{V_{cell}}{V_s} \tag{7}$$

When applying the BAY Model, a calibration process is required and a Mesh-Resolved VG Model can be used as reference for model calibration. Particularly for this model, the empirical constant C_{VG} of the Equation (7) is a relaxation parameter. This parameter is selected so that the simulated estimations for the lift force distribution fit the results corresponding to the computations of the mesh-resolved VG. Bender et al. [28] proposed that this Source Term Model can operate in two modes, the so-called asymptotic and linear ones. If V_s, defined as the total volume of the grid cells where the force term is applied, differs considerably from the volume of the vane type vortex generator, the model behaves in linear mode and is dependent on the constant C_{VG}. However, the model of the present work has been locally applied, such that the volume V_s is similar to the vortex generator volume. Therefore, the model operates in the asymptotic mode and is non sensitive to C_{VG}. Commonly values for this constant are $C_{VG} \cong 10$, see the study by Jirasek [30]. In the present work, instead of applying forces in all the subdomain cells as the BAY Model specifies, the forces are applied in cells placed just in the outline of the VG geometry, see Figure 4.

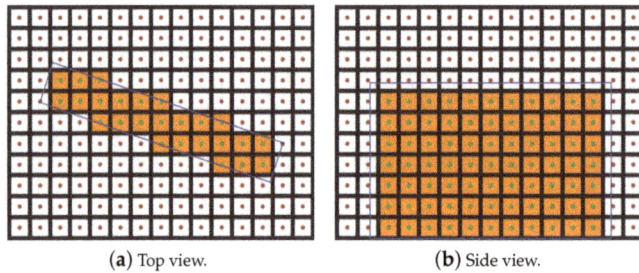

(a) Top view. (b) Side view.

Figure 4. Cells corresponding to the VG coloured in orange where the body forces are applied. VG dimensions are indicated by the blue line.

Taking into account that the Source Term Model was designed to be user-friendly, any vane size of height H and length L and angle of incidence β can be modelled. When using the OpenFOAM CFD code, only three parameters have to be specified to model the vortex generator (VG): the grid cells where the Source Term Model will be applied, the vortex generator area S_{VG} and the angle of incidence β of the main flow direction with respect to the VG orientation. In Figure 5, two examples of subdomains of the selected mesh cells which represent the two vortex generators (VGs) of heights $H_1 = 0.8\delta$ and $H_2 = 0.2\delta$ in the Source Term Model are shown.

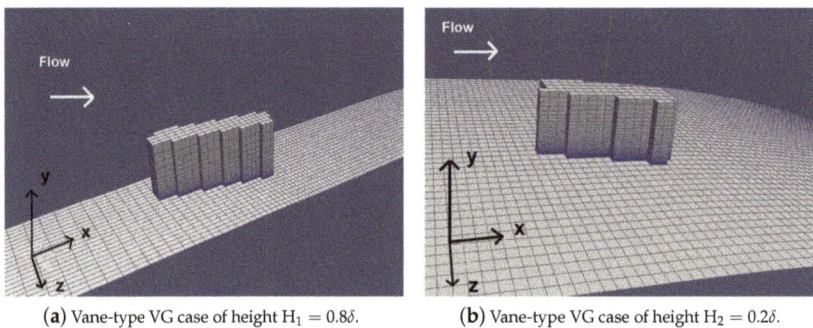

(a) Vane-type VG case of height $H_1 = 0.8\delta$. (b) Vane-type VG case of height $H_2 = 0.2\delta$.

Figure 5. Selected cells representing the two vortex generators (VGs) of heights H_1 and H_2 according to the Source Term Model and positioned on the flat plate upstream the backward-facing ramp.

The reader should finally note that the Source Term Model can replace the Mesh-Resolved VG Model when modelling VGs in certain cases. In a fully Mesh-Resolved VG Model, a fine mesh in the vicinity of the vortex generator (VG) must be designed in order to capture properly the VG effect on the domain flow. Thus, Fernández-Gamiz et al. [26] designed a fine mesh of approximately eight million cells for simulating the primary vortex induced by a rectangular sub-boundary layer VG on a flat plate in a fully meshed-resolved model. When using the BAY Model, a coarser mesh can be used to capture the VG influence on the flow saving computational time and resources as will be indicated later. However, attention must be paid when defining the mesh because the BAY Model accuracy can eventually be dependent on mesh resolution according to some authors, e.g., Florentie et al. [31].

3. Experimental Data

Experimental data for this study has been taken from the experiments carried out by Lin et al. [33]. The experiments were performed in the NASA Langley 20 by 28-inch Shear-Flow Control Tunnel and free stream velocity of $U_\infty = 40.23$ m·s^{-1}. Baseline flow separation was determined on a 25° sloped backward facing curved ramp of radius $R = 20.32$ cm with a behaviour similar to a classical two-dimensional flow. The ramp spanned the wind tunnel test section width. The local boundary layer thickness δ was approximately 32.51 mm and the Reynolds number was $Re_\theta = 9100$ at the upstream edge of the sloped backward facing curved ramp which was placed approximately at 1.98 m from the wind tunnel test section entrance. In the experiments, different types of passive flow control devices were analysed and between them vortex generators (VGs). Furthermore, sets of vortex generators (VGs) were simulated upstream of the backward-facing ramp. In the present study, only the pressure distribution data corresponding to a counter-rotating conventional vane-type vortex generator (Figure 2) of height $H_1 = 0.8\delta$ and a sub-boundary layer vane-type vortex generator of height $H_2 = 0.2\delta$ where δ represents the local boundary layer thickness at the upstream edge of the ramp will be analysed for comparison with the data obtained from the OpenFOAM simulations. The higher vortex generator of height H_1 was positioned at a distance of 5δ upstream of the ramp and the smaller vortex generator of height H_2 at a distance of 2.5δ.

4. Computational Configuration

Non-commercial open source code OpenFOAM [32] has been used for simulating the vortex and its effects in the present study. The open source CFD code is an object-oriented library package designed in C++ programming language to analyse systems of computational continuum mechanics.

The current computational domain (Figure 6) partially based on a previous simulation study by Konig et al. [34] consists of a flat plate with a backward-facing ramp and adverse pressure gradient where selected cells corresponding to a single rectangular vortex generator according to the Source Term Model namely jBAY previously presented in Jirasek [30] and positioned at a point where the VG height is 80% and 20% the local boundary layer thickness δ at the upstream edge of the ramp for the cases H_1 and H_2 respectively. The incidence angle of the vane with respect to the oncoming flow is $\beta = 15°$ for the $H_1 = 0.8\delta$ vane-type VG case and $\beta = 25°$ for he $H_2 = 0.2\delta$ vane-type VG case as shown in Figure 3. The modelled VGs have an aspect ratio (AR) defined as the relationship between the height H and the length L of AR = 0.5 (Table 1). A representation of the mesh subdomain and selected cells representing the vortex generator is shown in Figure 5. The origin of the computational domain was placed at the ramp center.

Table 1. Vortex generator geometry and parameters corresponding to the vane-type VG cases of height $H_1 = 0.8\delta$ and $H_2 = 0.2\delta$. In the present study: $\delta = 32.51$ mm.

VG Parameter	$H_1 = 0.8\delta$ VG Case	$H_2 = 0.2\delta$ VG Case
Height H	0.8δ	0.2δ
Length L	1.6δ	0.4δ
Incident angle β	15°	25°
Location from the ramp	5δ	2.5δ
Spacing λ	3.2δ	0.8δ

(**a**) Computational domain.

(**b**) Mesh around the ramp region.

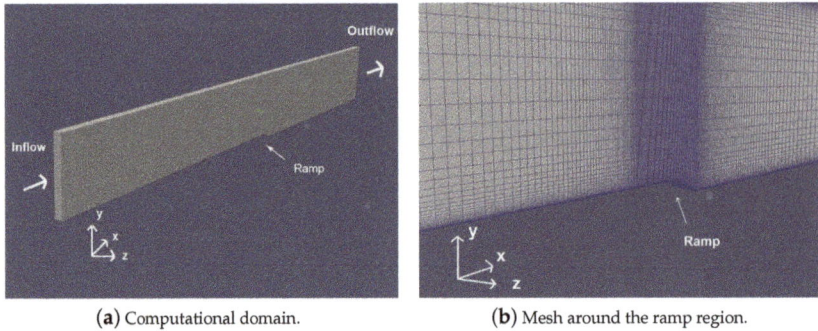

Figure 6. Computational domain and mesh detail of the VG positioned on the flat plate upstream the backward-facing ramp.

The variables of particular interest which require boundary conditions are the fluid quantities, velocity and pressure. Non-slip boundary conditions were applied to the flat plate and ramp of the computational domain to allow for boundary layer growth (see Figure 1). The sides of the computational domain have been defined as symmetry condition. The domain consists of only one vane and the symmetry assumption used in the present computations can be justified by previous studies as Gutierrez-Amo et al. [36] and Sørensen et al. [37]. The atmospheric conditions were selected in such a way that a boundary layer thickness of approximately $\delta = 32.51$ mm just upstream edge of the ramp according to Lin [33] was attained in the simulation of the test section of the wind tunnel. As the k-ω Shear Stress Transport (SST) turbulence model designed by Menter [35] was applied, a requirement when setting boundary conditions for the turbulent viscosity ν_t, the turbulent kinetic energy k and the dissipation rate ω was needed. Nearby the flat plate and ramp surfaces, wall functions were not used in the computations. Slip-wall boundary conditions were also implemented at the beginning and end of the flat plate (Figure 1) according to Konig et al. [34].

Two CFD solvers were applied in this three-dimensional simulation. The potentialFoam solver was firstly used to create initial fields to accelerate the convergence process and secondly the simpleFoam solver for the steady-state, incompressible and turbulent flow using the RANS (Reynolds Average Navier-Stokes) equations. This second solver is based on the k-ω SST turbulence model by Menter [35]. This turbulence model combines the Wilcox k-ω model for the close-to-wall regions and the k-ϵ model for the outer regions. The normal forces are applied in the selected cells of the computational domain according to the Source Term Model explained in Section 2 using the topoSet tool.

The computational domain analysed was discretized with a structured-type mesh and hexahedral faces of around 5×10^5 cells for the $H_1 = 0.8\delta$ vane-type VG case and 2×10^6 cells for the $H_2 = 0.2\delta$ vane-type VG case. Full second order linear-upwind schemes for the discretization were implemented for all the simulations. Two quality mesh parameters such as mesh orthogonality and mesh skewness for the two VG cases have been studied in order to analyse the mesh quality and the results are shown in Table 2. Mesh non-orthogonality is an indicator of how orthogonal the pairs of neighboring cells which share a face are and mesh skewness of how optimum the cell shape is in relation with the corner angles. The non-orthogonality parameter should be close to 0 degrees and the maximum skewness should not exceed 0.85 as suggested by Gutierrez-Amo et al. [36] for hexahedral cells to obtain an accurate solution.

Table 2. Quality meshing parameters corresponding to the mesh of the two vane-type VG cases.

VG Case	Non-Orthogonality (Average)	Maximum Skewness (-)
$H_1 = 0.8\delta$	5.44°	0.56
$H_2 = 0.2\delta$	5.32°	0.38

According to Richardson and Gaunt [38] Extrapolation Method, a grid dependency study was performed for three different mesh resolutions: coarse, medium and fine. The method was applied to a vortex parameter, namely the normalized peak vorticity $(\omega_{xmax}\delta)/U_\infty$ analysed in the next section at a plane normal to the streamwise direction located at $x/\delta = -3$ past the vane for the $H_1 = 0.8\delta$ VG case and $x/\delta = -1$ for the $H_2 = 0.2\delta$ VG case. Three parameters were calculated following the Richardson Extrapolation Method: the extrapolated solution RE, the order of accuracy p and the error ratio R. Table 3 exhibits the results obtained in the mesh independency study where a monotonic convergence was fulfilled. The higher resolution of the fine mesh was used in the computations for the two vane-type VG cases analysed. The iterative solution process was carried out in the simulations of the two VG cases until the residual errors dropped below 10^{-4} for the pressure p and 10^{-5} for the velocities U_x, U_y and U_z and the turbulence quantities used, e.g., the turbulent kinetic energy k and the specific dissipation rate ω. The values obtained for the dimensionless distance of the first wall cell for the meshes corresponding to the two VG cases is summarized in Table 4. In both cases the value of y+ is lower than 1 (y+ < 1) as required by the turbulence model adopted.

Table 3. Mesh dependency study based on the Richardson Extrapolation Method performed for the normalized peak vorticity $(\omega_{xmax}\delta)/U_\infty$ on three meshes of Coarse, Medium and Fine resolution.

VG Case	Plane Location	Peak Coarse	Vorticity Medium	$(\omega_{xmax}\delta)/U_\infty$ (-) Fine	RE	p	R
$H_1 = 0.8\delta$	$x/\delta = -3$	74.25	122.78	128.32	127.61	3.13	0.11
$H_2 = 0.2\delta$	$x/\delta = -1$	12.00	27.91	29.54	29.35	3.29	0.10

Table 4. Wall dimensionless distance y+ values for the two VG cases simulated.

VG Case	Minimum	Maximum	Average
$H_1 = 0.8\delta$	0.14	1.29	0.64
$H_2 = 0.2\delta$	0.06	0.71	0.35

5. Results

5.1. Vortex Visualization

Figure 7 shows the vortex visualization based on the velocity field distribution at four planes normal to the streamwise direction and located at normalized distances x/δ respect to the boundary layer thickness δ from the backward-facing ramp where the domain origin is placed. The scale in the y and z axis for the four snapshots in this figure is approximately 13 mm × 13 mm. Thus, the vortex can be clearly visualized for the $H_1 = 0.8\delta$ vane-type VG case as expected. Note that the larger the normalized distance x/δ is, the closer the vortex is to the VG trailing edge (TE). The four planes depicted are located between the selected cells representing the VG and the ramp.

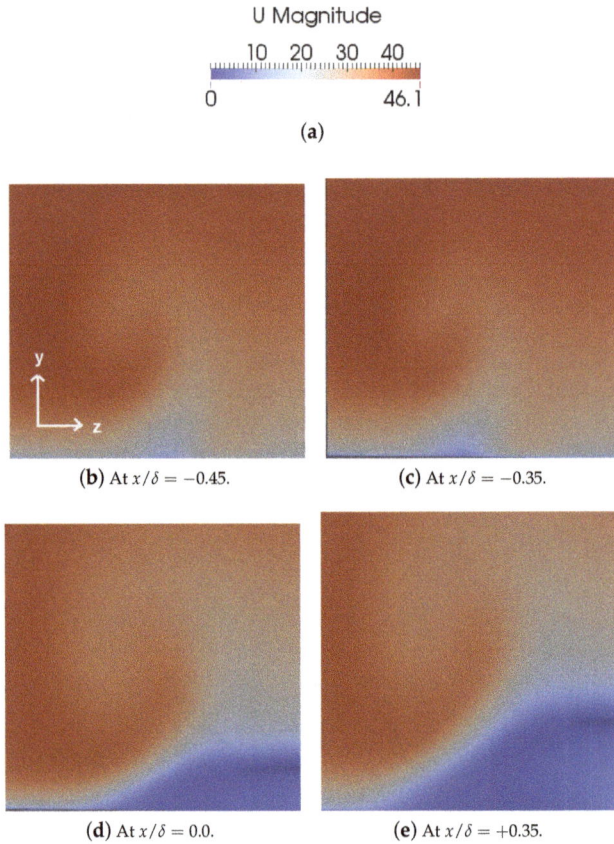

(a)

(b) At $x/\delta = -0.45$.   (c) At $x/\delta = -0.35$.

(d) At $x/\delta = 0.0$.   (e) At $x/\delta = +0.35$.

Figure 7. Velocity distribution (m·s^{-1}) at planes normal to the streamwise direction and located at normalized distances x/δ from the backward-facing ramp where the origin is placed. $H_1 = 0.8\delta$ vane-type VG case.

Figure 8 shows again the vortex visualization based on the velocity field distribution for the $H_2 = 0.2\delta$ vane-type VG case at four planes normal to the main flow direction. Now, the scale for the four snapshots is approximately 12.2 mm × 12.2 mm. The formation of the vortex is slightly observed but not as clear as it was observed in the previous $H_1 = 0.8\delta$ vane-type VG case shown in Figure 7. This seems to indicate that the vortex generated by the sub-layer boundary VG has a smaller strength than that corresponding to the conventional and higher VG. For instance, the vortex shape at plane $x/\delta = +0.35$ is not as easy to observe in the figure as the vortex shape is for the $H_1 = 0.8\delta$ vane-type VG case.

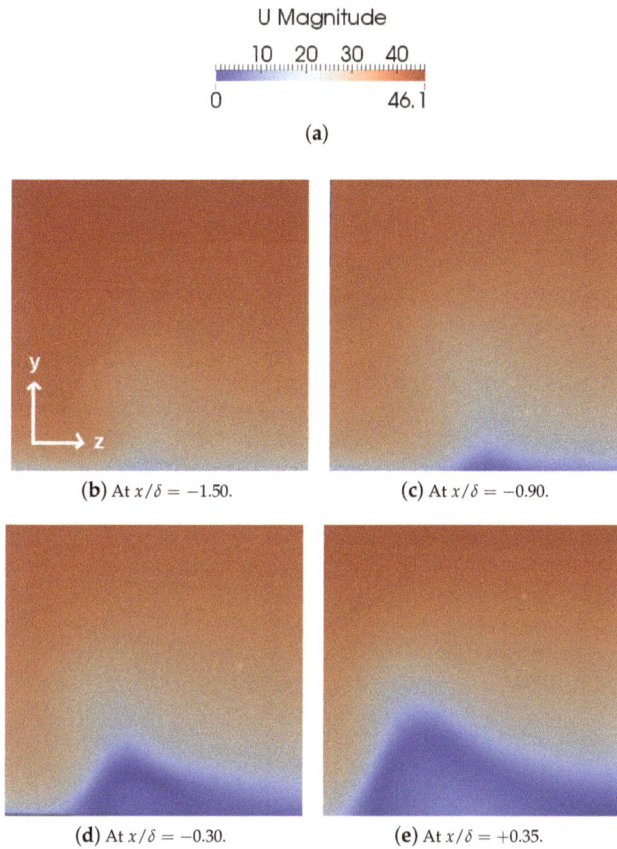

(a)

(b) At $x/\delta = -1.50$.

(c) At $x/\delta = -0.90$.

(d) At $x/\delta = -0.30$.

(e) At $x/\delta = +0.35$.

Figure 8. Velocity distribution (m·s^{-1}) at planes normal to the streamwise direction and located at distances x/δ from the backward-facing ramp. $H_2 = 0.2\delta$ vane-type VG case.

5.2. Pressure Distribution

The effect generated by the $H_1 = 0.8\delta$ vane-type VG placed at $x = -5\delta$ upstream the ramp on the streamwise pressure coefficient distribution c_p along the measurement lines A, B and C (Figure 2b) on the flat plate floor is appreciated in Figure 9. The three curves correspond to the pressure distribution (jBAY model, red rectangles) obtained from the CFD simulation with the jBAY Source Term Model implemented, the distribution data (Baseline, blue circles) obtained from the CFD simulation for a baseline case with no vortex generator positioned and the experimental pressure distribution data (EXP, green triangles) according to Lin et al. [33]. The error between the modelled (jBAY model) and the experimental (EXP) data is also shown in black vertical lines. The reader should note that the center of the backward-facing ramp is placed at the normalized distance $x/\delta = 0$ and the VG is positioned in a negative value of x/δ.

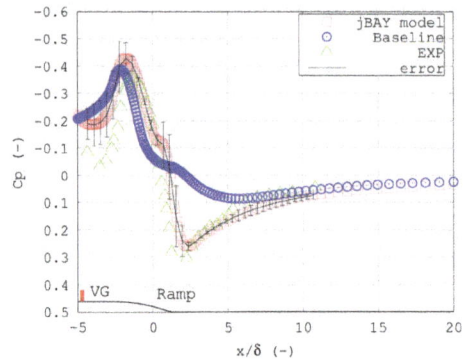

(**a**) Measurement line A.

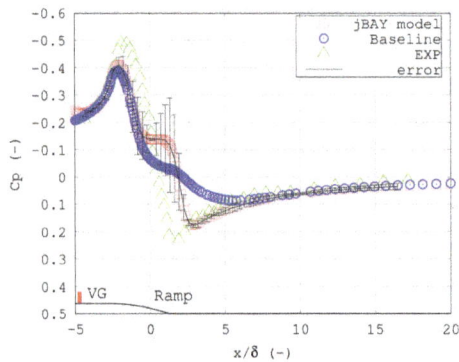

(**b**) Measurement line B.

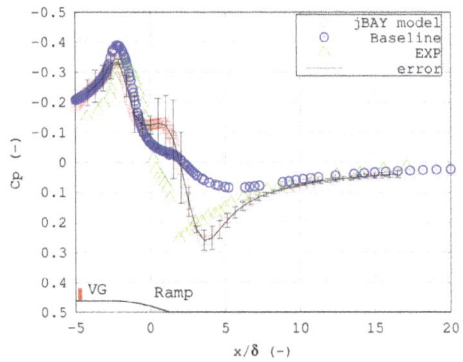

(**c**) Measurement line C.

Figure 9. Streamwise pressure distribution c_p along the measurement lines A, B and C on the flat plate floor. The red rectangles represent the simulated data (jBAY model), the blue circles the distribution data (Baseline) obtained from the baseline simulation and the green triangles the experimental (EXP) distribution data. Plots corresponding to the $H_1 = 0.8\delta$ vane-type VG case.

According to Lin et al. [33] and Konig et al. [34], the pressure distribution variations and the two peaks observed in the three plots at the beginning and end of the ramp correspond to accelerations and decelerations as expected. The influence of the VGs on the pressure distribution is meaningful when comparing with the pressure distribution for the baseline case. The results seem to be in agreement with the effects corresponding with those generated by vane-type VGs which considerably increase the pressure gradient between the beginning and the end of the ramp. As expected, the vortex generator is going to bring about variations in the flow structures. When observing the simulated pressure distribution along the measurement line A (Figure 9a), the initial suction peak at the beginning of the ramp is slightly overestimated and underestimated at the end of the ramp. The gradient of the pressure is quite well matched. Small deviations are appreciated between the experimental (EXP) and the simulated pressure distribution (jBAY model) before and behind the pressure peaks which could be related to the evidence that those regions are placed well within the three-dimensional flow structures induced by the VGs as indicated by Konig et al. [34]. When observing the pressure distribution along the measurement line B (Figure 9b) defined as the floor line starting from the VG trailing edge (TE) and parallel to the main flow (Figure 2b), the initial suction peak at the beginning of the ramp is underestimated although the gradient of the pressure as well as the pressure maximum at the end of the ramp are quite well matched. Similarly, Figure 9c shows that the simulated pressure distribution (jBAY model) along the measurement line C has a smaller magnitude between the two peaks at the beginning and end of the ramp than the amplitudes obtained along lines A and B. This distribution is shifted towards positive distances downstream the ramp.

Figure 10 depicts the effect generated by the $H_2 = 0.2\delta$ vane-type VG placed at $x = -2.5\delta$ upstream the ramp on the streamwise pressure distribution c_p along the measurement lines A, B and C on the flat plate floor. The error (black vertical lines) between the modelled (jBAY model) and the experimental (EXP) data is again indicated. The amplitude between the peaks is smaller than the amplitude obtained in the previous VG case of height H_1. On the other hand, the simulated pressure distribution (jBAY model, red rectangles) over the ramp as well as at the end of the ramp and for the three plots (Figure 10a–c) is slightly overestimated when comparing with the experimental distribution (EXP, green triangles).

In summary, the comparison of the simulated pressure distribution (jBAY model) with the experiment pressure distribution (EXP) for the $H_1 = 0.8\delta$ and $H_2 = 0.2\delta$ vane-type VG cases depicted in the two previous figures shows a relatively good agreement although relatively small deviations are appreciated, mostly for the $H_2 = 0.2\delta$ vane-type VG case. Thus, the mean absolute percentage error between the modelled and experimental data in the H_1 VG case is 7.57%, 8.10% and 7.10% for the three pressure distributions respectively measured in A, B and C lines. In the H_2 VG case, the mean absolute percentage error is 9.83%, 9.58% and 9.76%. The reason of these little bit larger deviations could be related to the fact that the VG height H_2 is within the sub-buffer zone of the local boundary layer where the shear and viscous stresses are predominant.

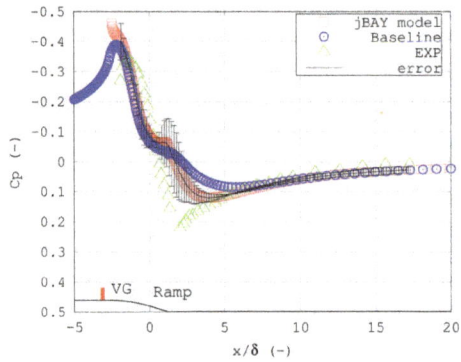

(**a**)Measurement line A.

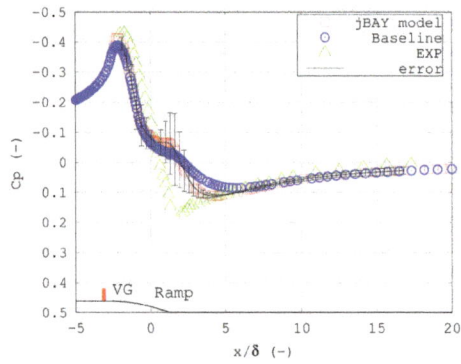

(**b**)Measurement line B.

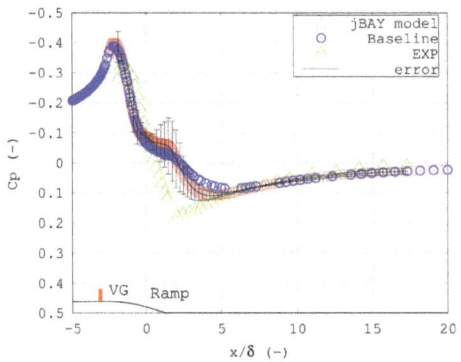

(**c**)Measurement line C.

Figure 10. Streamwise pressure distribution c_p along the measurement lines A, B and C on the flat plate floor. Plots corresponding to the $H_2 = 0.2\delta$ vane-type VG case.

5.3. Vortex Path

The path or trajectory of the vortex in the vertical (y) and lateral (z) directions can be determined by computing the location of the vortex center generated by the VG all along the downstream axis

x. The vortex center can be analytically defined as the point in a cross-stream plane which has the maximum value of the vorticity, the so-called peak vorticity ω_{max}. In Figure 11, a comparison between the vortex paths corresponding to the $H_1 = 0.8\delta$ and $H_2 = 0.2\delta$ vane-type VG cases is indicated. Streamwise, lateral and vertical coordinates are normalized by the local boundary layer thickness δ to show the effects of VG scaling downstream of the vane.

The vertical path corresponding to the conventional $H_1 = 0.8\delta$ vane-type VG case represented in Figure 11a tends to be parallel to the flat plate and the flow direction far from the backward-facing ramp placed at $x/\delta \cong 0$, but when the vortex approaches the ramp, it starts moving downward probably due to the adverse gradient of pressure created by the ramp. As observed in the same figure for the sub-boundary layer $H_2 = 0.2\delta$ vane-type VG case, the vertical path has a fast downward deviation probably due to the adverse gradient of pressure created by the ramp and its proximity to the ramp. The starting point for both cases is dependent on the position of the trailing edge (TE) of the vane on the flat plate, $x/\delta = -5$ and $x/\delta = -2.5$ respectively. The figure also shows that the vertical trajectories followed by the two vortices are nearly parallel to each other with a quite similar mean negative slope for both cases of approximately 0.009.

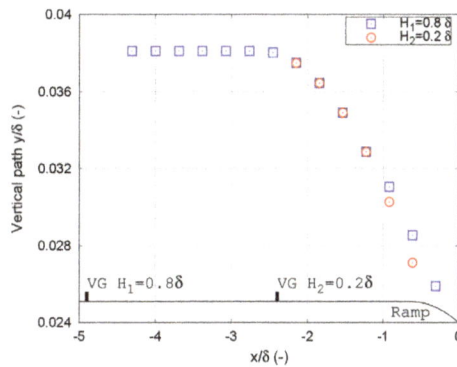

(**a**) Vertical path.

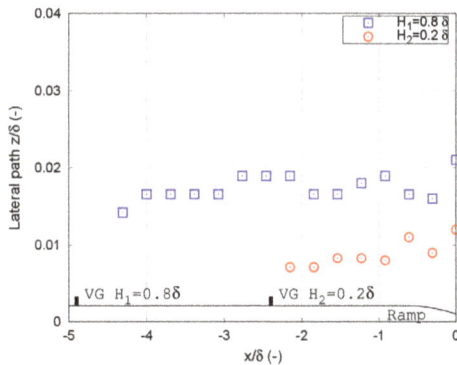

(**b**) Lateral path.

Figure 11. Normalized vertical (**a**) and lateral (**b**) paths function of the normalized distance x/δ from the backward-facing ramp for the $H_1 = 0.8\delta$ (blue rectangles) and $H_2 = 0.2\delta$ (red circles) vane-type VG cases. VGs, flat plate and ramp are also shown (not at scale).

The vortex paths for the $H_1 = 0.8\delta$ and $H_2 = 0.2\delta$ vane-type VG cases in the lateral direction are represented in Figure 11b. Both lateral trajectories seem to show a relatively small increasing trend in the same direction in which the vanes are pointed. As expected, the deviation of the lateral trajectory of the primary vortex described in the spanwise direction (z) by the highest VG is larger than the lateral deviation observed in the lowest VG. The reason for this different trajectory would lay on the fact that the lowest VG is within sub-buffer region of the boundary layer and consequently its effect on the oncoming flow is less significant. The starting point for both cases again depends on the location of the vane on the flat plate. This lateral increasing trend for the two trajectories is almost the same for both cases although for the $H_2 = 0.2\delta$ vane-type VG case the movement occurs in a lower streamwise distance. Thus, the mean positive slopes of the lateral trajectory for the H_1 and H_2 cases are around 0.005 and 0.006 respectively.

5.4. Vortex Decay

In order to illustrate how the vortex decays, the non-dimensional streamwise distribution of the normalized peak vorticity $(\omega_{xmax}\delta)/U_\infty$ can be studied. In Figure 12 the non-dimensional streamwise distribution of the normalized peak vorticity is plotted function of the normalized downstream distance x/δ for the two vane-type cases. The figure depicts the vortex decay in the streamwise direction and ratifies that the peak vorticity ω_{xmax} quite fast weakens past the VG for the two cases analysed. As observed, the maximum of the streamwise peak vorticity is obtained downstream the vane at the position $x/\delta = -2.8$ and $x/\delta = -1.8$ for the $H_1 = 0.8\delta$ and $H_2 = 0.2\delta$ VG cases respectively. After the maximum, the peak vorticity seems to decay in a decreasing exponential way to x/δ which is in concordance with Fernández-Gamiz et al. [39]. As expected, the peak vorticity magnitude decreases when the VG height is reduced.

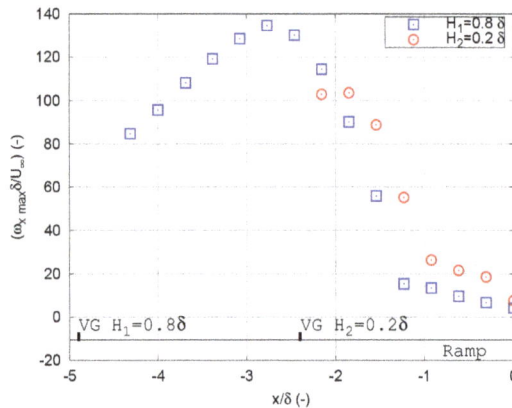

Figure 12. Non-dimensional streamwise peak vorticity $(\omega_{xmax}\delta)/U_\infty$ function of the normalized distance x/δ for the $H_1 = 0.8\delta$ (blue rectangles) and $H_2 = 0.2\delta$ (red circles) vane-type VG cases. VGs, flat plate and ramp also shown (not at scale).

5.5. Vortex Size

The vortex size is a key parameter when modelling vortices. A suitable way to study the vortex size is by means of the half-life radius $R_{0.5}$ defined as the radial distance from the vortex center to the point where the vorticity is half the peak vorticity ω_{max} captured in a cross-stream plane according to Bray [40]. At that point, the measurement errors when calculating the vortex size are negligible and the accuracy degree is high.

Figure 13 shows the vortex size evolution expressed in terms of the normalized half-life radius $R_{0.5}$ for the $H_1 = 0.8\delta$ and $H_2 = 0.2\delta$ vane-type VG cases at different locations x/δ past the VG. This figure shows a increasing trend in the vortex size when moving downstream for both VG cases. This relationship observed between both magnitudes could be formulated by a mathematical expression and this could eventually allow establishing a prediction model between the vortex size and the streamwise distance. As expected, the vortex size generated by the conventional VG of height $H_1 = 0.8\delta$ is larger than the vortex generated by the sub-boundary layer VG of height $H_2 = 0.2\delta$.

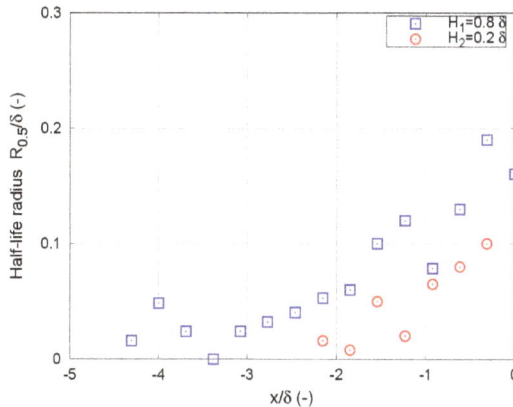

Figure 13. Vortex size evolution expressed in terms of the normalized half-life radius $R_{0.5}$ at difference normalized distances x/δ for the $H_1 = 0.8\delta$ (blue rectangles) and $H_2 = 0.2\delta$ (red circles) vane-type VG cases.

5.6. Wall Shear Stress

Figure 14 shows the wall shear stress distribution along the measurement line B on the flat plate floor for the two vane-type VG cases analysed in the present study. The wall shear stress distribution obtained from the implementation of the jBAY Source Term Model (red rectangles) is compared with that corresponding to the CFD simulation for the baseline case (blue circles). According to Godard and Stanislas [41], the implementation of VGs leads to an increase of the values of the wall shear stress downstream of the vanes in comparison with the cases where passive devices are not implemented. The results show a larger increment between the jBAY and baseline distributions for the $H_1 = 0.8\delta$ vane-type VG case than for the $H_2 = 0.2\delta$ vane-type VG case. This higher increase could be related to the fact that in the case of the conventional VG of height $H_1 = 0.8\delta$ the streamwise vortex induced is stronger and consequently its effects on the wall shear stress bigger than in the case of the sub-boundary layer VG of height $H_2 = 0.2\delta$ whose vortex induced is weaker and so are its effects, for instance, on the wall shear stress. However, these results obtained for the jBAY modelled wall shear stress seem to be again in accordance with those generated by rectangular VGs located on a flat plate which increase the wall shear stress downstream of the VG. According to Godard and Stanislas [41], this increase tends to delay the boundary layer detachment. Two wall shear stress peaks are appreciated at the beginning and end of the ramp which indicate the influence of the ramp on the shear stress. As observed, these two peaks are larger for the $H_1 = 0.8\delta$ vane-type VG case than for the $H_2 = 0.2\delta$ vane-type VG case.

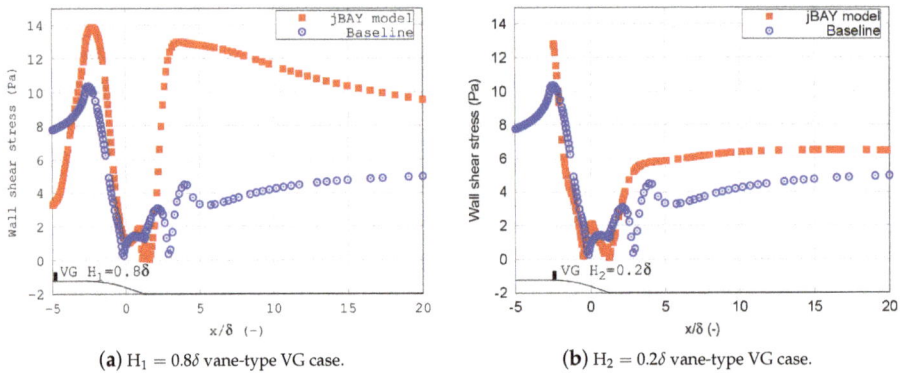

(**a**) $H_1 = 0.8\delta$ vane-type VG case.

(**b**) $H_2 = 0.2\delta$ vane-type VG case.

Figure 14. Wall shear stress distribution along the measurement line B. Red rectangles represent the data (jBAY model) obtained from the jBAY Source Term Model implemented and the blue circles the data (Baseline) obtained from the CFD simulation for the baseline case. The two vane-type VG cases are plotted. VGs, flat plate and ramp are also shown (not at scale).

6. Conclusions

The generation of vortices and their effects by a conventional vortex generator and a sub-boundary layer vortex generator positioned on a three-dimensional flat plate with a backward-facing ramp and adverse gradient pressure has been carried out by means of CFD simulations using the open-source code OpenFOAM. The influence of these two vane-type vortex generators (VGs) on the computational domain flow is implemented by using a source term in the corresponding Navier-Stokes equations according to the so-called jBAY Source Term Model. Steady-state, incompressible and turbulent flow is assumed and Reynolds Average Navier-Stokes (RANS) turbulence modelling is applied in the simulations.

The Source Term Model seems to simulate relatively well the streamwise pressure distributions along the floor of the flat plate for the two vane-type vortex generators studied of heights $H_1 = 80\%$ and $H_2 = 20\%$ the local boundary layer thickness δ at the upstream edge of the ramp. The results obtained for the jBAY modelled pressure distributions are in concordance with the experiments where the influence of these VG devices is measured. However, the streamwise pressure distribution has been slightly underestimated in certain regions far from the ramp and overestimated in other regions by the jBAY Model.

The jBAY modelled pressure distributions for the conventional vortex generator seem to be more accurate than those obtained for the sub-boundary layer vortex generator. The reason of this fact could be related to the height of the sub-boundary layer vortex generator which is within the sub-buffer zone of the local boundary layer where the shear and viscous stresses are predominant. Consequently, the results could potentially indicate that the jBAY Source Term Model presents more difficulties when modelling sub-boundary layer vortex generators.

Other parameters for vortex characterization analysed in the present study such as vortex path, vortex decay and vortex size obtained by means of the jBAY Source Term modelling, are in relatively good agreement with the results expected.

The implementation of the Source Term Model can represent an advantage over a fully Mesh-Resolved Vortex Generator Model for certain application cases due to a meaningful decrease in the cell number of the computational domain with the corresponding saving of computational time and resources. In the present work where the Source Term Model has been applied in the simulations of a conventional and a sub-boundary layer vortex generator of respective heights H_1 and H_2, the total cell number of the mesh could be eventually small when compared with the conventionally used

Energies **2019**, *12*, 605

fully Mesh-Resolved Vortex Generator Model. In addition, the jBAY Source Term Model could be quite helpful to reduce the meshing time.

Further research should be done in order to model and characterize vortex generators (VGs) of different dimensions and geometries positioned in surfaces of interest such as high-lift airfoils or ducts by implementing the Source Term Model or an optimized version of the Source Term Model into OpenFOAM or other CFD codes.

Author Contributions: I.E., U.F.-G., P.M.-F. and J.M.B. designed and were responsible for running the numerical simulations, carrying out the post-processing and writing the manuscript.

Funding: This research was partially funded by the Government of the Basque Country through the ELKARTEK 16/24.

Acknowledgments: The authors are grateful to the University of the Basque Country UPV/EHU through the EHU 12/26 research program.

Conflicts of Interest: The authors declare no conflict of interest.

References

1. Johnson, S.J.; Baker, J.P.; van Dam, C.P.; Berg, D. An overview of active load control techniques for wind turbines with an emphasis on microtabs. *Wind Energy* **2009**, *13*, 239–253. [CrossRef]
2. Gad-el-Hak, M. *Flow Control: Passive, Active, and Reactive Flow Management*; Cambridge University Press: Cambridge, UK, 2007; ISBN 9780521036719.
3. Taylor, H.D. *Summary Report on Vortex Generators*; Research Department Report No. R-05280-9; Research Department, United Aircraft Corporation: Moscow, Russia, 1947.
4. Liu, Y.; Sun, J.; Tang, Y.; Lu, L. Effect of slot at blade root on compressor cascade performance under different aerodynamic parameters. *Appl. Sci.* **2016**, *6*, 421. [CrossRef]
5. Wood, R.M. Discussion of aerodynamic control effectors concepts (ACEs) for future unmanned air vehicles (UAVs). In Proceedings of the AIAA 1st Technical Conference and Workshop on Unmanned Aerospace Vehicle, Systems, Technologies and Operations, Portsmouth, VA, USA, 20–23 May 2002.
6. Aramendia, I.; Fernández-Gamiz, U.; Ramos-Hernanz, J.A.; Sancho, J.; Lopez-Guede, J.M.; Zulueta, E. Flow Control Devices for Wind Turbines. *Energy Harvest. Energy Effic. Technol. Methods Appl.* **2017**, *37*, 629–655. [CrossRef]
7. Becker, R.; Garwon, M.; Gutknecht, C.; Barwolff, G.; King, R. Robust control of separated shear flows in simulation and experiment. *J. Process Control.* **2005**, *15*, 691–700. [CrossRef]
8. Rao, D.M.; Kariya, T.T. Boundary-layer submerged vortex generators for separation control, an exploratory study. In Proceedings of the AIAA/ASME/SIAM/APS 1st National Fluid Dynamics Congress, Cincinnati, OH, USA, 25–28 July1998.
9. Lin, J.C.; Robinson, S.K.; McGhee, R.J.; Valarezo, W.O. Separation Control on high-lift airfoils via micro vortex generators. *J. Aircr.* **1994**, *31*, 1317–1323. [CrossRef]
10. Doerffer, P.; Barakos, G.N.; Luczak, M.M. Recent Progress in Flow Control for Practical Flows. In *Results of the STADYWICO and IMESCON Projects*; Springer International Publishing AG: Cham, Switzerland, 2017; ISBN 978-3-319-50568-8.
11. Steijl, R.; Barakos, G.; Badcock, K. A framework for CFD analysis of helicopter rotors in hover and forward flight. *Int. J. Numer. Methods Fluids* **2006**, *51*, 819–847. [CrossRef]
12. Velte, C.M.; Hansen, M.O.L. Investigation of flow behind vortex generators by stereo particle image velocimetry on a thick airfoil near stall. *Wind Energy* **2013**, *16*, 775–785. [CrossRef]
13. Velte, C.M.; Hansen, M.O.L.; Okulov, V.L. Helical structure of longitudinal vortices embedded in turbulent wall-bounded flow. *J. Fluid Mech.* **2009**, *619*, 167–177. [CrossRef]
14. Fernández-Gamiz, U.; Velte, C.M.; Réthoré, P.; Sørensen, N.N.; Egusquiza, E. Testing of self-similarity and helical symmetry in vortex generator flow simulations. *Wind Energy* **2016**, *19*, 1043–1052. [CrossRef]
15. Urkiola, A.; Fernández-Gamiz, U.; Errasti, I.; Zulueta, E. Computational characterization of the vortex generated by a vortex generator on a flat plate for different vane angles. *Aerosp. Sci. Technol.* **2017**, *65*, 18–25. [CrossRef]

16. Gao, L.; Zhang, H.; Liu, Y.; Han, S. Effects of vortex generators on a blunt trailing-edge airfoil for wind turbines. *Renew. Energy* **2015**, *76*, 303–311. [CrossRef]

17. Baldacchino, D.; Manolesos, M.; Ferreira, C.; Gonzalez Salcedo, A.; Aparicio, M.; Chaviaropoulos, T.; Diakakis, K.; Florentie, L.; Garcia, N.R.; Papadakis, G.; et al. Experimental benchmark and code validation for airfoils equipped with passive vortex generators. *J. Phys.* **2016**, *753*, 022002. [CrossRef]

18. Keuthe, A.M. Effect of streamwise vortices on wake properties associated with sound generation. *J. Aircr.* **1972**, *9*, 715–719. [CrossRef]

19. Holmes, A.; Hickey, P.; Murphy, W.; Hilton, D. The application of sub-boundary layer vortex generators to reduce canopy "Mach rumble" interior noise on the Gulfstream III. In Proceedings of the 25th AIAA Aerospace Sciences Meeting, Aerospace Sciences Meetings, Reno, NV, USA, 24–26 March 1987 [CrossRef].

20. Lin, J.C.; Howard, F.G.; Selby, G.V. Small submerged vortex generators for turbulent-flow separation control. *J. Spacecr. Rocket.* **1990**, *27*, 503–507. [CrossRef]

21. Martinez-Filgueira, P.; Fernández-Gamiz, U.; Zulueta, E.; Errasti, I.; Fernández-Gauna, B. Parametric study of low-profile vortex generators. *Int. J. Hydrogen Energy* **2017**, *42*, 17700–17712. [CrossRef]

22. Lin, J. Review of research on low-profile vortex generators to control boundary-layer separation. *Prog. Aerosp. Sci.* **2002**, *38*, 389–420. [CrossRef]

23. Kenning, O.C.; Kaynes, I.W.; Miller, J.V. The potential application of flow control to helicopter rotor blades. In Proceedings of the 30th European Rotorcraft Forum, Marseille, France, 14–16 September 2004; 14p.

24. Schubauer, G.B.; Spangenber, W.G. Forced mixing in boundary layers. *J. Fluid Mech.* **1960**, *8*, 10–32. [CrossRef]

25. Bragg, M.B.; Gregorek, G.M. Experimental study of airfoil performance with vortex generators. *J. Aircr.* **1987**, *24*, 305–309. [CrossRef]

26. Fernández-Gamiz, U.; Errasti, I.; Gutierrez-Amo, R.; Boyano, A.; Barambones, O. Computational modelling of rectangular sub-boundary-layer vortex generators. *Appl. Sci.* **2018**, *8*, 138. [CrossRef]

27. Troldborg, N.; Zahle, F.; Sorensen, N.N. Simulation of a MW rotor equipped with vortex generators using CFD and an actuator shape model. In Proceedings of the 53th AIAA Aerospace Sciences Meeting, Kissimmee, FL, USA, 5–9 January 2015; 10p.

28. Bender, E.E.; Anderson, B.H.; Yagle, P.J. Vortex Generator Modelling for Navier–Stokes Codes. In Proceedings of the 3rd ASME/JSME Joint Fluids Engineering Conference, San Francisco, CA, USA, 18–23 July 1999.

29. Dudek, J.C. Modeling Vortex Generators in a Navier–Stokes Code. *AIAA J.* **2011**, *49*, 748–759. [CrossRef]

30. Jirasek, A. Vortex-Generator Model and Its Application to Flow Control. *J. Aircr.* **2005**, *42*, 1486–1491. [CrossRef]

31. Florentie, L.; Hulshoff, S.J.; van Zuijlen, A.H. Adjoint-based optimization of a source-term representation of vortex generators. *Comput. Fluids* **2018**, *162*, 139–151. [CrossRef]

32. OpenFOAM. Available online: https://openfoam.org (accessed on 1 October 2018).

33. Lin, J.C. Control of Turbulent Boundary-Layer Separation using Micro-Vortex Generators. *Am. Inst. Aeronaut. Astronaut.* **1999**, 1–16. [CrossRef]

34. Konig, B.; Fares, E.; Nolting, S. Fully-Resolved Lattice-Boltzmann Simulation of Vane-Type Vortex Generators. In Proceedings of the 7th AIAA Flow Control Conference, Atlanta, GA, USA, 16–20 June 2014; pp. 2795–3008. [CrossRef]

35. Menter, F.R. Zonal two equation k-w turbulence model for aerodynamic flows. *AIAA J.* **1993**, *93*, 2906. [CrossRef]

36. Gutierrez-Amo, R.; Fernández-Gamiz, U.; Errasti, I.; Zulueta, E. Computational modelling of three different sub-boundary layer vortex generators on a flat plate. *Energies* **2018**, *11*, 3107. [CrossRef]

37. Sørensen, N.; Zahle, F.; Bak, C.; Vronsky, T. Prediction of the Effect of Vortex Generators on Airfoil Performance. *Phys. Conf. Ser.* **2014**, *524*, 012019. [CrossRef]

38. Richardson, L.F.; Gaunt, J.A. The deferred approach to the limit. Part I. Single lattice. Part II. Interpenetrating lattices. *Philos. Trans. R. Soc. Lond. Ser. A* **1927**, *226*, 299–361. [CrossRef]

39. Fernández-Gamiz, U.; Zamorano, G.; Zulueta, E. Computational study of the vortex path variation with the VG height. *J. Phys. Conf. Ser.* **2014**, *524*, 012024. [CrossRef]

40. Bray, T.P. A Parametric Study of Vane and Air-Jet Vortex Generators. Ph.D. Thesis, College of Aeronautics, Cranfield University, Cranfield, UK, 1998.
41. Godard, G.; Stanislas, M. Control of a decelerating boundary layer. Part 1. Optimization of passive vortex generators. *Aerosp. Sci. Technol.* **2006**, *10*, 181–191. [CrossRef]

energies

MDPI

Article

Numerical Study on the Effect of Tunnel Aspect Ratio on Evacuation with Unsteady Heat Release Rate Due to Fire in the Case of Two Vehicles

Younggi Park, Youngman Lee, Junyoung Na and Hong Sun Ryou *

School of Mechanical Engineering, Chung-Ang University,82, Heukseok-ro, Dongjak-gu, Seoul 06974, Korea; pyg0511@cau.ac.kr (Y.P.); ymlee@alllitelife.com (Y.L.); junyoung628@naver.com (J.N.)
* Correspondence: cfdmec@cau.ac.kr; Tel.: +82-2-820-5280

Received: 30 November 2018; Accepted: 26 December 2018; Published: 1 January 2019

Abstract: In this study, the characteristics of fires in case of two vehicles in a tunnel are analyzed by Computational Fluid Dynamics analysis for varying tunnel aspect ratios. Unsteady heat release rates over time are set as the input conditions of fire sources considering real phenomena. Unsteady heat release rate values are obtained from experiments. As a result, the smoke velocities above the fire source appear faster in the case of tunnels with a large aspect ratio because the higher the height of the tunnel, the faster the smoke velocity caused by buoyancy forces. The smoke velocity in the longitudinal direction increases quickly. However, the temperature distribution in the vicinity of the ceiling is low when the tunnel aspect ratio is large because the height of the tunnel is not directly affected by the flames. Also, the higher the height of the tunnel, the lower the visibility distance due to the heat and smoke coming down along the wall surface. However, in the tunnels represented in this study, it is considered that the visibility of evacuees is sufficiently secured.

Keywords: aspect ratio; evacuation; fire propagation; tunnel vehicle fire; unsteady heat release rate

---

1. Introduction

Tunnels have been continuously constructed to overcome topographical barriers in mountainous terrains and urban areas, as well as to reduce waste of time. Generally, fires in tunnels have a complex flow structure because it is a physical phenomenon affected by the tunnel geometry, ventilation system, and fan location, as well as chemical reactions and heat and mass transfer. Therefore, it is one of the most challenging research topics. When there are fires in tunnels, there are often critical casualties because smoke spreads rapidly along the longitudinal direction of the tunnel. According to the National Fire Protection Association (NFPA) fire statistics report, more than 70% of all tunnel fire casualties are caused by suffocation due to smoke and toxic gas inhalation [1]. Therefore, lots of studies regarding fires in tunnels have been undertaken using experimental and numerical approaches.

In the case of experimental studies, real-scale or reduced-scale tunnel experiments with pool fires have normally been conducted to figure out the heat release rate (HRR), temperature, smoke velocity and so forth. Some studies have conducted experiments with real vehicles, not in tunnels but rather in large scale experimental rooms. By changing the initial ignition location of the fire Katsuhiro et al. figured out that the temperature distribution and maximum heat release rate reached 3 MW. They found that as soon as a window breaks, the fire suddenly diffused due to the inflow of exterior air [2]. Most experimental studies were conducted considering single vehicle situations [3,4], but in reality, vehicle accidents normally occur as collisions between two vehicles, however, there are only a few such studies. Many researchers have conducted pool fire experiments and the pool size was calculated using the scaling law [5], and based on the type of vehicles considering maximum Heat Release Rate (HRR). However, lots of reduced-scale experimental studies have been conducted, even if there are different temperature and HRR distributions between vehicles and pool fires due to material properties. Lee et al. [6] conducted small-scale experimental studies to investigate the change of fire characteristics such as ceiling jet flow and critical velocity for varying tunnel aspect ratios. They applied new dimensionless velocity and heat release rates to suggest modified critical velocities.

Numerical methods have been used many studies because the results of experimental studies can be compared with numerical studies and fire phenomena in tunnel fires possibly verified under various conditions [7–9]. In particular, the aspect ratio of the tunnel is the one of the most interesting factors in tunnel smoke propagation studies. The heat flow generated in a fire is different for varying tunnel aspect ratios, and this directly affects the evacuation result. In tunnels with the same hydraulic diameter, the critical velocity to prevent the smoke spreading for evacuees is different for various aspect ratios. Therefore, investigating the fire characteristics according to different aspect ratios is an important parameter when designing smoke control systems [10]. In previously studies, model tunnels have been constructed in which one can change the tunnel aspect ratio and determine the effect on mass loss rate, maximum smoke temperature and temperature distribution. It was found that aspect ratio is an important factor that affects the smoke temperature distribution. As can be seen from related papers regarding the effect of aspect ratio with numerical study, aspect ratio is a decisive parameter for designing of smoke control systems. Figure 1 presents a schematic diagram of the basic concepts about varying aspects ratio on tunnels. Therefore, it is very important to investigate the fire characteristics in case of two vehicle fires in real situations. Thus, in this study, we numerically investigated the effect of tunnel aspect ratios on unsteady heat release rates.

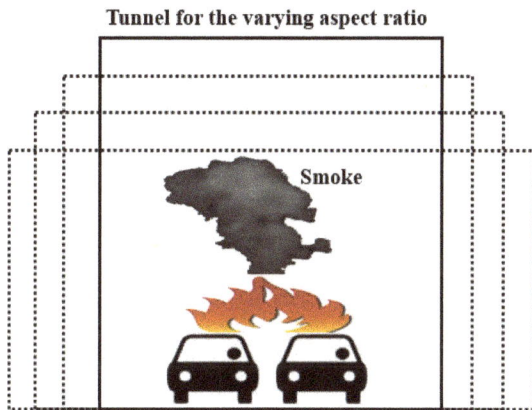

Figure 1. Schematic diagram of the basic concepts about varying aspect ratios of tunnels.

2. Numerical Method

The values of heat release rate obtained from a large scale cone calorimeter experimental apparatus are set as the fire input conditions for the numerical analysis. Park et al. have presented detailed information about this experimental study, currently under review for the journal *Fire Technology* [11]. Figure 2 presents whether the heat release rate shows the same result values when applied under the conditions of a fire source. In addition, unlike the pool fire experiments, in real vehicle experiments, there are some sections where the heat release rate decreases or stabilizes for the various points at which the fire source is propagated. Section 1 indicates that the fire spreads from the passenger seat to the driver's seat and rear seat. In addition, Section 2 is the time when the fire propagated from the vehicle where the initial fire occurred to the vehicle next to it, and Section 3 represents the time to propagate from indoors to outside vehicle components such as fuel tank, engine room and bumpers. Section 4 was a period for comparing Sections 1–3 that represent other trends. This is because, unlike the usual pool fire experiment, the heat release rates does not increase continuously but are steady or decrease. In this study, temperature, smoke velocities and visibility and so forth changes caused by the aspect ratio at the point of reaching the maximum heat release rate in four sections are analyzed using Fire Dynamic Simulator (FDS) numerical analysis program [12].

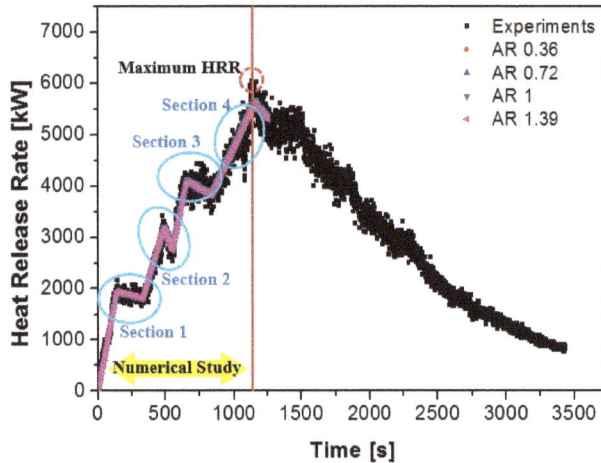

Figure 2. Heat release rate (HRR) results from experiments and numerical studies.

Figure 3a presents the computational domain of a tunnel with 100 m of longitudinal direction without any ventilation system. Vehicle sizes are 1.8 m (W) × 4.7 m (L) × 1.4 (H) and distance between vehicles is 0.6 m. Figure 3b represents the location of thermocouples and analysis planes. It depends on tunnel aspect ratio, however, the thermocouples are attached 0.2 m below the ceiling and 0.6 m above the vehicles. The interval between each thermocouple is 0.6 m. Figure 3c presents four difference tunnel aspect ratios but the total cross section area is almost constant. The aspect ratio is defined as the height divided by the width. Adiabatic conditions are set on the tunnel wall, and pressure outlet conditions at the tunnel exits. Figure 3d represents the three various flow directions occurring in a tunnel. The smoke velocity above fire source, velocities against the ceiling, and ceiling jet flows are affected by the tunnel's aspect ratio caused by heat gases and temperatures inside the tunnel. These flows are represented as flow 1, flow 2 and flow 3, respectively. The number of grids was approximately 770,000 and applied to differentiate the number of regions to improve the accuracy and efficiency of analysis. The size of the grid was 0.1, 0.2 and 0.4 m from the above of fire source to the exit of the tunnel, respectively. To distinguish the optimum grid size, the size of the minimum grid required for the flow modeling is determined by the characteristic flame diameter taking into account the heat release rate of the fire source [13]. Also, in order to improve the accuracy and efficiency of the analysis, the grid size is applied differently considering the flow characteristics:

$$D^* = \left(\frac{\dot{Q}}{\rho_\infty C_p T_\infty \sqrt{g}} \right)^{2/5} \tag{1}$$

where ρ_∞ [kg/m^3], C_p [kJ/(kg·K)], T_∞ [°C] represent the density, specific heat and temperature for ambient air (20 °C, 1 atm), respectively. For simulations involving buoyant plumes, the flow field is given by non-dimensional expression $D^*/\delta x$, where D^* is the characteristic fire diameter and δx is the nominal size of a mesh cell. The quantity, $\dot{Q}$, is the total heat release rate of the fire. The quantity $D^*/\delta x$ can be thought of as the number of computational cells spanning the characteristic diameter of the fire. Generally, a reasonable numerical results can be obtained at a grid size of $0.05 < \bar{\Delta}/D^* < 0.10$ [14] where, $\bar{\Delta}$ represents the average grid size. In this study, the average grid size is 0.35 m. Therefore, an average grid size of approximately 770,000 is considered when considering characteristic flame diameters.

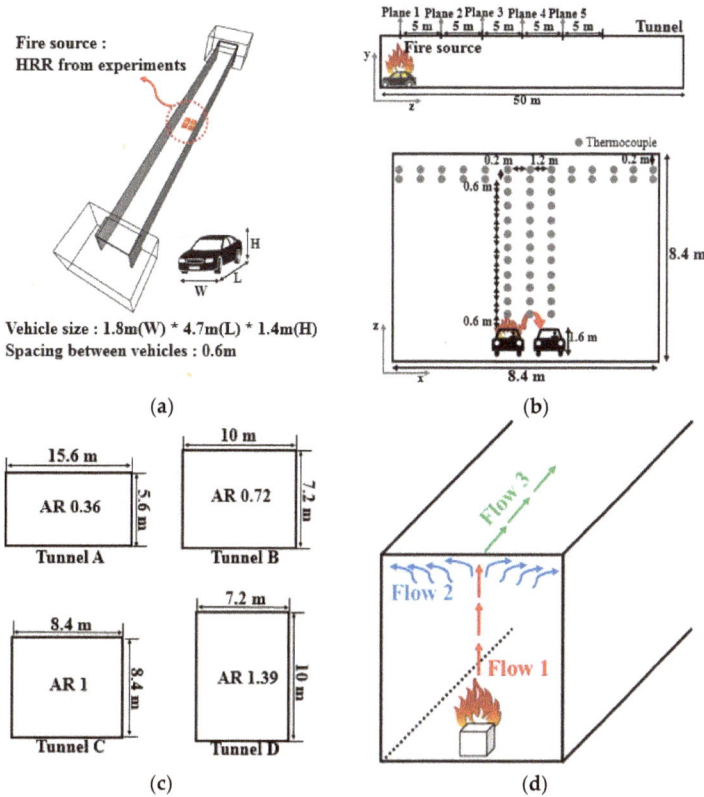

Figure 3. Numerical study: (**a**) computational domain for tunnel, (**b**) location of thermocouples and analysis planes, (**c**) cross-section of tunnel for varying aspect ratio and (**d**) main flow in case of tunnel fire.

3. Results and Discussion

3.1. Effect of Aspect Ratio on Tunnel

Figure 4 presents the temperature and smoke velocity contours for varying tunnel aspect ratios at 1500 s where it concerns the maximum heat release rate. When the aspect ratio of the tunnel is small, the temperature distribution above the fire source and temperature distribution in the horizontal direction are high. The lower the height of the tunnel, the more affected by the flame. Conversely, in case of smoke velocity, the higher the tunnel height, the more the momentum forces on the fire source increase, so that the smoke velocity appears faster. Also, the higher the height of the tunnel, the stronger the buoyancy force applied. Figure 5 presents the values on the change of temperature and smoke velocities for varying aspect ratios measured 0.2 m below the ceiling at 1500 s underpinning the effects of tunnel height on smoke velocities. To compensate the severe fluctuations, the value on 1500 s was calculated as an average value before the 10 seconds and after the 10 seconds at the specific time. Because of the different tunnel heights, the results shown are obtained by normalizing the tunnel height.

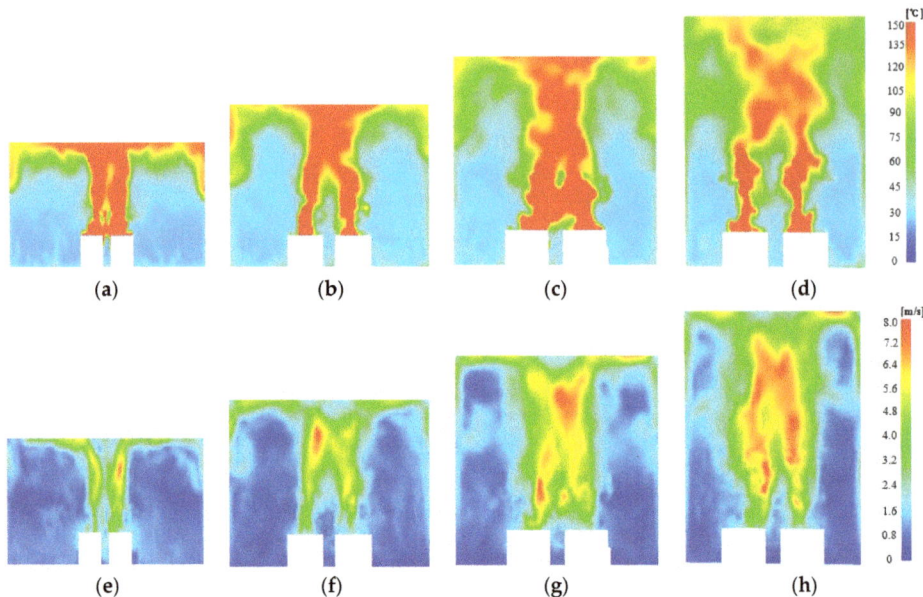

Figure 4. Temperature and smoke velocity contours at 1500 s; (**a–d**) represents the temperature distribution where the tunnel aspect ratio is 0.36, 0.72, 1 and 1.39, respectively, (**e–h**) represents the smoke velocities.

The value of 0 on the Y axis represents the temperature just above the fire source and the value of 1 represents at a point 0.2 m beneath from the ceiling. When the maximum heat release rate occurs, the temperature above the fire source is from about 120 to 200°C, and the temperature at the ceiling is from about 150 to 450°C. It varied by almost two- and three-fold depending on the location and tunnel aspect ratio. The smoke velocity is the fastest near the center of the tunnel, with a speed difference about 1.4 times higher. Furthermore, the smoke velocity is the fastest and has a tendency to decrease at the middle of the tunnel. This is due to buoyancy forces, which accelerate the smoke velocity over time, but near the tunnel wall, the smoke velocity is lowered due to no-slip conditions, friction and the normalized tunnel height to analyze the tendency of the tunnel aspect ratio.

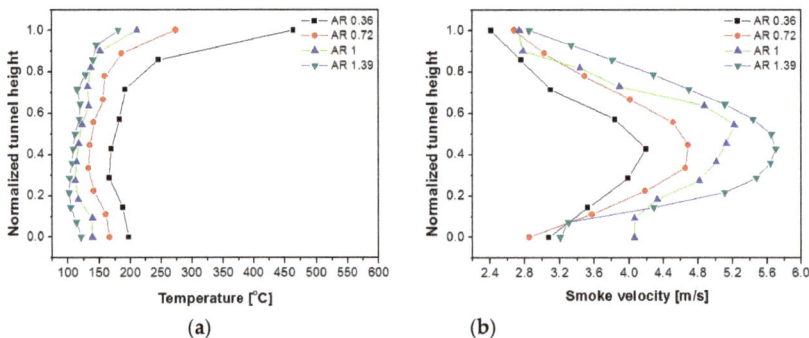

Figure 5. Temperature and smoke velocity distributions where it is located below 0.2 m of ceiling for varying aspect ratio at 1500 s. (**a**) Temperature distribution, (**b**) Smoke velocities.

Figure 6a presents the temperature changes 0.2 m below the tunnel ceiling for varying aspect ratios and distances. In the case of the tunnel with the smallest aspect ratio, the temperature above the fire source is the highest. In addition, the slope of the temperature distribution along the tunnel distances show that the temperature drop rate greatly changes when the aspect ratio is low. This is because when the height of the tunnel is high, the width of the tunnel is relatively small and the heat loss to the tunnel ceiling is reduced. Figure 6b presents the smoke velocities at the same point with various aspect ratios. It appears that the temperature distribution below the ceiling at the point 5 m away from the fire source is faster than the smoke velocity above the fire source, and the smoke velocity declines as the distance increases. This is due to the fact that the smoke velocity above the fire source is suddenly changed by the fluctuation and the smoke spreads along longitudinal direction before reaching 0.2 m from the ceiling. This is underpinned in Figure 7 by a vector in the longitudinal direction.

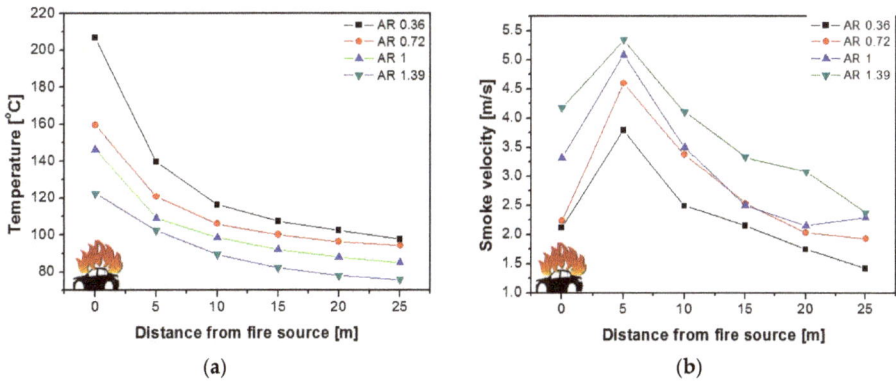

Figure 6. Temperatures and smoke velocities below 0.2 m of the tunnel ceiling for varying aspect ratio and distances from fire source at 1500 s. (**a**) Temperature distribution, (**b**) Smoke velocity distribution.

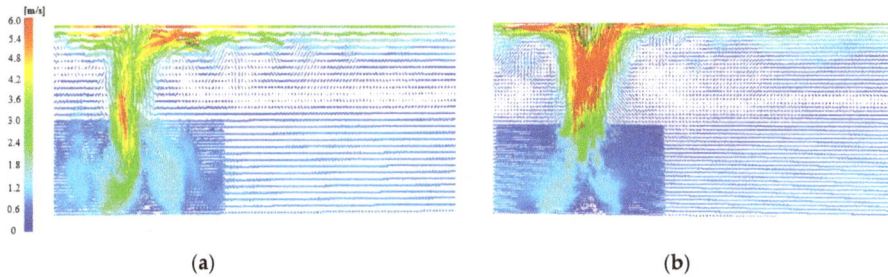

Figure 7. Velocity vector on longitudinal direction at 1500 s, (**a**,**b**) represent the tunnel aspect ratio of 0.36 and 1.39, respectively.

Also, in the case of the smoke velocity distribution according to distance, when the tunnel is high, buoyancy forces are have a great influence and it seems that the momentum force of the ceiling jet flow spreading in the longitudinal direction acts stronger. Therefore, when the height of the tunnel is high, the exit time to the tunnel exit is also fastest and this is represented in Figure 8. The time for the initial smoke to exit the tunnel was different as 90 s, 86 s, 79 s and 75 s, respectively, for the various aspect ratios. The smaller the aspect ratio, the longer it took for smoke to be emitted. The lower the height of the tunnel, the higher the temperature is because the flame directly reaches the ceiling, and the smoke temperature spreads in the longitudinal and lateral directions. However, the smoke velocity at which the smoke is spreading increased as the height is increased by the buoyancy forces.

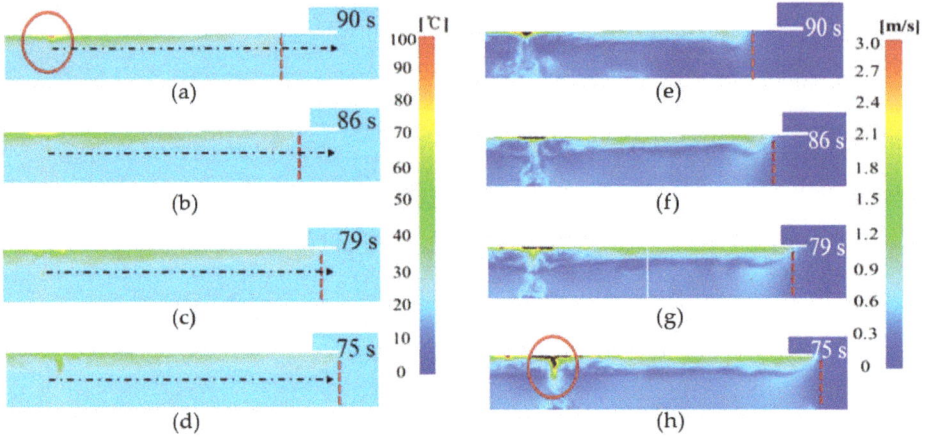

Figure 8. Temperature and smoke velocity distributions according to the initial time when the smoke exits the tunnel, (**a–d**) represents the various aspect ratio of 0.36, 0.72, 1 and 1.39, respectively. (**e–h**) represents the smoke velocity for varying aspect ratio of 0.36, 0.72, 1 and 1.39, respectively.

The temperature of the upper part of the tunnel increases continuously as time goes by. However, the temperature presents almost constant values because external cold air continuously flows into the lower part of the tunnel. Therefore, in order to secure the evacuation route of the evacuees, a flow is introduced in the opposite direction of the ceiling jet flow when the fire occurs. As the ceiling jet flow changes according to the aspect ratio, Lee et al. [9] suggested the formula considering aspect ratio as follows:

$$V_c = 0.73 AR_{tu}{}^{0.2}\sqrt{g\overline{H}}\left(\frac{Q}{\rho_0 c_{p,0} T_0 \sqrt{AR_{tu}g\overline{H}^5}}\right)^{1/3} \tag{2}$$

where, AR represents the tunnel aspect ratio and ρ_∞, C_p, T_∞ represent the density, specific heat and temperature for the ambient air (20 °C, 1 atm) and $\overline{H}$ is the hydraulic diameter of the tunnel. Based on the formula presented by Lee, the critical velocities for varying aspect ratios are simply 1.74, 1, 0.77 and 0.59, respectively. However, from an energy efficiency point of view, it is essential to calculate the appropriate air flow rate over time. Therefore, the critical velocity considering the heat release over time is shown in Figure 9. The critical velocity should be introduced considering the specific shape of the tunnel.

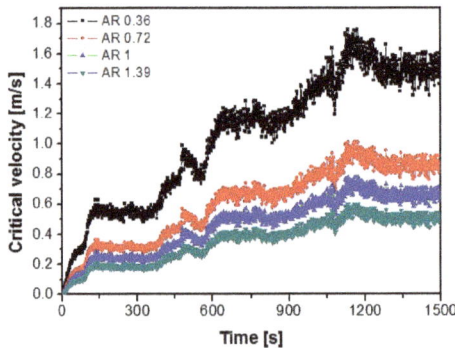

Figure 9. Critical velocity considering the aspect ratio of a tunnel with unsteady heat release rate.

3.2. Visibility

Figure 10 shows how much visible distance the fire can reach at a person's average eye level of 1.7 m. It shows the contour at the point of maximum heat release rate based on the FDS setting value having a maximum visible distance of 30 m. A value of 30 means that people can be visually identified up to 30 meters away. As can be seen from A in Figure 10 showing the fire source surface, the visibilities near the tunnel wall are low in a narrow tunnel width. On the other hand, in wide tunnels, the visual distance is relatively clear. This is because as the tunnel width becomes narrower, the smoke spreads in the lateral direction, and the smoke bumping against the wall generates a vortex and descends downward. If the tunnel width is wide, less vortex is generated. Visibility at 5 m and 10 m from the fire source is similar, but visibility is lower for a narrow tunnel width. Furthermore, velocity vectors are represented in Figure 11 to underpin the vortex for varying aspect ratios. As can be seen in Figure 11, the vortex size can be found to be large in high tunnels. It is because heat flow from the ceiling on the wall quickly flows into the bottom of the tunnel and fresh air from the outside quickly flows into the top of the tunnel, resulting in a larger vortex and it results in reduced visibility for evacuees. However, although the visibility varies depending on the aspect ratio of the tunnel, the visibility at a point where the average height of a person is 1.7 m seems to be clear.

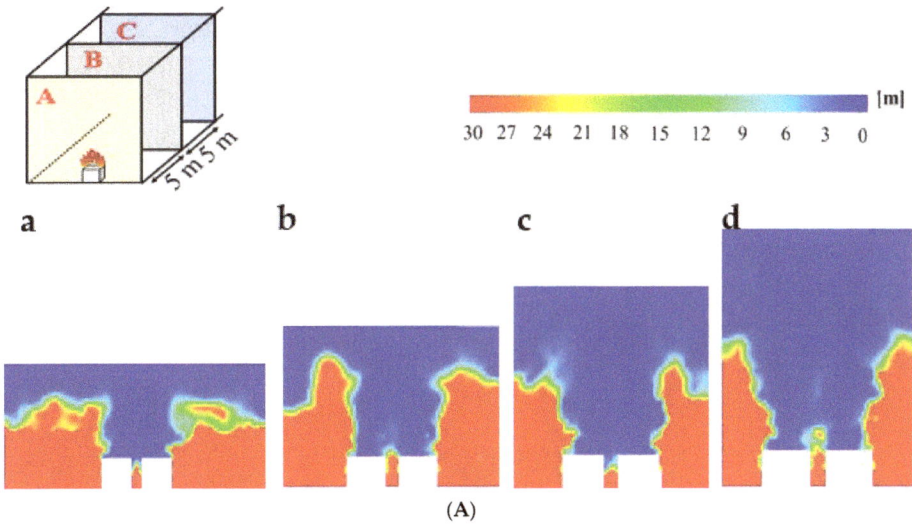

(A)

Figure 10. *Cont.*

(B)

(C)

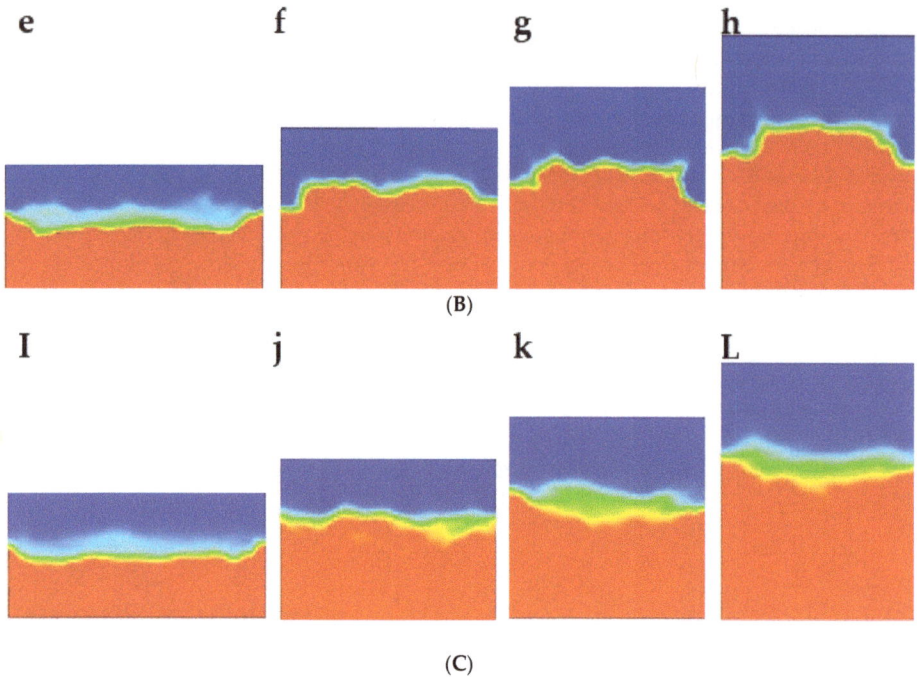

Figure 10. Visibility depending on distances, (**A**) represents on Fire plane, (**B,C**) represent the planes of 5 m and 10 m away from the fire sources, respectively. The aspect ratios of a, e, and I are 0.36, b, f, and j are aspect ratios of 0.72, c, g, and k are aspect ratios of 1, and d, b, and L are aspect ratios of 1.39.

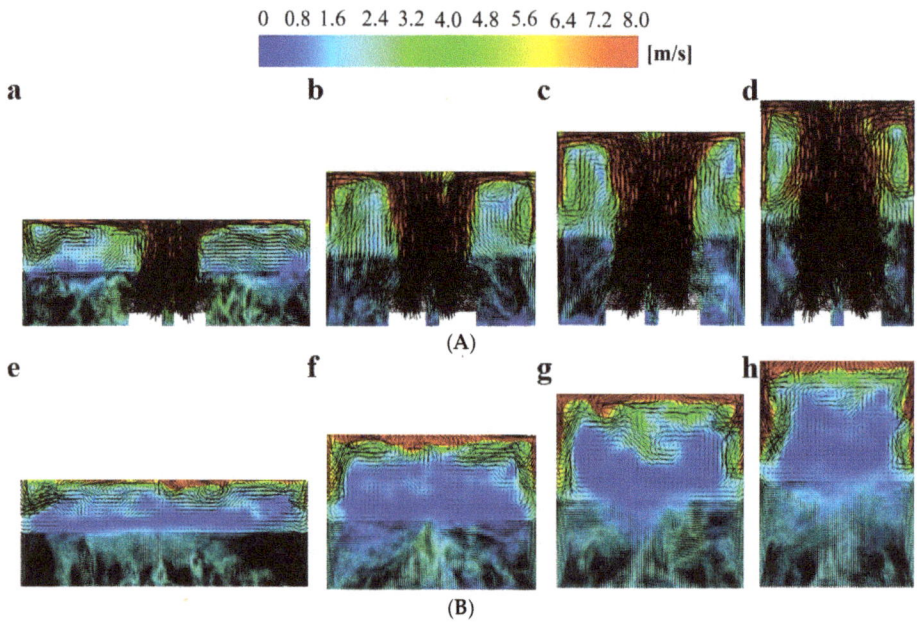

(A)

(B)

Figure 11. *Cont.*

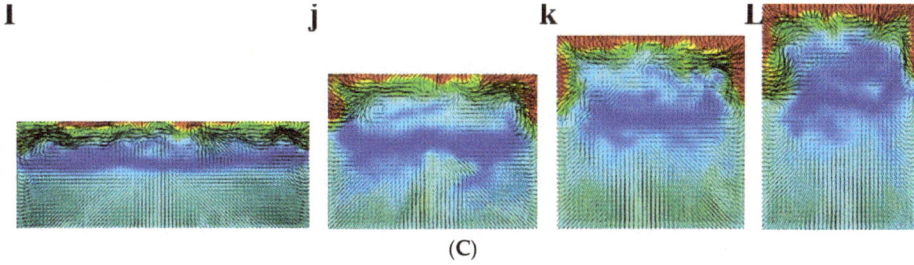

(C)

Figure 11. Velocity vector to figure out vortex depending on distances and aspect ratio, (**A**) represents on Fire plane, (**B,C**) represent the planes of 5 m and 10 m away from the fire sources, respectively. The aspect ratios of a, e, and I are 0.36, b, f, and j are aspect ratios of 0.72, c, g, and k are aspect ratios of 1, and d, b, and L are aspect ratios of 1.39.

Figure 12 presents the temperature distributions over time at the height of 1.7 m to consider an average human average height. It shows the same temperature change irrespective of the aspect ratio. This shows the temperature distribution in the same vicinity as the location of the fire source, so the effect of the tunnel width and length is not shown. Although it is difficult to set the evacuation time considering only the temperature, the evacuation must be within completed 450 s (7 min 30 s) before reaching 60 °C in order to reduce deaths due to fire in a tunnel.

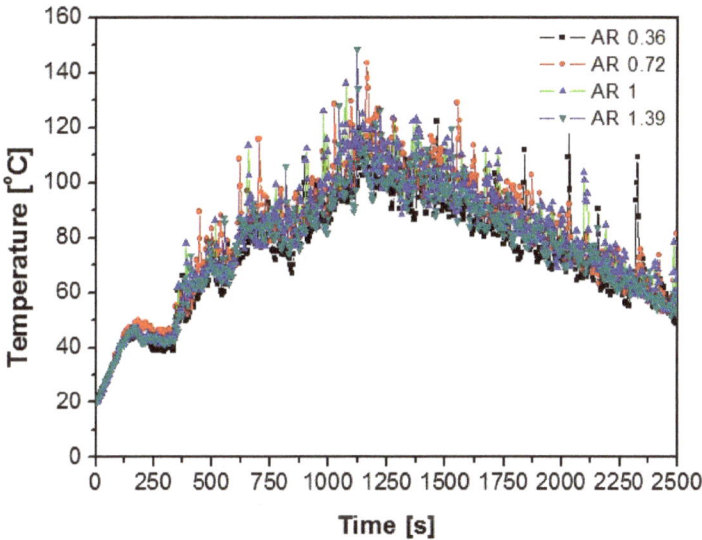

Figure 12. Temperature changes over time at the height of 1.7 m.

Figure 13 shows the temperature distribution at the specific section for the various aspect ratios in Sections 1–4, 0.2 m below the tunnel ceiling. Sections 1–3 are the sections where the heat release rate is stable or decreases, whereas Section 4 represents the section where the heat release rate increases radically in order to compare with other sections. As presented in Figure 13, the temperature increases and decreases are relatively stable. However, the temperature distribution in the region where the heat release rate increases rapidly differs from that in other regions.

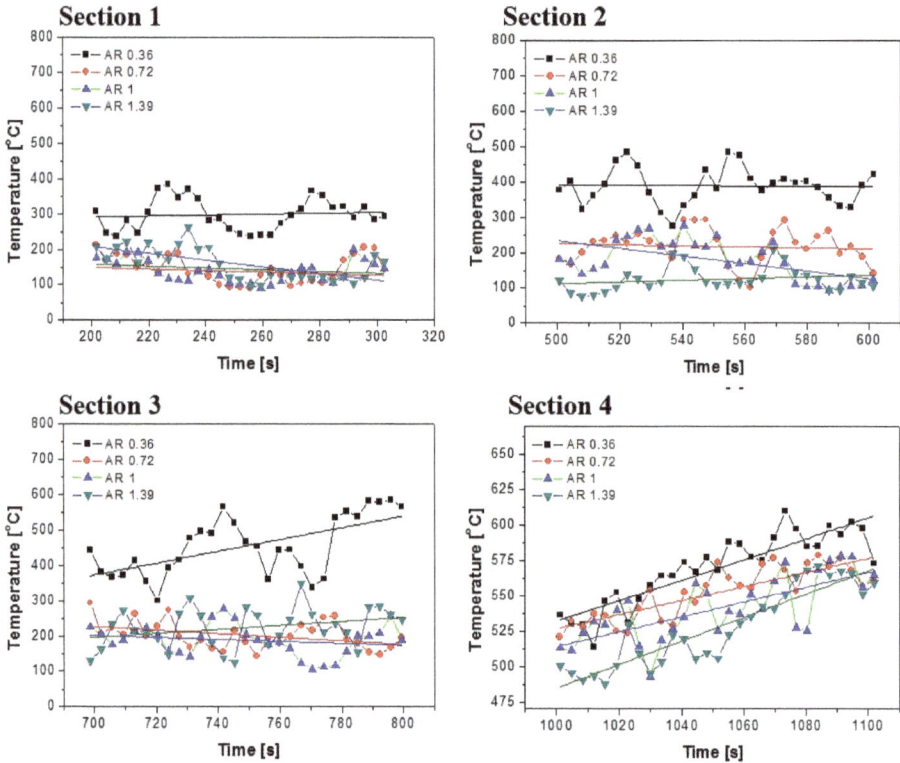

Figure 13. Temperature distribution at the specific times for the various aspect ratio at the point of where below 0.2 m of ceiling, (**Section 1**) represents the temperature values from 200 s to 300 s, (**Section 2**) represents the temperature values from 500 s to 600 s, (**Section 3**) represents the temperature values from 700 s to 800 s and (**Section 4**) represents the temperature values from 1000 s to 1100 s to compared to other sections.

4. Conclusions

The aim of this study was to investigate the fire characteristics of the tunnel vehicle fires for varying the aspect ratio of the tunnel cross-section using FDS. For this purpose, the heat release rates obtained from experimental study in case of two vehicles were measured. The following conclusions were obtained:

1) In a tunnel with a large aspect ratio, the smoke velocity above the fire source was generated rapidly. The buoyancy force and momentum force which affected the smoke velocity in the direction of longitudinal is greater in the case of a small aspect ratio tunnel.

2) As the aspect ratio becomes smaller, the temperature distribution in the cross-section area of the tunnel was higher because the flame directly affects the below of the ceiling but the flame was relatively small when the tunnel aspect ratio was large.

3) When the tunnel width was narrow, the vortex generated by the wall was generated strongly, and the smoke falls along the wall surface, so it is difficult to secure visibility.

4) The temperature changes in the section where the heat release rate was constant are insufficient, but the temperature change in the section where the heat release rate increases was apparent.

Author Contributions: Conceptualization, methodology, investigation: Y.P. and H.S.R.; software, validation, writing—original draft preparation: Y.P.; formal analysis, investigation: J.N. and Y.L.; writing—review and editing, supervision, project administration, funding acquisition: H.S.R.

Acknowledgments: This research was supported by the Chung-Ang University research grant in 2018.

Conflicts of Interest: The authors declare no conflict of interest.

References

1. *NFPA 92A: Recommended Practice for Smoke control Systems*; Natl Fire Protection Assn: Quincy, MA, USA, 1988.
2. Katsuhiro, O.; Norimichi, W.; Yasuaki, H.; Yasuaki, H.; Tadaomi, C.; Ryoji, M.; Hitoshi, M.; Satoshi, O.; Hideki, S.; Yohsuke, T.; et al. Burning behavior of sedan passenger cars. *Fire Saf. J.* **2009**, *44*, 301–310. [CrossRef]
3. Mangs, J.; Keski-Rahkonen, O. Characterization of the Fire Behaviour of a Burning Passenger Car. Part1: Car Fire Experiments. *Fire Saf. J.* **1994**, *23*, 17–35. [CrossRef]
4. Katsuhiro, O.; Takuma, O.; Hiroki, M.; Masakatsu, H.; Norimichi, W. Burning behavior of minivan passenger cars. *Fire Saf. J.* **2013**, *62*, 272–280. [CrossRef]
5. Quintiere, J.G. Scaling Applications in Fire Research. *Fire Saf. J.* **1989**, *15*, 3–29. [CrossRef]
6. Lee, S.R.; Ryou, H.S. An Experimental Study of the Effect of the Aspect Ratio on the Critical Velocity in Longitudinal Ventilation Tunnel Fires. *Fire Sci. J.* **2005**, *23*, 119–138. [CrossRef]
7. Jain, S.; Kumar, S.; Kumar, S.; Sharma, T.P. Numerical simulation of fire in a tunnel: Comparative study of CFAST and CFX predictions. *Tunn. Undergr. Space Technol.* **2008**, *23*, 160–170. [CrossRef]
8. Banjac, M.; Nikolic, B. Numerical Study of Smoke Flow Control in Tunnel Fires Using Ventilation Systems. *FME Trans J.* **2008**, *36*, 145–150.
9. Lee, S.R.; Ryou, H.S. A numerical study on smoke movement in longitudinal ventilation tunnel fires for different aspect ratio. *Build. Environ. J.* **2006**, *41*, 719–725. [CrossRef]
10. Ji, J.; Bi, Y.; Venkatasubbaiah, K.; Li, K. Influence of aspect ratio of tunnel on smoke temperature distribution under ceiling in near field of fire source. *Appl. Therm. Eng.* **2016**, *106*, 1094–1102. [CrossRef]
11. Park, Y.; Kim, J.; Ryou, H.S. Experimental Study on the Fire-Spreading Characteristics and Heat Release Rates of Burning Vehicles using a Large-Scale Cone Calorimeter. *Fire Technol. J.* **2018**. submitted.
12. McGrattan, K.; Hostikka, S.; McDermott, R.; Floyd, J.; Weinschenk, C.; Overholt, K. Fire Dynamics Simulator Technical Reference Guide Volume 1: Mathematical Model. *NIST Publ.* **2017**. [CrossRef]
13. Bounagui, A.; Benichou, N.; McCarteny, C.; Kashef, A. Optimizing the grid size used in CFD sumulations to evaluate fire safety in houses. In Proceedings of the 3rd NRC Symposium on Computational Fluid Dynamics, High performance Computing Virtual Reality, Ottawa, ON, Canada, 4 December 2003.
14. McGrattan, K.B.; Jason, F.; Forney, G.P.; Baum, H.R.; Hoskikka, S. Improved radiation and combustion routines for a large eddy simulation fire model. *Fire Saf. Sci.* **2003**, *7*, 827–883. [CrossRef]

energies

MDPI

Article

Numerical Analysis on the Effect of the Tunnel Slope on the Plug-Holing Phenomena

Ji Tae Kim, Ki-Bae Hong and Hong Sun Ryou *

School of Mechanical Engineering, Chung-Ang University, Seoul 06974, Korea; sdd322@naver.com (J.T.K.); gbhong@ut.ac.kr (K.-B.H.)
* Correspondence: cfdmec@cau.ac.kr; Tel.: +82-02-813-3669

Received: 30 November 2018; Accepted: 21 December 2018; Published: 25 December 2018

Abstract: Preventing the plug-holing phenomena of a natural ventilation system in a shallow underground tunnel is important for improving the ventilation performance, and the tunnel slope has a significant influence on the smoke flow. In this study, the effect of the tunnel slope on plug-holing in a shallow underground tunnel was analyzed by numerical method. The tunnel slope was increased by 0–8 degrees and the fire source was assumed to be 5 MW, which is equivalent to one sedan vehicle. As a result, the possibility of plug-holing decreased as the tunnel slope increased. However, when the tunnel slope is more than 4°, the fresh air from the entrance of the tunnel and smoke are diluted before reaching the shaft, so the flow temperature passing through the shaft is lowered, and the ventilation performance begins to decrease. In particular, plug-holing does not occur at the tunnel slopes of 6 and 8°, but the ventilation performance is expected to decrease because the temperature of the smoke discharged to the shaft is much lower than the general smoke temperature. Therefore, it is necessary to design the natural ventilation system considering the influence of the tunnel slope.

Keywords: plug-holing; tunnel slope; fire; natural ventilation; ventilation performance

---

1. Introduction

Statistics and reports by NFPA (National Fire Protection Association in USA) have shown that smoke inhalation is a major fatal factor in fire accidents [1]. Therefore, it is very important to ensure the good operation of the smoke ventilation system to reduce human casualties.

Generally, the vertical natural ventilation system is widely used in the shallow road tunnel due to low installation and maintenance costs [2]. In order to design an accurate vertical natural ventilation system, it is necessary to understand the tunnel fire phenomena in order to effectively exhaust the smoke, which is generated by the fire.

The plug-holing phenomena is one of the important factors to vertical natural ventilation systems. The plug holing phenomena is defined as: The fresh air beneath the smoke layer is exhausted through the shaft with smoke. Therefore, understanding the plug-holing phenomena, and preventing the plug-holing phenomena, is an important factor in improving the performance of the natural exhaust system.

Some researchers have shown that the horizontal momentum force of the ceiling jet and the buoyancy force of the vertical shaft play an important role in the plug-holing phenomena, and suggests the criterion is the modified Froude number for the effective ventilation system design without plug-holing. Hinkley et al. [3] introduced the plug-holing phenomena and presented a modified Froude number, which is defined as the ratio of the momentum force of the smoke to the buoyancy force in the shaft and the height of the tunnel. J. Ji et al. [4] analyzed the effect of the heat release rate on plug-holing, and proposed a modified Richardson number as the criterion for plug-holing. Additionally, J. Ji et al. [5] shows the influence of the cross-section area and aspect ratio of the shaft.

D. Baek et al. [6,7] presented, through numerical analysis, a modified Froude number with the effect of the hydraulic diameter and tunnel aspect ratio on the plug-holing phenomena.

As reported in previous studies, the effect of geometry is an important factor affecting the ventilation performance of the vertical natural ventilation system. Particularly, it is known that the slope of the tunnel greatly affects the buoyancy force of the smoke flow, the critical velocity, and the back layering length [8–10].

Therefore, the slope of the tunnel relates to the plug-holing phenomena, which reduces the ventilation performance in the vertical natural ventilation system. However, the effect of the tunnel slope on the plug-holing phenomena has not been clearly understood yet.

Therefore, in this study, a numerical study was carried out to analyze the effect of the tunnel slope on the plug-holing phenomena in a shallow underground tunnel.

2. Numerical Details

2.1. Geometry and Boundary Conditions

Figure 1 presents the geometry of the tunnel for analyzing the plug-holing phenomena according to the tunnel slope. Generally, the tunnel slope of the shallow underground tunnel is constructed less than 8 degrees below the urban center to relieve traffic congestion in urban areas. In the shallow underground tunnel, the vertical shaft vents are installed at intervals of about 100 to 180 m [9].

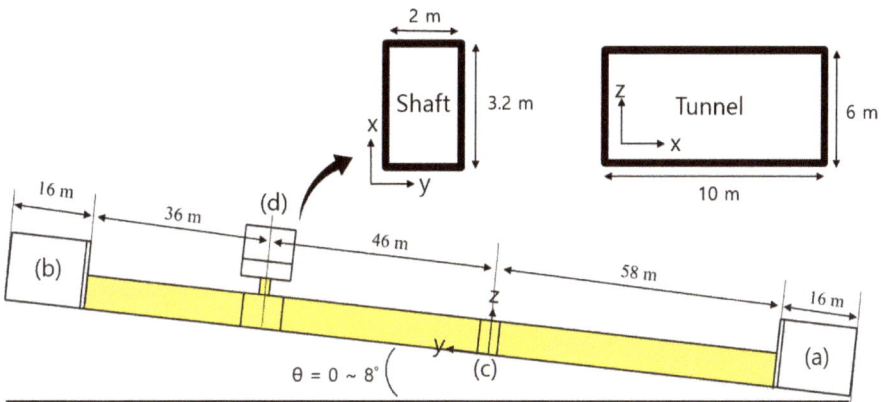

Figure 1. Geometry and boundary conditions of the tunnel.

Therefore, in this study, five tunnel slopes were selected at 0, 2, 4, 6, and 8 degrees. The tunnel is assumed to be a two-lane tunnel, and the width of the tunnel was set to 10 m and the height of the tunnel to 6 m. In addition, the length of the tunnel was set to be installed at intervals of 140 m with repeated shafts. The center of the vertical shaft was located at a distance of 58 m from the entrance of the tunnel, as shown in Figure 1d. The width and depth of the vertical shaft were 3.2 and 2 m, respectively, and the height was 3 m. The walls of the tunnels were assumed to be adiabatic, and zero-gauge pressure was applied to the atmospheric conditions at the exit of both ends of the tunnel and the outlet of the vertical shaft, as shown in Figure 1a,b,d, respectively. The fire source in tunnels is assumed to occur in passenger cars. According to previous studies, the maximum heat release rate (HRR) is known to be about 3–5 MW [11,12]. Thus, in this study the fire source was located at $y = 0$, as shown in Figure 1a and the maximum HRR was set to 5 MW. The size of the fire was $x = 1.2$ m, $y = 1.2$ m, and $z = 1$ m, and heat release rate applied to the surface was $z = 1$ m.

2.2. Numerical Methods

The fire dynamics simulator (FDS), developed by the National Institute of Standards and Technology (NIST), was used to analyze the fire-driven fluid flow. NIST performed the verification and validation of the models used in the FDS, and provided validated values such as turbulent Prandtl number, turbulent Schmidt number, convective heat transfer coefficient, and radiation fraction from fire source.

FDS solves the low Mach number Navier—Stokes equation (Ma < 0.3). The large eddy simulation (LES) model was used to calculate turbulence for the fire driven flow. The Deardorff model was used for turbulent viscous model [5]. The gray gas model was used to analyze the radiative heat transfer. Radiation fraction of fire was set to 0.35 and gravity was applied to 9.81 m/s in the −z direction. The atmospheric temperature was set at 25 °C. The grid for numerical analysis was generated at 0.2 m, by grid independency test between 0.1 and 0.4 m in grid size, based on equation (1), which was recommended in the FDS user guide. Where D^* is grid size, $\dot{Q}$ is heat release rate, ρ_∞ is ambient density, c_P is heat capacity of the inflow at constant pressure, T_∞ is ambient temperature, and g is gravity.

Also, near the fire source and the shaft, the grid size was formed at 0.1 m to improve the accuracy of the numerical analysis.

$$D^* = \left(\frac{\dot{Q}}{\rho_\infty c_P T_\infty \sqrt{g}} \right)^{\frac{2}{5}} \tag{1}$$

3. Results and Discussion

Figure 2 shows the temperature contour at x = 0 for the average value about 20 s after the flow due to the quasi-steady state. The tunnel slope was 0 degrees, the high-temperature combustion products generated from the fire source formed a smoke layer along the tunnel ceiling. The smoke layer was formed with a similar temperature field in both directions of the tunnel. The slope of the tunnel was 2 degrees, the flow of the smoke layer was relatively shifted toward the +y direction of the tunnel. Moreover, the tunnel slope was 4 degrees or more, and the flow of the smoke layer formed toward the entrance of the tunnel with a back layering of about 2 m.

Also, the slope of the tunnel was 0 degrees, fresh air flowed in both directions of the tunnel, and a flow field of about 25 °C was formed below the smoke layer. The angle of the tunnel angle was 2 degrees, the temperature of the flow from the fire source to the shaft was about 45 °C near the tunnel floor, as shown in the Figure 2b. The temperature field of 0 and 2 degrees were deeply related with the flow field. The slope of the tunnel was 0 degrees, the flow of the high temperature smoke products spread in both directions from the fire source, and the fresh air flow below the smoke layer was formed from both tunnel ends to the fire source. However, as shown in the Figure 3b, the slope of the tunnel was 2 degrees, and only the flow was formed in the +y direction from the fire source, with some very low velocity flows. Thus, the low velocity flow increased the temperature of the flow near the tunnel bottom.

In addition, as shown in the Figure 2c–e, when the slope of the tunnel was more than 4 degrees, it can be seen that the temperature below the smoke layer was about 30 °C or more, which was higher than the fresh air by more than 5 °C.

As shown in the Figure 3c–e, the flow velocity near the bottom of the tunnel was relatively higher, from the tunnel slope of 4 degrees. Because, the fresh air flow into from the −y direction was heated by the fire source and flows only in the +y direction by buoyancy.

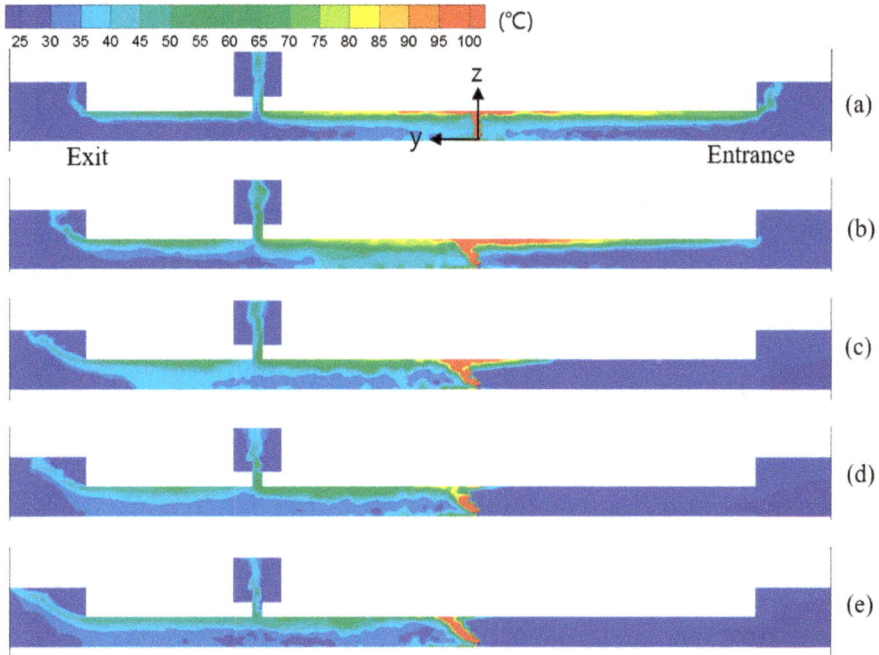

Figure 2. Temperature contour at x = 0: (**a**) 0°, (**b**) 2°, (**c**) 4°, (**d**) 6°, and (**e**) 8°.

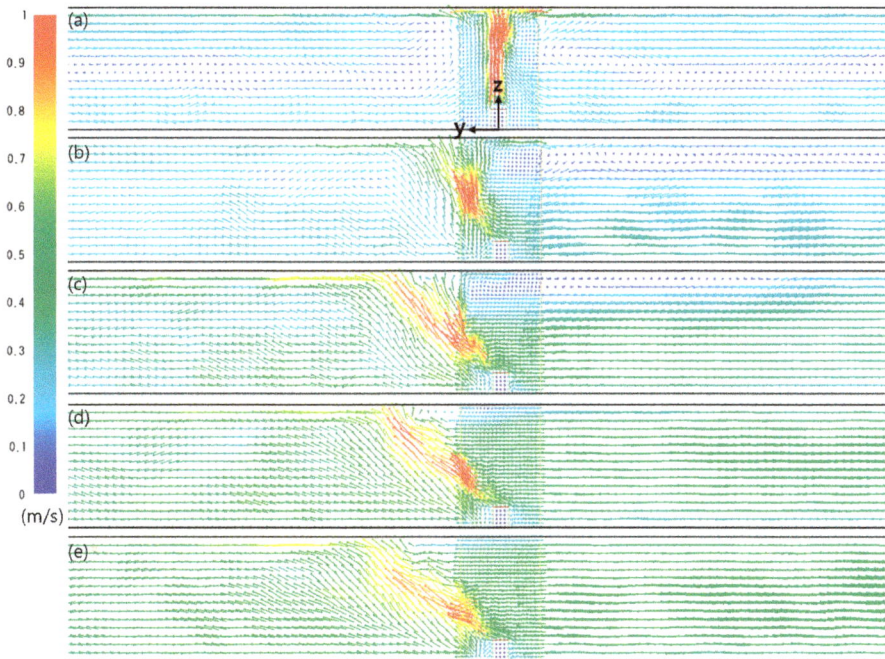

Figure 3. Velocity vector near the fire source at x = 0: (**a**) 0°, (**b**) 2°, (**c**) 4°, (**d**) 6°, (**e**) 8°.

Figure 4 represents the temperature distribution of the flow inside of the shaft. The temperature distribution was obtained from x = 0, y = 0, the dimensionless distance 0 was the starting location of the shaft, and the end location in the y direction was the dimensionless distance 100. The internal temperature of the shaft according to the tunnel slope at the dimensionless distance zero was similar to the smoke layer temperature at the shaft start location.

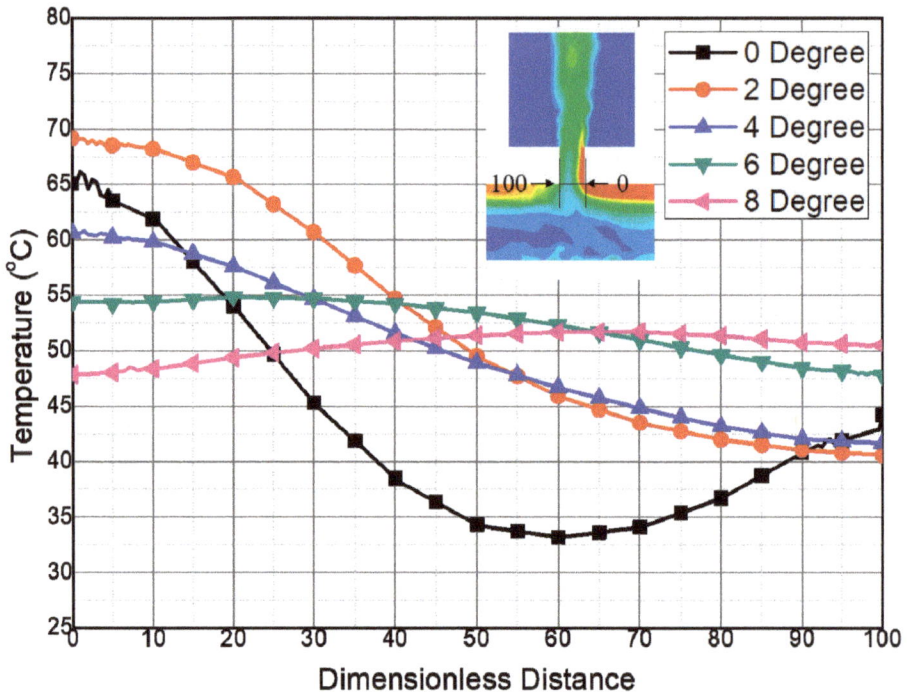

Figure 4. Temperature distributions inside of the shaft.

Generally, when plug-holing occurs, a flow of fresh air that has a lower temperature than the smoke layer flows into the shaft. When the tunnel slope was 0 degrees, the flow temperature inside the shaft was 33.5 °C at the dimensionless distance 60, which means that a flow lower than the temperature of the smoke layer flows into the shaft, which means the possibility of plug-holing is high.

However, the possibility of plug-holing will decrease at the tunnel slope of 2 and 4° because of the temperature at the shaft being higher than the fresh air temperature. In particular, when the tunnel slope exceeds 6 degrees, the maximum temperature deviation inside the shaft is significantly decreased about 8 °C.

These temperature characteristics inside the shaft were deeply related to flow field in the tunnel according to the tunnel slope.

As shown in Figure 5, the smoke layer that flows in the direction of the tunnel exit was exhausted vertically at the shaft. The smoke and the fresh air flows into the shaft and recirculation flow occurs at the region where these two flows collide.

However, as the tunnel slope increases, the velocity of the flow increases with the longitudinal direction of the tunnel. Therefore, as can be seen from the vector near the inlet of the shaft, the flow is oriented longitudinally, to the vertical direction of the shaft, at the tunnel slope of more than 4 degrees.

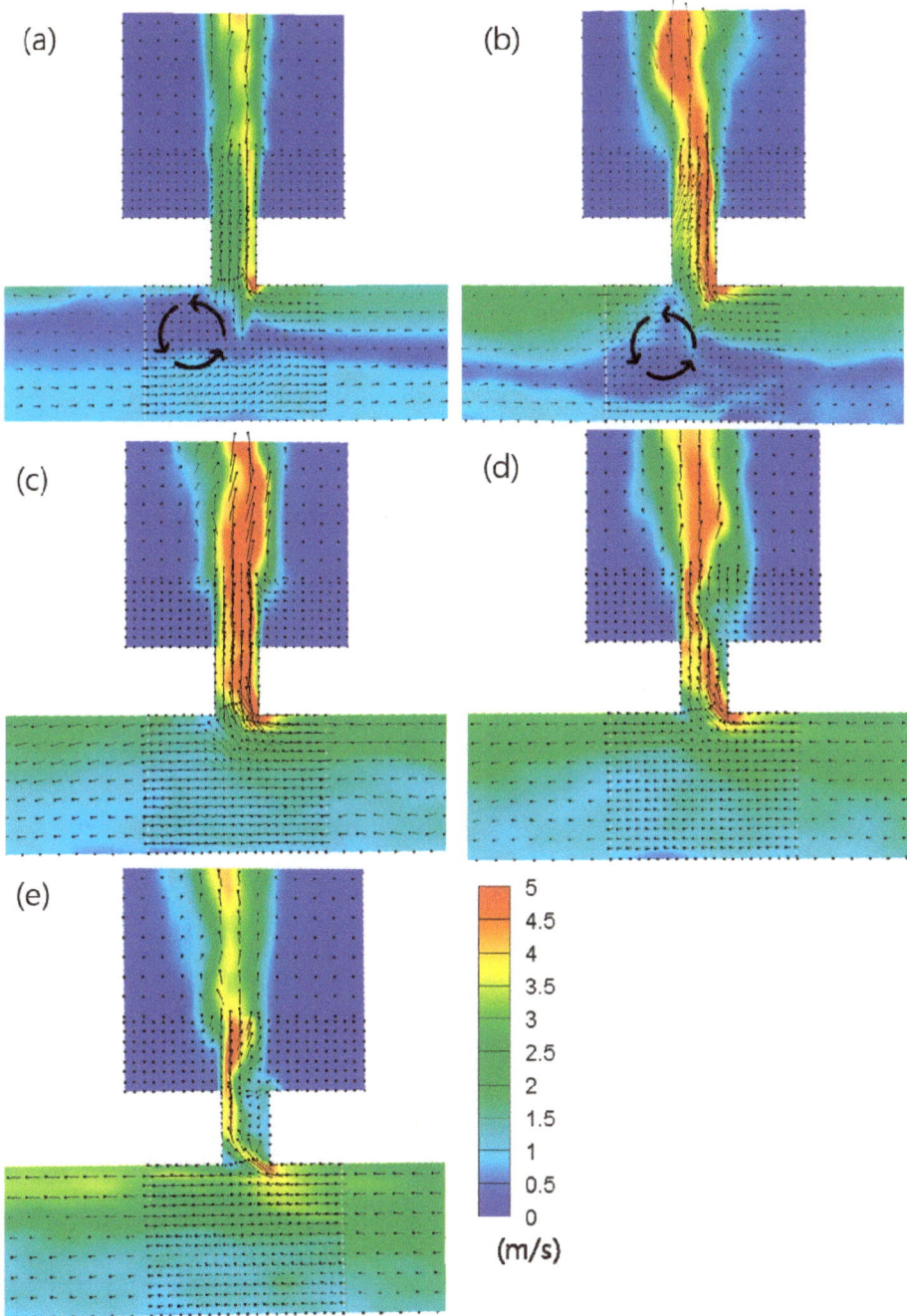

Figure 5. Velocity vector contour (x = 0): (**a**) 0°, (**b**) 2°, (**c**) 4°, (**d**) 6°, and (**e**) 8°.

Figure 6 represents the velocity distribution inside the shaft. The velocity of the shaft is strongly related to buoyancy force, which is influenced by the smoke layer temperature. As shown in Figure 4, the velocity inside of the shaft increases the slope of the tunnel to 4 degrees, and the temperature distribution inside of the shaft decreases by greater than 6 degrees.

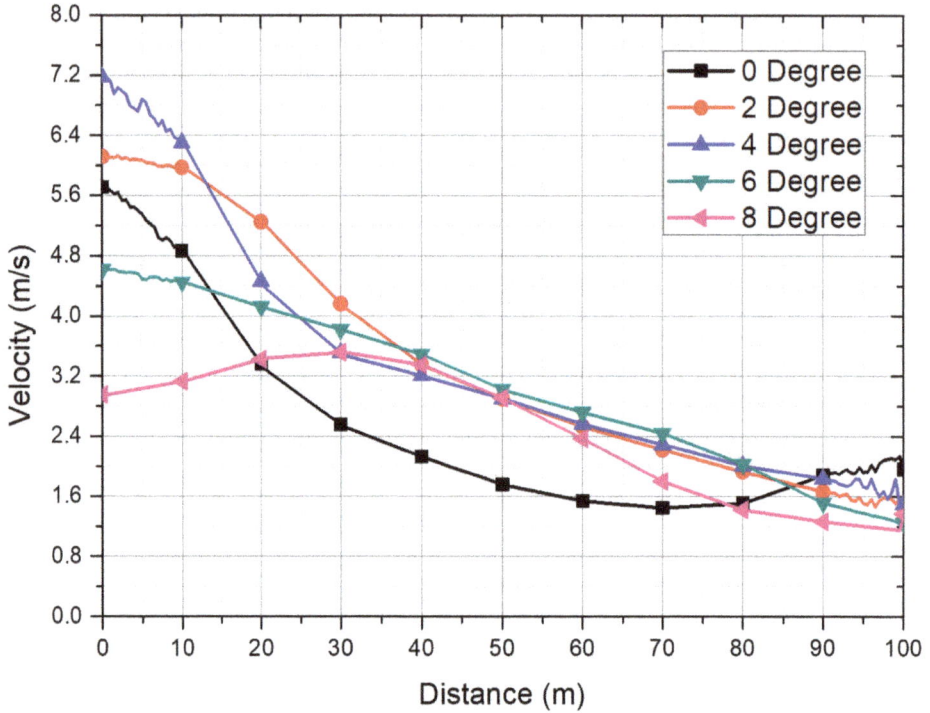

Figure 6. Velocity distributions inside of the shaft.

As shown in Figures 4 and 5, the plug-holing phenomena will not occur even if the slope of the tunnel exceeds 2 degrees. Additionally, the ventilation performance of the shaft is expected to increase because the flow temperature and flow velocity of the shaft are higher than the tunnel slope of 0 degrees, and until the slope of the tunnel reaches 4 degrees. However, above the tunnel slope of 6 degrees the ventilation performance is expected to decrease, because the internal velocity of the shaft is lower than the tunnel slope of 4 degrees.

4. Conclusions

The plug-holing phenomenon and the ventilation performance of the tunnel are affected by the tunnel slope. The possibility of plug-holing decreases as the slope of the tunnel increases. Moreover, when the tunnel slope is 4 degrees, the flow and temperature of the flow exhausted through the shaft are the highest. Therefore, the ventilation performance is the best when the tunnel slope is 4 degrees.

However, when the tunnel slope is greater than 4 degrees, fresh air from the entrance of the tunnel is diluted with smoke before reaching the shaft, thus the possibility of plug-holing is very low above 6 degrees and the ventilation performance begins to decrease. Therefore, it is necessary to design the shaft for the natural ventilation system considering the influence of the tunnel slope.

Author Contributions: Investigation, J.T.K.; Writing-Original Draft Preparation, J.T.K.; Writing-Review & Editing, J.T.K.; Supervision, H.S.R.; Funding Acquisition, K.-B.H.

Funding: This research was supported by the Fire Fighting Safety and 119 Rescue Technology Research and Development Program, funded by the Ministry of Public Safety and Security (MPSS-Fire Fiting-2015-80).

Conflicts of Interest: The funders had no role in the design of the study; in the collection, analyses, or interpretation of data; in the writing of the manuscript, and in the decision to publish the results.

References

1. Alarie, Y. The toxicity of smoke from polymeric materials during thermal decomposition. *Annu. Rev. Pharmacol. Toxicol.* **1985**, *25*, 325–347. [CrossRef] [PubMed]
2. Heskestad, G. Smoke distributions from fire plumes in uniform downdraft from a ceiling. *Fire Saf. J.* **2004**, *39*, 358–374. [CrossRef]
3. Hinkley, P.L. The flow of hot gases along an enclosed shopping mall a tentative theory. *Fire Saf. Sci.* **1970**, *807*, 1–17.
4. Ji, J.; Gao, Z.H.; Fan, C.G.; Zhong, W.; Sun, J.H. A study of the effect of plug-holing and boundary layer separation on natural ventilation with vertical shaft in urban road tunnel fires. *Int. J. Heat Mass Transf.* **2012**, *55*, 6032–6041. [CrossRef]
5. Ji, J.; Han, J.Y.; Fan, C.G.; Gao, Z.H.; Sun, J.H. Influence of cross-sectional area and aspect ratio of shaft on natural ventilation in urban road tunnel. *Int. J. Heat Mass Transf.* **2013**, *67*, 420–431. [CrossRef]
6. Baek, D.; Bae, S.; Ryou, H.S. A numerical study on the effect of the hydraulic diameter of tunnels on the plug-holing phenomena in shallow underground tunnels. *J. Mech. Sci. Technol.* **2017**, *31*, 2331–2338. [CrossRef]
7. Baek, D.; Sung, K.H.; Ryou, H.S. Experimental study on the effect of heat release rate and aspect ratio of tunnel on the plug-holing phenomena in shallow underground tunnels. *Int. J. Heat Mass Transf.* **2017**, *113*, 1135–1141. [CrossRef]
8. Ko, G.H.; Kim, S.R.; Ryou, H.S. An experimental study on the effect of slope on the critical velocity in tunnel fires. *J. Fire Sci.* **2010**, *28*, 27–47.
9. Chow, W.K.; Wong, K.Y.; Chung, W.Y. Longitudinal ventilation for smoke control in a tilted tunnel by scale modeling. *Tunn. Undergr. Space Technol.* **2010**, *25*, 122–128. [CrossRef]
10. Atkinson, G.T.; Wu, Y. Smoke control in sloping tunnels. *Fire Saf. J.* **1996**, *27*, 335–341. [CrossRef]
11. Okamoto, K.; Watanabe, N.; Hagimoto, Y.; Chigira, T.; Masano, R.; Miura, H.; Ochiai, S.; Satoh, H.; Tamura, Y.; Hayano, K.; et al. Burning behavior of sedan passenger cars. *Fire Saf. J.* **2009**, *44*, 301–310. [CrossRef]
12. Shintani, Y.; Kakae, N.; Harada, K.; Masuda, H.; Takahash, W. Experimental investigation of burning behavior of automobiles. *Fire Saf. Sci.* **2004**, *6*, 1–13.

energies

MDPI

Article

Disc Thickness and Spacing Distance Impacts on Flow Characteristics of Multichannel Tesla Turbines

Wenjiao Qi, Qinghua Deng *, Yu Jiang, Qi Yuan and Zhenping Feng

Shaanxi Engineering Laboratory of Turbomachinery and Power Equipment, Institute of Turbomachinery, School of Energy and Power Engineering, Xi'an Jiaotong University, Xi'an 710049, China; qiwenjiao@stu.xjtu.edu.cn (W.Q.); guaiily@stu.xjtu.edu.cn (Y.J.); qyuan@mail.xjtu.edu.cn (Q.Y.); zpfeng@mail.xjtu.edu.cn (Z.F.)
* Correspondence: qhdeng@mail.xjtu.edu.cn

Received: 9 November 2018; Accepted: 20 December 2018; Published: 24 December 2018

Abstract: Tesla turbines are a kind of unconventional bladeless turbines, which utilize the viscosity of working fluid to rotate the rotor and realize energy conversion. They offer an attractive substitution for small and micro conventional bladed turbines due to two major advantages. In this study, the effects of two influential geometrical parameters, disc thickness and disc spacing distance, on the aerodynamic performance and flow characteristics for two kinds of multichannel Tesla turbines (one-to-one turbine and one-to-many turbine) were investigated and analyzed numerically. The results show that, with increasing disc thickness, the isentropic efficiency of the one-to-one turbine decreases a little and that of the one-to-many turbine reduces significantly. For example, for turbine cases with 0.5 mm disc spacing distance, the former drops less than 7% and the latter decreases by about 45% of their original values as disc thickness increases from 1 mm to 2 mm. With increasing disc spacing distance, the isentropic efficiency of both kinds of turbines increases first and then decreases, and an optimal value and a high efficiency range exist to make the isentropic efficiency reach its maximum and maintain at a high level, respectively. The optimal disc spacing distance for the one-to-one turbine is less than that for the one-to-many turbine (0.5 mm and 1 mm, respectively, for turbine cases with disc thickness of 1 mm). To sum up, for designing a multichannel Tesla turbine, the disc spacing distance should be among its high efficiency range, and the determination of disc thickness should be balanced between its impacts on the aerodynamic performance and mechanical stress.

Keywords: Tesla turbine; fluid dynamics; disc thickness; disc spacing distance; isentropic efficiency

1. Introduction

In the last two decades, one of the interests is to develop small-scale turbomachinery due to market demands driving and manufacture techniques making progress [1], which is mainly applied to small power generation unit, such as distributed energy systems using low-grade energy and increasing energy utilization efficiency. In addition, small-scale turbomachinery is also applied as the mobile or small power device, which can be used in unmanned aircraft, miniaturized radio-controlled vehicles [2,3], and so on.

Several research groups designed, manufactured and experimented some small or micro gas turbines to study their feasibility and effectiveness [4–9]. These studies show that two major problems are faced when conventional turbines are scaled down. One is the rapidly increasing rotational speed, and the other is the significantly decreasing flow efficiency. Moreover, Brayton cycle system cannot be realized when the flow efficiency of conventional turbines goes down to a certain amount due to their microminiaturization [10]. However, a Tesla turbine is considered as one of the best alternatives [11].

In 1913, the first Tesla turbine was designed, manufactured and patented by the famous scholar Nikola Tesla [12]. It is an unconventional bladeless turbomachinery, which uses the viscous stress acting on rotating disc walls to drag the rotor to rotate, as shown in Figure 1. The rotor of Tesla turbines is composed by several flat, parallel, rigid, co-rotating discs, which are placed closely and mounted on a central shaft. The narrow spaces between the adjacent discs are disc channels. In Tesla turbines, the working fluid accelerates in nozzles or volute and obtains its highest flow velocity at the stator outlet; then injects into the disc channels nearly tangentially; and finally flows spirally towards the disc holes or slots around the shaft.

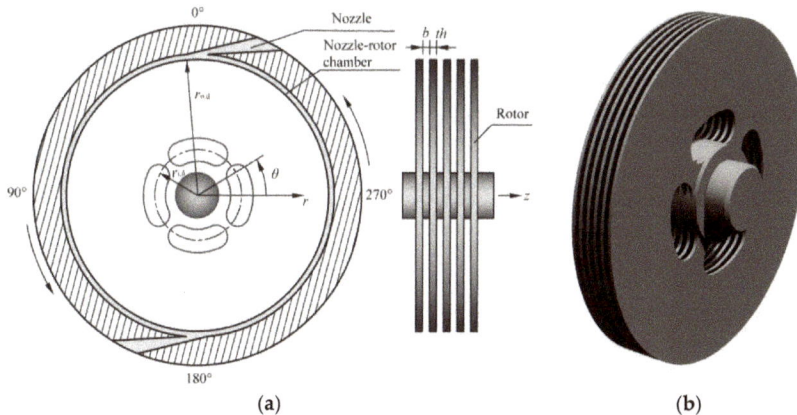

Figure 1. Tesla turbine schematic: (**a**) 2-D sketch; and (**b**) 3-D rotor.

During 1913–1950, little research was conducted on Tesla turbines due to the invention of gas turbines and its high efficiency. Thereafter, the theoretical and experimental analysis on Tesla turbines has been restarted due to its advantages; for example, it is simple to manufacture and maintain with low cost and it can use kinds of working fluid [13–16]. However, the study progress of Tesla turbines was still slow. In the last two decades, much research on Tesla turbines has been reported using not only theoretical and experimental methods but also numerical simulations due to the rapid development of computational fluid dynamics (CFD) and advanced computer techniques.

According to the inlet geometry, Tesla turbines are classified into two types: voluted Tesla turbine and nozzled Tesla turbine. There are few reports on the voluted Tesla turbine. The flow characteristics and loss mechanism of voluted Tesla turbines were investigated using experimental and numerical methods [17]. The results show that the theoretical limit of the isentropic efficiency is about 40%, if the parasitic losses, mainly including bearing loss, viscous loss on end walls, and eddy loss in the volute, can be minimized.

The main focus of Tesla turbines is on investigating flow characteristics in nozzled Tesla turbines. A one-dimensional analysis method of flow field in the disc channel is proposed with some hypotheses, and the model agrees well with previous experimental performance data [18]. Sengupta and Guha formulated a mathematical theory by simplifying the Navier–Stokes equations using a magnitude analysis method, which can assess the turbine performance [19–21]. Talluri et al. developed a new method for designing of a Tesla turbine for Organic Rankine Cycle (ORC) applications, in which almost all losses are considered using real gas physical properties [22]. It can calculate the aerodynamic performance of the multichannel Tesla turbine; however, the results have not been validated by numerical and experimental analysis. To sum up, most theoretical analysis can only be applied to one channel Tesla turbines, in which the influence of disc thickness has not been taken into account yet.

Some researchers investigated the total aerodynamic performance and detailed flow fields using experimental method. The research group of Guha set up a test rig of Tesla turbines, and put forward several methods for measuring power [23]. In addition, they improved the inlet and nozzle to enhance the turbine efficiency [24]. Schosser et al. designed and set up a test rig to reveal the detailed flow field in the disc spacing of one channel Tesla turbine using 3D tomographic Particle Image Velocimetry (PIV) and Particle Tracking Velocimetry (PTV) measurements, however its rotational speed is restricted because of the mechanical stress [25]. From the above research, the detailed flow field in multichannel Tesla turbines cannot be investigated experimentally.

The complicated turbine model cannot be analyzed theoretically, although it can be simulated by CFD method to reveal the detailed flow characteristics. The effects of some parameters, such as the rotational speed and nozzle number, have been studied for a Tesla turbine with a low-boiling medium [11,26]. The effects of disc spacing distance and rotational speed on a Tesla turbine were analyzed numerically by our research group, and the results show that an optimal value of both parameters exists for the turbine to obtain its highest efficiency, respectively [27–29]. The flow fields in the disc channel of one channel Tesla turbine (no stator) with different inlet conditions, including flow coefficient and inlet geometries, were simulated numerically; the efficiency decreases dramatically with increasing flow coefficient, and the inlet non-uniformity is also a factor that makes turbine efficiency decrease [30]. Sengupta et al. studied the influence of four parameters, namely the nozzle number, disc thickness, rotational speed and nozzle-rotor radial clearance, on the performance of a Tesla turbine, and the results show that in the turbine design, thin discs with flat disc tip edge, more nuzzles and an optimal radial clearance are suggested [31].

In the authors' previous study, the nozzled multichannel Tesla turbine is classified into two categories according to nozzle geometry: one nozzle channel to one disc channel (one-to-one Tesla turbines) and one nozzle channel to several disc channels (one-to-many Tesla turbines), as shown in Figure 2 [29]. The one-to-one Tesla turbine and the one-to-many turbine with the same geometry operating under the same conditions were studied numerically and the only difference between two kinds of Tesla turbines is their nozzle geometries. The objective was to reveal and compare their fluid dynamic, and the results show that the flow mechanism of the two Tesla turbines are totally different.

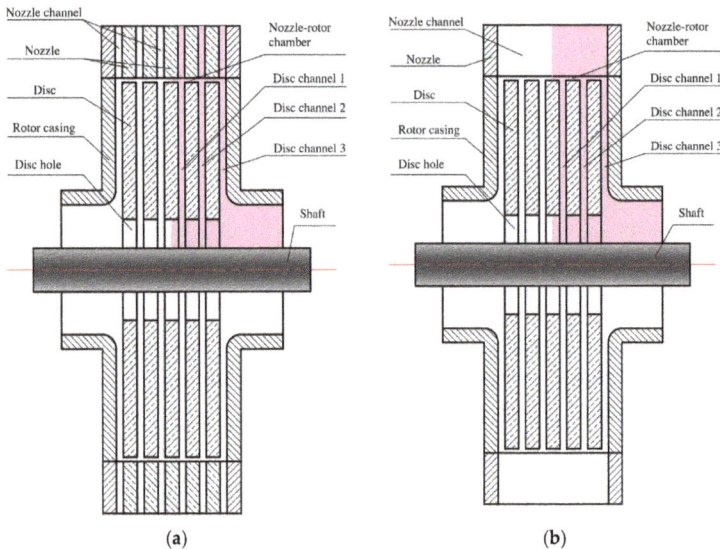

Figure 2. Sketch maps of multichannel Tesla turbines: (**a**) one-to-one Tesla turbine; and (**b**) one-to-many Tesla turbine.

For multichannel Tesla turbines, the affecting factors include the geometrical parameters (such as nozzle number, nozzle geometry, disc outer diameter, disc spacing distance and disc thickness) and operating parameters (such as rotational speed mass flow rate and turbine pressure ratio). The above reference review shows that the influence of most affecting parameters has been investigated, although some research is on one channel Tesla turbine. However, up to now, few studies have been conducted on the impacts of disc thickness and disc spacing distance on the performance of multichannel Tesla turbines. The effect of disc thickness is studied in Ref. [31], but only the flow in the nozzle-rotor chamber and the rotor are included in numerical calculations. In addition, it has been analyzed experimentally for the one-to-many turbine, while, for the one-to-one turbine, its influence and the fluid mechanism have not been investigated [32].

The impacts of disc spacing distance on the one channel Tesla turbine were analyzed by our group, and the results show that an optimal disc spacing distance exists to enable the Tesla turbine to obtain its best performance [28]. In the turbine with narrower disc spacing distance, two boundary layers on two disc walls overlap, resulting in a decrease in frictional force and torque; in the turbine with wider disc spacing distance, more working fluid outside the boundary layer flows out of the disc channels resulting in less momentum exchange to drive a rotor. Because disc thickness has to be taken into consideration in multichannel Tesla turbines, the influence of disc spacing distance on flow fields must be different from that in one channel Tesla turbines.

In this paper, two kinds of multichannel Tesla turbines (one-to-one turbines and one-to-many turbines) with different disc spacing distance and disc thickness are simulated numerically to investigate their influence on the aerodynamic performance and flow characteristics. This paper provides theoretical and engineering reference for designing a multichannel Tesla turbine.

2. Numerical Approach

2.1. Geometry Model and Boundary Conditions

In this study, two groups of multichannel Tesla turbines were calculated numerically. One group is the one-to-one Tesla turbine, and the other is the one-to-many Tesla turbine. Each group has six cases of Tesla turbines, and the parameters of disc spacing distance and disc thickness are given in Table 1, in which the case is named as "disc thickness-disc spacing distance". In this research, the working fluid is compressed air.

Table 1. Geometrical parameters of each case.

Group 1 (One-To-One Multichannel Tesla Turbines)			
Case	*th* (mm)	*b* (mm)	*b/th* (-)
Case 1-0.3	1	0.3	0.3
Case 1-0.5	1	0.5	0.5
Case 1-1	1	1	1
Case 2-0.3	2	0.3	0.15
Case 2-0.5	2	0.5	0.25
Case 2-1	2	1	0.5
Group 2 (One-To-Many Multichannel Tesla Turbines)			
Case	*th* (mm)	*b* (mm)	*b/th* (-)
Case 1-0.3	1	0.3	0.3
Case 1-0.5	1	0.5	0.5
Case 1-1	1	1	1
Case 2-0.5	2	0.5	0.25
Case 2-1	2	1	0.5
Case 2-2	2	2	1

Each turbine case consists of five discs and six disc channels. The other geometrical parameters and aerodynamic parameters are all the same for the two groups of the Tesla turbine, which are the same as those in the authors' previous study [29] (Table 2).

To save computing time and resources, the calculation domain was reduced based on a rotational symmetry and a symmetry about a middle plane that is normal to the axis, which finally became a quarter of the whole turbine. It is shown in Figure 2 with red color.

Table 2. Geometrical and aerodynamic parameters.

Symbol	Unit	Value
N_n	(-)	2
$d_{o,d}$	(mm)	100
$d_{i,d}$	(mm)	38.4
c	(mm)	0.25
N_d	(-)	5
N_{dc}	(-)	6
α	(°)	10
p_{nt}/p_i	(-)	3.42
T_{nt}	(K)	373

The boundary conditions are given in Table 2. In addition, because of the calculation domain reduction, the symmetry and rotational periodicity boundary conditions were set up at corresponding surfaces. The adiabatic and no-slip boundary condition was adopted for all walls. The frozen rotor method was applied to deal with the rotor-stator interface.

2.2. Numerical Solver and Mesh Sensitivity

In this research, all numerical calculations were conducted using commercial software ANSYS CFX (ANSYS Inc., Canonsburg, PA, USA), in which the Reynolds Averaged Navier–Stokes (RANS) equations in the calculation domain were solved for turbulent flow. The RANS equation groups are as follows.

Continuity equation:

$$\frac{\partial \rho}{\partial t} + \frac{\partial}{\partial x_j}(\rho U_j) = 0 \tag{1}$$

Momentum equations:

$$\frac{\partial \rho U_i}{\partial t} + \frac{\partial}{\partial x_j}(\rho U_i U_j) = -\frac{\partial p}{\partial x_i} + \frac{\partial}{\partial x_j}(\tau_{ij} - \rho \overline{u_i u_j}) + S_M \tag{2}$$

Energy equations:

$$\frac{\partial(\rho h_{tot})}{\partial t} - \frac{\partial p}{\partial t} + \frac{\partial}{\partial x_j}(\rho U_j h_{tot}) = \frac{\partial}{\partial x_j}(\lambda \frac{\partial T}{\partial x_j} - \rho \overline{u_j h}) + \frac{\partial}{\partial x_j}[U_i(\tau_{ij} - \rho \overline{u_i u_j})] + S_E \tag{3}$$

in which x_j are coordinates in three directions, U_j are average velocity in three directions, p is the static pressure, T is the temperature, ρ is the density, and τ is the molecular stress tensor.

In Equations (2) and (3), S_M and S_E are source terms, $\rho \overline{u_i u_j}$ are the Reynolds stresses, $\rho \overline{u_j h}$ is an additional turbulence flux term, $\frac{\partial}{\partial x_j}[U_i(\tau_{ij} - \rho \overline{u_i u_j})]$ represents the viscous work term. h_{tot} is the mean total enthalpy, h is the static enthalpy, and $h_{tot} = h + U_i U_j/2 + k$, where k is the turbulent kinetic energy and $k = \overline{u_i^2}/2$.

In addition, the density can be solved according to the state equation for ideal air,

$$\rho = p_{abs}/RT \tag{4}$$

where p_{abs} is the absolute pressure and R is the specific gas constant of air.

According to the above discussion, it can be found that RANS equations are not closed due to two additional terms, which are Reynolds stress and Reynolds flux. Therefore, some turbulence models in ANSYS CFX are provided to calculate turbulent flow.

For Tesla turbines, the detailed experimental data cannot be found in public references, therefore the flow model verification could not be conducted. The critical Reynolds number for the plane Poiseuille flow is about 1000 [33], and is defined by the half disc spacing distance and the average velocity in Tesla turbines. In most Tesla turbines, the Reynolds number is always above the critical values, and the minimum Reynolds number was about 1500 in our computational cases. Therefore, the Shear Stress Transport (SST) turbulence model was adopted in this research, which is also used in numerical research by other researchers [11,25,26].

SST turbulence model is a kind of eddy viscosity model, and, according to eddy viscosity hypothesis, the Reynolds stresses are considered to be proportional to mean velocity gradients. They can be expressed as follows.

$$-\rho\overline{u_i u_j} = \mu_t\left(\frac{\partial U_i}{\partial x_j} + \frac{\partial U_j}{\partial x_i}\right) - \frac{2}{3}\delta_{ij}\left(\rho k + \mu_t\frac{\partial U_k}{\partial x_k}\right) \tag{5}$$

where μ_t is the eddy viscosity.

In addition, the Reynolds fluxes of a scalar is stated as a linear relationship to the mean scalar gradient based on the eddy diffusivity hypothesis:

$$-\rho\overline{u_i\phi} = \Gamma_t\frac{\partial \Phi}{\partial x_i} \tag{6}$$

where $\rho\overline{u_i\phi}$ is the Reynolds flux of a scalar Φ, Γ_t is the eddy diffusivity, written as $\Gamma_t = \mu_t/\mathrm{Pr}_t$, and Pr_t is the turbulent Prandtl number.

Equations (5) and (6) give expressions of turbulent fluctuations as functions of the mean variables when the turbulent viscosity is known. In ANSYS CFX solver, SST turbulence model, which derives from the $k - \varepsilon$ and $k - \omega$ turbulence models, uses this variable and expresses the turbulent viscosity as,

$$\mu_t = \rho\frac{k}{\omega} \tag{7}$$

where ω is turbulent frequency.

The Reynolds averaged transport equations for k and ω in SST turbulence model are as follows.

$$\frac{\partial(\rho\omega)}{\partial t} + \frac{\partial}{\partial x_j}(\rho U_j\omega) = \frac{\partial}{\partial x_j}\left[\left(\mu + \frac{\mu_t}{\sigma_{\omega3}}\right)\frac{\partial\omega}{\partial x_j}\right] + (1 - F1)2\rho\frac{1}{\sigma_{\omega2}\omega}\frac{\partial k}{\partial x_j}\frac{\partial\omega}{\partial x_j} + \alpha_3\frac{\omega}{k}P_k - \beta_3\rho\omega^2 + P_{kb} \tag{8}$$

$$\frac{\partial(\rho k)}{\partial t} + \frac{\partial}{\partial x_j}(\rho U_j k) = \frac{\partial}{\partial x_j}\left[\left(\mu + \frac{\mu_t}{\sigma_{k2}}\right)\frac{\partial k}{\partial x_j}\right] + P_k - \beta'\rho k\omega + P_{kb} \tag{9}$$

For the detailed coefficients in Equations (8) and (9), refer to Ref. [34]. Up to now, the RANS equations are closed, and can be solved to obtain the turbulent flow fields.

For all numerical simulations in this research, spatial discretization for the above flow governing equations was conducted with high-resolution second-order central difference scheme and time discretization was with second-order backward Euler scheme. In addition, an empirical automatic wall function was adopted to guarantee the solution accuracy. The accuracy of second order was obtained in all numerical simulations. In addition, the y+ should be less than 2 for all walls as the SST turbulence model requires.

In this study, a structured grid was generated for fluid domain, using ANSYS mesh generation software ICEM CFD (ANSYS Inc., Canonsburg, PA, USA). The mesh independence analysis for the one-to-many turbine Case 1-0.5 was conducted to ensure the accuracy of the results.

Three sets of mesh for this turbine were generated and the detailed information are given in Table 3. The N-R chamber stands for the chamber between the nozzle and the rotor.

Table 3. Mesh information.

Case No.	Stator (Nozzle/N-R Chamber)		Rotor (Each Disc Channel)	
	Number of Nodes (r,θ,z Directions)	Total Node Number	Number of Nodes (r,θ,z Directions)	Total Node Number
Case 1	$(55/13) \times (36/269) \times 99$	526,516	$65 \times 288 \times 23$	400,660
Case 2	$(67/17) \times (45/335) \times 107$	923,517	$81 \times 333 \times 29$	782,217
Case 3	$(87/21) \times (57/417) \times 135$	1,830,306	$102 \times 417 \times 37$	1,581,306

The aerodynamic performance parameters and their relative variation values (based on the results of Case 3) are given in Table 4. The isentropic efficiency η is defined as the actual shaft power P divided by the isentropic power across the whole turbine $m\Delta h_{\text{isen}}$, as follows:

$$\eta = \frac{P}{m\Delta h_{\text{isen}}} = \frac{M\Omega}{mc_p T_{\text{nt}}\left[1 - (p_i/p_{\text{nt}})^{(\gamma-1)/\gamma}\right]} \tag{10}$$

where P equals the whole torque M multiplied by the angular speed of the rotor Ω. c_p and γ are the physical properties of the working fluid, which are specific heat at constant pressure and the specific heat ratio. The subscript "nt" represent the total parameters at the turbine inlet, while "i" indicates the static parameter at the turbine outlet.

Figure 3 presents the Mach number contours on the mid-section of DC1 ("DC" is short for "disc channel") for the one-to-many turbine Case 1-0.5 with three different node cases. The detailed locations of the three DCs are indicated in Figure 2. Note that DC1 and DC2 each consists of two rotating disc walls, and DC3 has one rotating disc wall and one motionless casing wall.

Table 4 indicates that the variation value of each aerodynamic performance decreases with node number. Figure 3 shows that the Mach number distribution for each case is similar, but the difference between Cases 2 and 3 is much less than that between Cases 1 and 2. To sum up, the results of Case 2 fulfill the requirements for mesh independence, therefore the mesh of Case 2 was applied for the this turbine in numerical simulations. Moreover, for other turbine cases with different geometry, the grid distribution changes accordingly to get a reliable result.

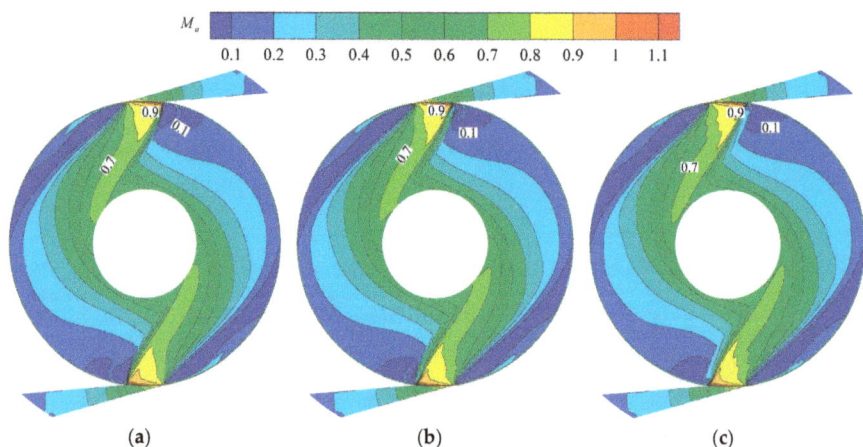

Figure 3. Mach number contours on the mid-sections of DC1 for the one-to-many turbine Case 1-0.5 with three different node cases: (a) Case 1; (b) Case 2; and (c) Case 3.

Table 4. Mesh independence.

Case No.	Node Number (million)	m (kg/s)	δm(%)	P (kW)	δP (%)	η (-)	$\delta \eta$ (%)
Case 1	1.72	0.03576	0.619	0.5960	1.568	0.1504	0.940
Case 2	3.27	0.03562	0.225	0.5886	0.307	0.1491	0.067
Case 3	6.57	0.03554	0	0.5868	0	0.1490	0

3. Results and Discussions

3.1. Total Aerodynamic Performance of Two Kinds of Multichannel Tesla Turbines

Two kinds of multichannel Tesla turbines with disc thickness of 1 mm and 2 mm were calculated numerically, respectively. It is obvious that, if the disc thickness is too large, the air flowing into disc channels will be more difficult, therefore, the aerodynamic performance becomes worse. However, the disc thickness should not be too small, which is mainly confined to its mechanical stress and processing problem. As disc thickness decreases, the stiffness of the discs decreases rapidly, and accordingly the mechanical stress increases, which should be lower than material allowable stress. Previous study on the computational solid mechanical (CSM) analysis indicates that the disc thickness should not be less than 1 mm for the Tesla turbine with disc outer diameter of 100 mm.

3.1.1. Isentropic Efficiency

Figure 4 shows the variation curves of the isentropic efficiency for each turbine case with the rotational speed for the two kinds of multichannel Tesla turbines. In Figure 4, note that the negative isentropic efficiency has no physical meaning and only indicates that under this condition the Tesla turbine consumes external energy without power outputting. It can be seen that the isentropic efficiency of the one-to-one turbine is much higher than that of the one-to-many Tesla turbine, and it changes much faster with rotational speed.

For the one-to-one turbine, the isentropic efficiency of the turbine cases with same disc spacing distance but different disc thickness is almost equal, although the turbine cases with smaller disc thickness are slightly more efficient, which is more obvious at higher rotational speeds. For example, the isentropic efficiency of Case 2-0.5 has a less than 7% reduction of that of Case 1-0.5 at all rotational speeds lower than 45,000 r/min. The turbine case with disc spacing distance of 0.5 mm obtains the highest efficiency at lower rotational speeds. That is, an optimal disc spacing distance for the one-to-one multichannel Tesla turbine exists to obtain its best performance, which is in accordance with that for one channel Tesla turbines [28]. In fact, a high efficiency range of disc spacing distance exists, among which the isentropic efficiency can remain at a high level. However, the turbine with disc spacing distance of 1 mm performs best at higher rotational speeds, because the isentropic efficiency changes more rapidly with rotational speed for the turbine case with narrower disc spacing distance. It can be predicted that an optimal disc spacing distance must exist at higher rotational speeds, which must be higher than 0.5 mm.

For the one-to-many turbine, obviously, the isentropic efficiency of the turbine cases with greater disc thickness is much lower than that of the cases with smaller disc thickness, which agrees well with the experimental results in Ref. [32]. In detail, the turbine isentropic efficiency of Case 2-0.5 is about 55% of that of Case 1-0.5. Meanwhile, the one-to-many turbine also has an optimal disc spacing distance.

To sum up, for designing multichannel Tesla turbines, a one-to-one Tesla turbine is recommended. Moreover, the disc spacing distance should be in its high efficiency range, and the disc thickness should be as small as possible while satisfying the requirement of allowable material stress. Usually, the disc thickness can be determined by the results of CSM, and it should be larger than 1 mm for discs of 100 mm diameter.

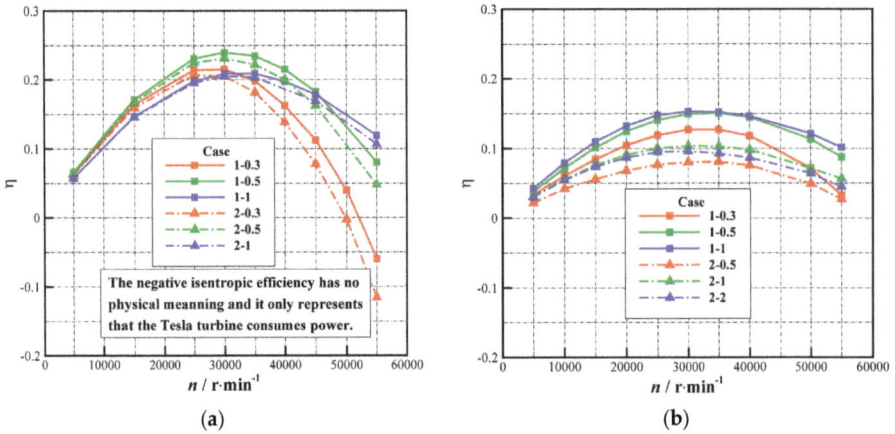

Figure 4. Isentropic efficiency versus rotational speed of each turbine case for two kinds of multichannel Tesla turbines: (**a**) one-to-one Tesla turbine; and (**b**) one-to-many Tesla turbine.

3.1.2. Flow Coefficient

Figure 5 presents the curves of the flow coefficient for each turbine case versus the rotational speed. The flow coefficient C_m is defined as:

$$C_m = \frac{v_{o,d}}{\Omega r_{o,d}} = \frac{m}{2\pi \rho_{o,d} \Omega b r_{o,d}^2} \tag{11}$$

where $v_{o,d}$ and $\omega r_{o,d}$ represent the average radial velocity and the disc rotational linear velocity at the rotor inlet, respectively.

The flow coefficient of the one-to-one Tesla turbine is much less than that of the one-to-many turbine with same disc spacing distance and disc thickness (Figure 5). In detail, for turbine Case 1-0.5, that of the one-to-many is nearly double that of the one-to-one turbine. Compared to the one-to-one turbine, when the operating parameters and other geometrical parameters are given, the mass flow rate of the one-to-many turbine is much higher due to much larger nozzle throat area (for Case 1-0.5, the one-to-many turbine is about 2.6 times); the density at the rotor inlet changes much less with the multichannel Tesla turbine type (for Case 1-0.5, the one-to-many turbine is about 1.3 times) than the mass flow rate; therefore, the flow coefficient of the one-to-many turbine is much higher.

For the one-to-one turbine, the flow coefficient of the turbine cases with same disc spacing distance is almost the same (the relative variation between Cases 1-0.5 and 2-0.5 is less than 2%), although it has a little bit increase with disc thickness, especially at higher rotational speeds. Meanwhile, this coefficient of the turbine cases with disc spacing distance of 0.3 mm and 0.5 mm is almost the same, and clearly higher than that of 1 mm, which is because the air density at the rotor inlet goes up with disc spacing distance.

For the one-to-many turbine, the variation relationships of the flow coefficient with disc thickness and spacing distance become complicated, and these two geometrical parameters affect significantly on this coefficient. As shown in Figure 2, the axial length ratio of nozzle channel to disc channels of the one-to-many Tesla turbine is larger than 1 and it goes up with increasing disc thickness or decreasing disc spacing distance, thus the flow coefficient increases as disc thickness goes up and disc spacing distance goes down, respectively. Same with the one-to-one turbine, the air density at the rotor inlet mainly depends on disc spacing distance, and increases with it, which also makes the coefficient decrease as disc spacing distance increases.

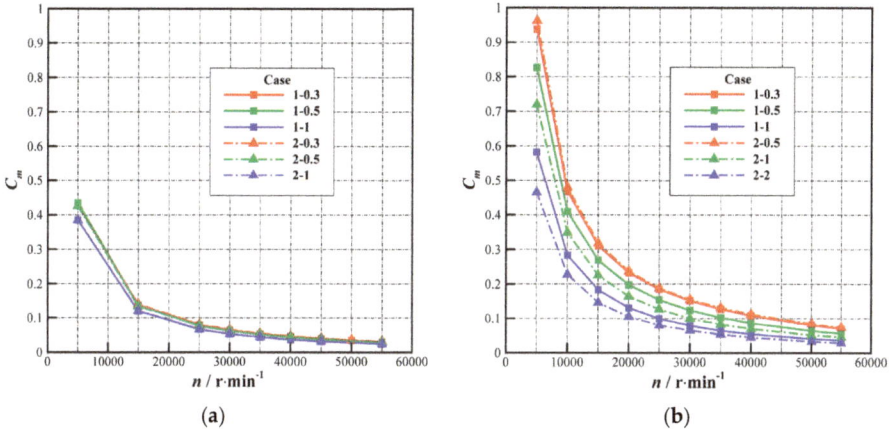

Figure 5. Flow coefficient versus rotational speed of each case for two kinds of multichannel Tesla turbines: (**a**) one-to-one Tesla turbine; and (**b**) one-to-many Tesla turbine.

3.1.3. Percentages of Mass Flow Rate and Torque in Disc Channels

Figure 6 shows the mass flow rate and torque percentages in three DCs for all the turbine cases at 30,000 r/min. In Figure 6, both percentages in DCs 1 and 2 are nearly equal for both kinds of turbines, which indicates that the flow fields in DCs 1 and 2 are similar. In addition, DCs 1 and 2 have more air than DC3, while generating more torque.

For the one-to-one turbine, as disc thickness changes, both percentages in each disc channel are almost the same for turbine cases with same disc spacing distance, respectively. The torque percentage of DC3 for the turbine cases with 0.5 mm in disc spacing distance are almost the same, about 17%, while that for turbine with 0.3 mm in disc spacing distance decreases to 12%. All these flow phenomena are explained in the next sections.

For the one-to-many turbine, the variation relationship of mass flow rate percentage with disc spacing distance is not coincident. For the turbine cases with disc thickness of 1 mm, those in DCs 1 and 2 decrease as disc spacing distance increases, while for the turbine cases with disc thickness of 2 mm, they decrease first and then increase. In detail, those of Cases 1-1 and 2-1 reach the minimum values respectively. Moreover, disc thickness and spacing distance have few effects on the torque percentage in each disc channel.

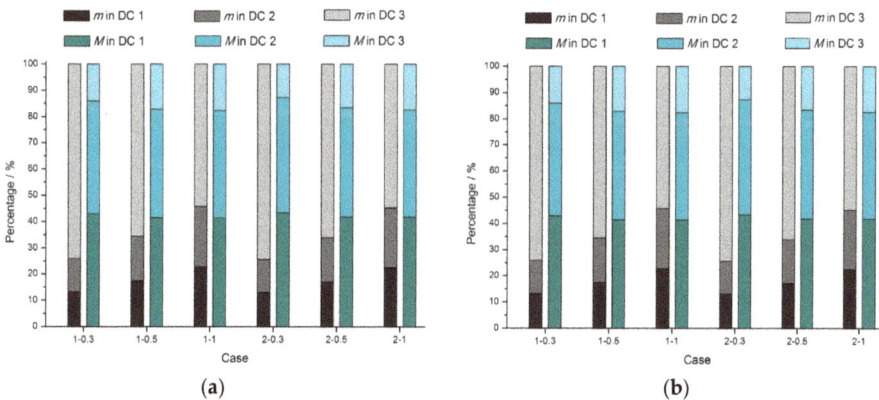

Figure 6. Mass flow rate and torque percentages in three DCs for two kinds of Tesla turbines at 30,000 r/min: (**a**) one-to-one Tesla turbine; and (**b**) one-to-many Tesla turbine.

To be noted, the calculation domain has five disc rotating walls and the average torque percentage is 20%. However, for all the turbine cases, the torque percentage in DC3 is less than 20%. Meanwhile, DC3 flows more working fluid than DCs 1 and 2. That means DC3 uses much more working fluid and generates less torque and power. Thus, the disc spacing distance of DC3 should be smaller to get less working fluid. To sum up, disc thickness slightly influences the aerodynamic performance of the one-to-one turbine. The isentropic efficiency of the turbine cases with disc thickness of 2 mm is a little lower than that of 1 mm. Specifically, for turbine cases with 0.5 mm in disc spacing distance, it decreases by only 7% at most rotational speeds. Moreover, the disc spacing distance affects greatly the aerodynamic performance of the one-to-one turbine, and an optimal disc spacing distance exists to make the turbine obtain the highest isentropic efficiency.

Both the disc thickness and spacing distance have significant impacts on the aerodynamic performance of the one-to-many turbine. The one-to-many turbine also has an optimal disc spacing distance, which is larger than that for the one-to-one turbine (1 mm and 0.5 mm, respectively, for turbine with disc thickness of 1 mm). In addition, the performance becomes remarkably worse with an increase in disc thickness (a reduction of about 45%).

3.2. Flow Status of One-To-One Multichannel Tesla Turbines

From the above discussion, the isentropic efficiency of the one-to-one Tesla turbine drops a little as disc thickness increases. It increases first and then decreases with increasing disc spacing distance, and reaches its maximum value for the turbine case with disc spacing distance of 0.5 mm at lower rotational speeds including the optimal rotational speed of 30,000 r/min. To reveal the reasons that the disc thickness and spacing distance affect the turbine performance, the flow characteristics in the Tesla turbine were analyzed.

3.2.1. One-To-One Turbine with Different Disc Thickness

Figures 7 and 8 show Mach number contours and streamlines and on the mid-sections of three DCs for the one-to-one turbine Cases 1-0.5 and 2-0.5 at 30,000 r/min, respectively. The streamlines and Mach number contours of DCs 1 and 2 are almost the same, but different from that of DC3, which agrees well with the above analysis. Therefore, in the next analysis, only the flow fields of DCs 1 and 3 are given.

Obviously, in flow channels 1 and 2, part of the air that just injects nearly tangentially into the disc channels flows into the N-R chamber due to centrifugal force, and finally flows into DC3. In flow channel 3, the air from the nozzle together with that from the N-R chamber flows into DC3 at a larger flow angle (relative to the tangential direction). Thus, it is easy to know that for the one-to-one turbine less air flows through DCs 1 and 2 than that through DC3. In addition, it can be observed that the Mach number on the mid-sections of DCs 1 and 2 is much higher than that of DC3, and the flow angle is much less than that in DC3, which leads to much higher relative tangential velocity and longer path lines in DCs 1 and 2. Therefore, the torque percentage in DC3 is lower than its average value of 20%.

Comparing Figures 7 and 8, the flow fields in each disc channel of the turbine cases with same disc spacing distance but different disc thickness have little difference. This indicates that their aerodynamic performance also has little difference.

Figure 9 illustrates the contours of relative tangential velocity and vector distributions on the radial cross-section of 0°, whose location is marked in Figure 1. In this figure, only a small part of the disc channels is presented and the axial size is doubled to show the flow field clearly. The reference vector is given near the text of "rotor".

In the disc channels, the contour represents the relative tangential velocity, and the vector indicates the radial and axial velocities. In the stator, the contour shows the velocity normal to this section, and the vector describes the axial velocity and another velocity on this section. In the disc channels, the positive relative tangential velocity indicates that the air at this region generates usable torque; otherwise, it consumes external mechanical energy, most of which appears near the casing wall.

Figure 9 indicates that on this section the axial velocity is close to zero, and the radial velocity is much lower than the relative tangential velocity. In addition, the air with high velocity flows from the nozzle through the N-R chamber and finally into the disc channels rapidly with little flow deflection to the axial direction. Some vortex generates at some region of the N-R chamber.

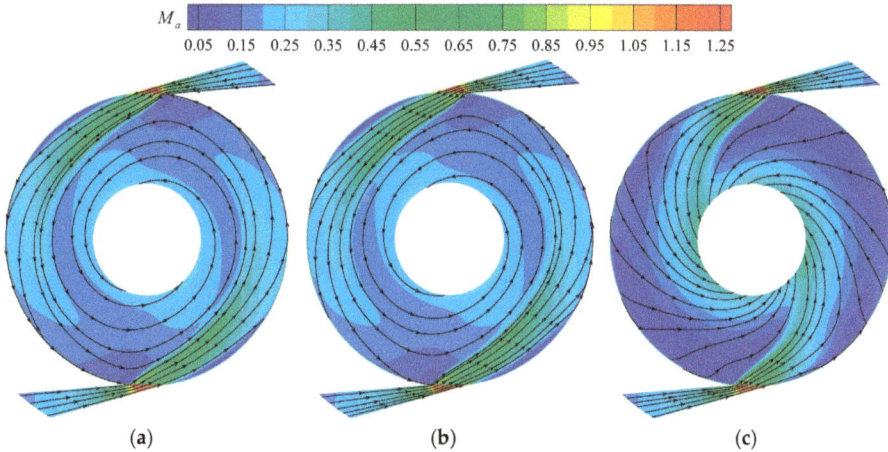

Figure 7. Mach number contours and streamlines on the mid-sections of three DCs for the one-to-one turbine Case 1-0.5 at 30,000 r/min: (**a**) DC1; (**b**) DC2; and (**c**) DC3.

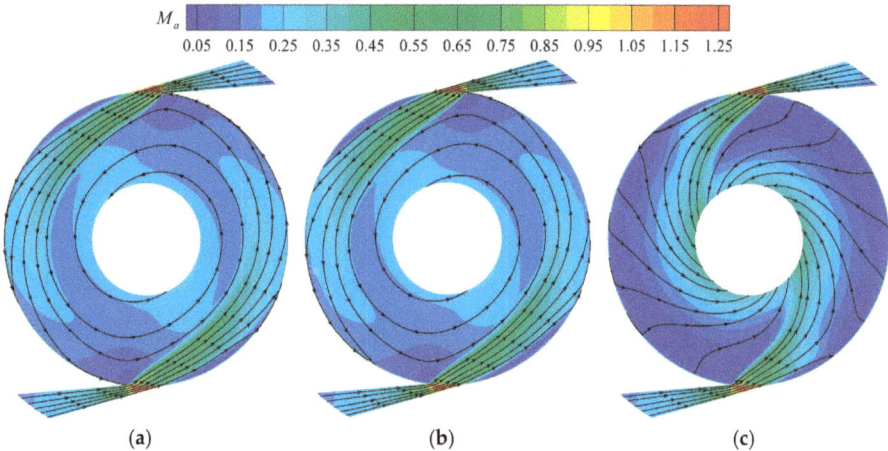

Figure 8. Mach number contours and streamlines on the mid-sections of three DCs for the one-to-one turbine Case 2-0.5 at 30,000 r/min: (**a**) DC1; (**b**) DC2; and (**c**) DC3.

The flow fields for turbine Cases 1-0.5 and 2-0.5 are almost the same in most regions on the radial cross-section. However, the flow velocity of the working fluid in the N-R chamber for turbine Case 2-0.5 is much lower than that for Case 1-0.5, which consumes more extra energy to move them. Moreover, it can be observed from the relative tangential velocity contours that the tangential velocity gradient near the rotating disc walls for Case 2-0.5 is a little less than that for Case 1-0.5, which leads to lower torque. All these factors lead to the isentropic efficiency of Case 2-0.5 a little lower than that of 1-0.5 (23.92% and 23.04%, respectively).

To sum up, the flow field and aerodynamic performance of the one-to-one Tesla turbine changes a little with disc thickness. In detail, the isentropic efficiency of the turbine with the thicker disc is little lower, which results from lower velocity in the N-R chamber and a little lower tangential velocity gradient close to the disc walls.

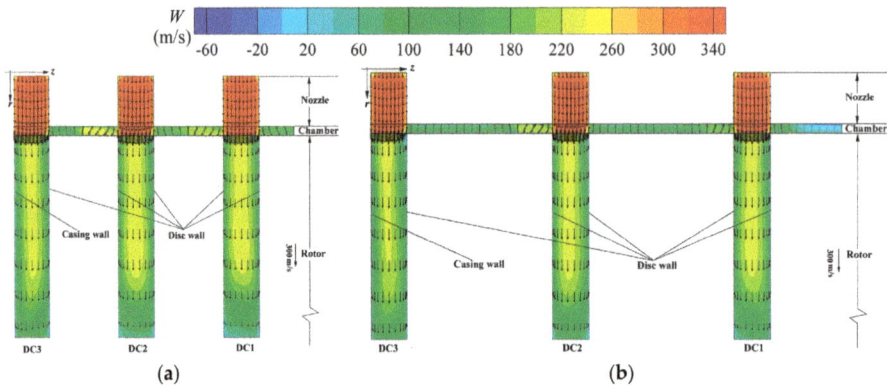

Figure 9. Contours of relative tangential velocity and vector distributions on radial cross-sections of 0° for the one-to-one Tesla turbine cases at 30,000 r/min: (**a**) Case 1-0.5; and (**b**) Case 2-0.5.

3.2.2. One-To-One Turbine with Different Disc Spacing Distance

To reveal the detailed flow characteristics in the disc channels of the one-to-one turbine with different disc spacing distance, Figure 10 shows the variation curves of circumferential mass flow average velocity, including relative tangential velocity and radial velocity, versus the radius ratio (local radius to rotor inlet radius) for different one-to-one turbine cases. In this figure, the abscissa values of 1 and 0.384 represent the rotor inlet and outlet, respectively.

As shown in Figure 10a, for all turbine cases, the average relative tangential velocity in DC1 obtains the highest value at the rotor inlet, and decreases with radius ratio due to the contribution to the output power. For turbine Cases 1-0.5 and 1-1, it then increases as radius ratio continues to decrease (the air flows towards the turbine outlet), and the stationary point of velocity occurs much earlier with increasing disc spacing distance. In addition, the average relative tangential velocity decreases at the region near the rotor inlet and increases at the region near the rotor outlet with increasing disc spacing distance. Similarly, the average relative tangential velocity in DC3 for all turbine cases decreases first and then increases as the air flows towards the turbine outlet. However, it is much higher for the turbine cases with larger disc spacing distance. In general, the average relative tangential velocity difference between DCs 1 and 3 decreases with increasing disc spacing distance.

As shown in Figure 10b, for all turbine cases, the average radial velocity in DC1 decreases with radius ratio, which seems to violate the continuity equation of working fluid. The flow area (cylindrical surface of the disc channel) decreases as the air flows through the disc channels to the rotor outlet (radius ratio decreases), while the density also decreases, thus the radial velocity should increase based on the continuity equation. However, it is completely opposite to the results in Figure 10, which is due to the partial admission of the nozzled Tesla turbine. These discrete nozzles are installed symmetrically at the disc tip, and most air is injected into DCs 1 and 2 by the nozzles. Although the flow area decreases as radius ratio decreases, the effective flow area increases due to the air diffusion when the air flows from the rotor inlet to outlet, which leads to a decrease in average radial velocity in DC1.

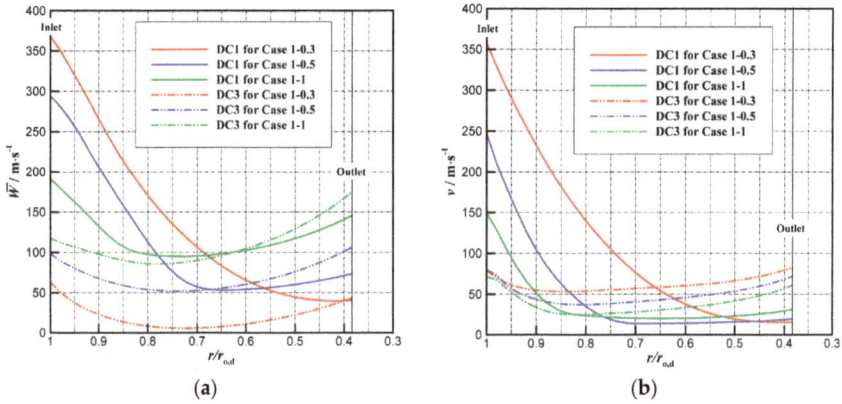

Figure 10. Variation curves of circumferential mass flow average velocity versus the radius ratio for three one-to-one turbine cases: (**a**) relative tangential velocity; and (**b**) radial velocity.

In addition, the average radial velocity in DC3 decreases first and then increases; however, it has much less variation than that in DC1. Some air from the N-R chamber flows into DC3, and the effective flow area of DC3 is much larger than that of DC1. However, the flow velocity at the rotor inlet from the N-R chamber is much less, and the average radial velocity still decreases slightly due to air diffusion. As the air flows towards the rotor outlet, the average radial velocity has a slight increase due to the pressure drop the rotor and the decrease in effective flow area. Moreover, the average radial velocity in DCs 1 and 3 both decreases with increasing disc spacing distance, and the average radial velocity at the rotor inlet in DC1 is much higher than that in DC3.

In summary, the relative tangential velocity, determining the torque and power, decreases with radius ratio due to outputting power, and then increases slightly, resulting from the decrease in disc rotational linear velocity and the pressure in the drop. Furthermore, its difference between DCs 1 and 3 decreases with increasing disc spacing distance. Moreover, the radial velocity, depending on the mass flow rate, has a strong relationship with the effective flow area, and it decreases with increasing disc spacing distance.

Figure 11 shows Mach number contours and streamlines and on the mid-sections of DCs 1 and 3 for the one-to-one turbine cases with different disc spacing distance at 30,000 r/min.

As shown in Figure 11, the Mach number at the nozzle outlet decreases as disc spacing distance increases because of the higher pressure drop in the disc channels of the turbine cases with wider disc spacing. With increasing flow velocity at the nozzle outlet, the flow angle at the rotor inlet should decrease if the nozzle outlet flow angle of different turbine cases is the same. However, it is totally opposite to the flow fields in Figure 11, which is because the nozzle outlet flow angle changes with disc spacing distance despite their same nozzle outlet geometrical angles. The air in the scarfed part of the nozzle tends to speed up and changes its direction to the side without the nozzle wall, when the pressure ratio of the nozzle (nozzle outlet pressure to nozzle inlet pressure) is under the critical pressure ratio of air. The deflection angle goes up with an increase in difference value between the nozzle pressure ratio and the critical pressure ratio, as shown in turbine Cases 1-0.3 and 1-0.5 in Figure 11. When the nozzle pressure ratio is over the critical pressure ratio, the flow will not deflect in the scarfed part and the flow angle at the rotor inlet is almost equal to the nozzle outlet geometrical angle, shown in turbine Case 1-1.

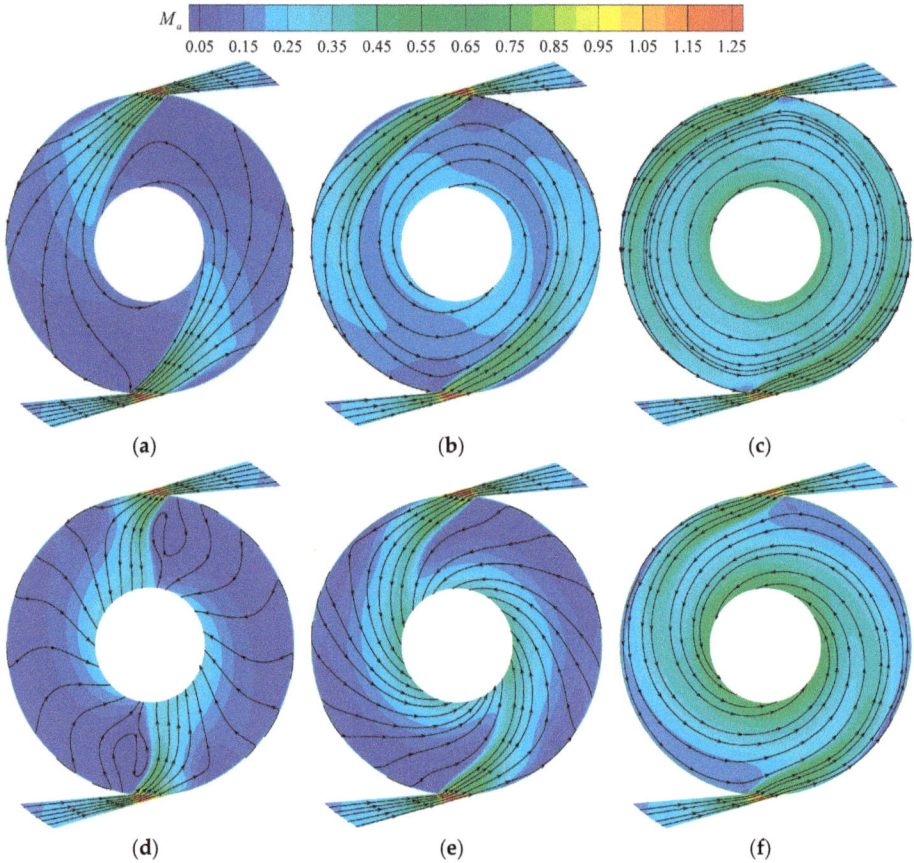

Figure 11. Mach number contours and streamlines on the mid-sections of DCs 1 and 3 for the one-to-one turbine cases with different disc spacing distance at 30,000 r/min: (**a**) DC1 for Case 1-0.3; (**b**) DC1 for Case 1-0.5; (**c**) DC1 for Case 1-1; (**d**) DC3 for Case 1-0.3; (**e**) DC3 for Case 1-0.5; and (**f**) DC3 for Case 1-1.

Comparing the flow fields in DCs 1 and 3, it can be found that more air flows into DC3 from both the nozzle and the N-R chamber, and the velocity of the working fluid from the N-R chamber is much lower. Therefore, the average relative tangential velocity and radial velocity at the rotor inlet of DC3 are under those of DC1. Apart from DC1 for turbine Case 1-0.3, the flow velocity in other disc channels decreases first and then increases as the radius decreases. Moreover, the flow field difference between DCs 1 and 3 decreases with increasing disc spacing distance. All the above flow phenomena are the same as those in Figure 10.

Based on the above discussion, the isentropic efficiency of the one-to-one turbine goes up first and then down with increasing disc spacing distance, that is to say, an optimal disc spacing distance exists to allow a Tesla turbine to obtain the highest isentropic efficiency. To explain this phenomenon, Table 5 gives the coefficients of component energy loss for the one-to-one turbine cases at 30,000 r/min, including the nozzle loss, the disc loss and the leaving-velocity loss, which are defined as each component energy loss divided by the isentropic enthalpy drop across the whole turbine.

Table 5. Coefficients of component energy loss for one-to-one Tesla turbine cases.

Case Name	Nozzle Loss Coefficient	Disc Loss Coefficient	Leaving-Velocity Loss Coefficient	Isentropic Efficiency
Case 1-0.3	0.2857	0.4196	0.0760	0.2187
Case 1-0.5	0.1849	0.4405	0.1303	0.2443
Case 1-1	0.1085	0.4358	0.2440	0.2117

With increasing disc spacing distance, the nozzle loss coefficient goes down and the leaving velocity loss coefficient goes up remarkably. The disc loss coefficient increases first and then decreases, and its variation values are relatively lower than those of the nozzle loss and the leaving-velocity loss.

The variation rules of the component energy loss for the one-to-one turbine with disc spacing distance are analyzed in detail based on the flow fields. In the multichannel Tesla turbine, the nozzle energy loss includes the loss resulting from the viscous friction (called "friction loss in the nozzle"), and the loss occurring in the nozzle and the N-R chamber caused by the sudden variation of flow area, when the air flows through the nozzle and the N-R chamber and finally to the disc channels (called "local loss in the nozzle"). For the one-to-one turbine, the friction loss in the nozzle decreases with increasing disc spacing distance, caused by a decrease in proportion of the boundary layer in the nozzle channel and a decrease in flow velocity (Figure 11). In addition, most air flows out of the nozzle to the disc channels quickly (Figure 9), and the N-R chamber has little influence on the flow field. Thus, the local loss in the nozzle slightly influences the nozzle loss. In detail, the flow area ratio of the N-R chamber to the disc channels decreases with increasing disc spacing distance, therefore, the local loss for the one-to-one turbine decreases. As a result, the nozzle loss, including the friction loss and the local loss, decreases remarkably with increasing disc spacing distance.

With an increase in disc spacing distance, more air outside the boundary layer has no use in momentum exchange, which indicates more energy loss in the rotor. This is in agreement with the result in one channel Tesla turbines [28]. The local loss occurring in the disc channels due to the flow area variation has a decrease with increasing disc spacing distance, although the local loss also has little influence on the disc loss. Therefore, the combined actions of the two factors lead to increasing first and then decreasing disc loss as disc spacing distance increases. The Mach number at the turbine outlet rises with disc spacing distance, resulting in increasing leaving-velocity loss coefficient, as shown in Figure 11. These results are the same as those in Table 5.

3.3. Flow Status of One-To-Many Multichannel Tesla Turbines

As discussed in the above section, the isentropic efficiency of the one-to-many turbine goes up first and then down with increasing disc spacing distance. It decreases remarkably as disc thickness increases. Table 6 gives the coefficients of component energy loss for the one-to-many turbine cases at 30,000 r/min. With increasing disc spacing distance, the nozzle loss coefficient and disc loss coefficient decrease, and the leaving-velocity loss coefficient rises. With increasing disc thickness, the nozzle loss has a slight increase, the disc loss decreases, and the leaving-velocity loss increases remarkably. The variation rules of these coefficients are explained in the following text.

Table 6. Coefficients of component energy loss for one-to-many Tesla turbine cases.

Case Name	Nozzle Loss Coefficient	Disc Loss Coefficient	Leaving-Velocity Loss Coefficient	Isentropic Efficiency
Case 1-0.3	0.0724	0.5539	0.2387	0.1350
Case 1-0.5	0.0578	0.4574	0.3310	0.1538
Case 1-1	0.0417	0.3936	0.4055	0.1592
Case 2-0.5	0.0699	0.4324	0.4040	0.0937
Case 2-1	0.0471	0.3504	0.4925	0.1100
Case 2-2	0.0375	0.3083	0.5561	0.0981

3.3.1. One-To-Many Turbine with Different Disc Thickness

To explain the variation rules of the aerodynamic performance of the one-to-many turbine with disc thickness, Figure 12 presents Mach number contours and streamlines on the mid-sections of DCs 1 and 3 for turbine Cases 1-0.5 and 2-0.5.

The Mach number at the nozzle outlet for Case 1-0.5 is slightly higher than that for Case 2-0.5. Therefore, the flow angle at the rotor inlet is a little lower. Moreover, the flow angle in most regions of the disc channels for Case 1-0.5 is lower than that for Case 2-0.5, leading to longer path lines and more momentum exchange. Different from the one-to-one turbine, Mach number increases suddenly at the rotor inlet directly facing the nozzle outlet, which is due to a sudden increase in flow area when the air passes the N-R chamber to the disc channels. The flow velocity at that place increases more significantly for Case 2-0.5 than that for Case 1-0.5.

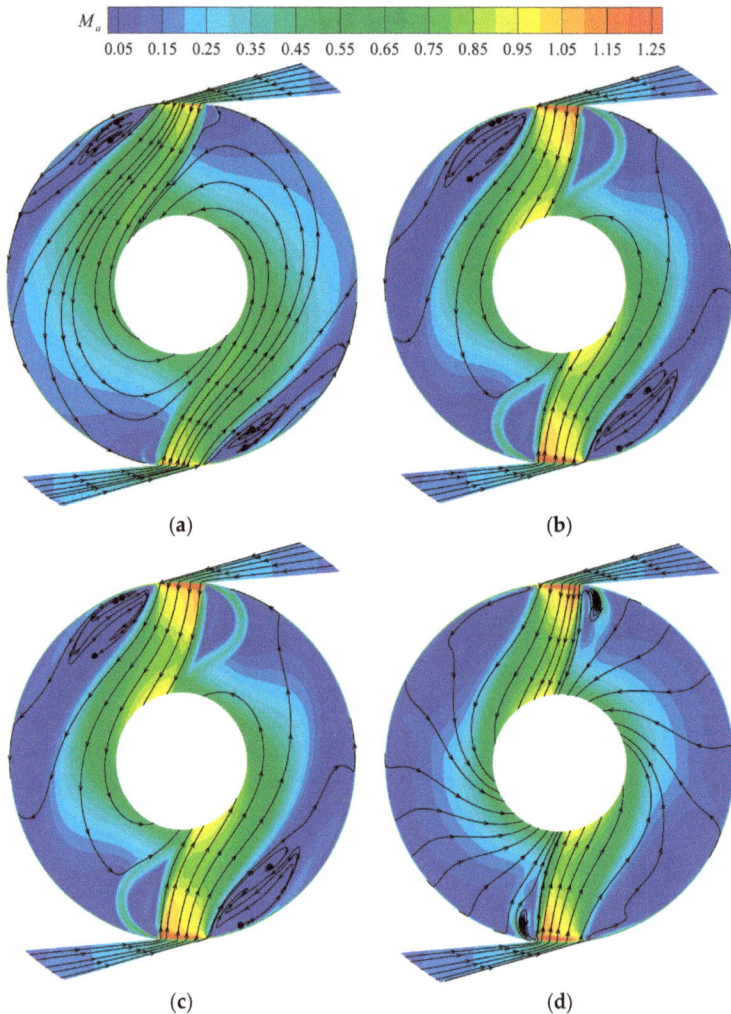

Figure 12. Mach number contours and streamlines on the mid-sections of DCs 1 and 3 for the one-to-many turbine cases with different disc thickness at 30,000 r/min: (**a**) DC1 for Case 1-0.5; (**b**) DC1 for Case 2-0.5; (**c**) DC3 for Case 1-0.5; and (**d**) DC3 for Case 2-0.5.

Figure 13 presents the contours of the pressure ratio on the mid-sections of DC1 for the one-to-many turbine cases, and the pressure ratio equals the ratio of the local pressure to the turbine inlet total pressure. The pressure ratio at the nozzle outlet for Case 1-0.5 is lower than that for Case 2-0.5, due to lower flow area ratio of the nozzle channel to the disc channels. It indicates a higher pressure drop in the nozzle and higher flow velocity at the nozzle outlet for Case 1-0.5 (Figures 12 and 13). A sudden decrease in pressure ratio also occurs at the same region of the sudden increase in Mach number. The pressure ratio decreases more remarkably for Case 2-0.5 than that for Case 1-0.5, because of much more decrease in flow area when the air flows through the N-R chamber to the disc channels.

The following is the detailed analysis of the component energy loss for the one-to-many turbine cases with different disc thickness based on the flow fields. With increasing disc thickness, the proportion of the boundary layer in the nozzle channel decreases and the flow velocity also decreases, which leads to lower friction loss in the nozzle. Actually, compared with the one-to-one turbine, the nozzle friction loss of the one-to-many turbine is much lower due to much less proportion of the boundary layer in the nozzle channel. It should be quite little and has a little influence on the nozzle loss. The local loss in the nozzle for the turbine case with thicker discs is much higher than that with thinner discs because the flow area reduces more when the air flows out of the N-R chamber to the disc channels They lead to a little increase in nozzle loss coefficient with increasing disc thickness (see Table 6).

As disc thickness increases, the working fluid becomes much more due to the larger nozzle throat area, and the flow velocity in the disc channels becomes much higher (see Figure 12). This leads to less boundary layer thickness, leading to less energy loss in the disc channels. The local loss at the inlet of the disc channels goes up with disc thickness, due to the higher flow area ratio of the N-R chamber to the disc channels. As a result, the disc loss decreases with increasing disc thickness (Table 6). The Mach number at the turbine outlet is much higher for Case 2-0.5 than that for Case 1-0.5 (Figure 12), which leads to higher leaving-velocity loss.

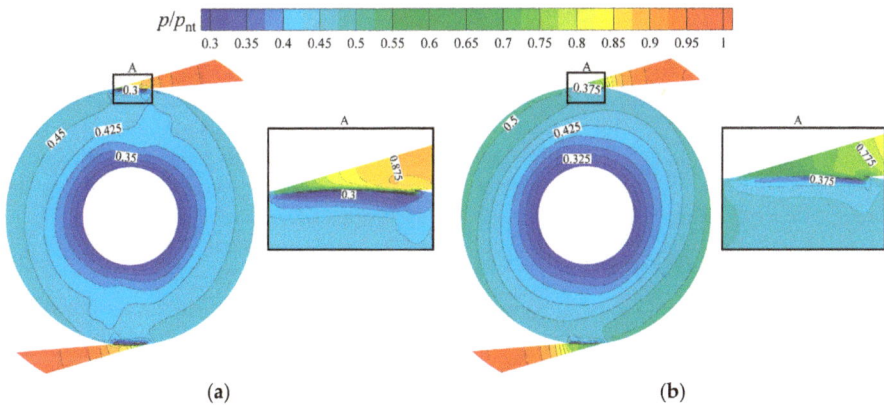

Figure 13. Contours of pressure ratio on the mid-sections of DC1 for the one-to-many Tesla turbine cases with different disc thickness at 30,000 r/min: (**a**) Case 1-0.5; and (**b**) Case 2-0.5.

Figure 14 gives contours of relative tangential velocity and vector distributions on the radial cross-sections of 0° for the one-to-many Tesla turbine cases. Unlike the one-to-one turbine, the air at the nozzle outlet in the one-to-many turbine bends to the axial direction to flow into the disc channels, especially at the nozzle of which downstream is a wall. The vortex occurs on the downstream wall of the N-R chamber and the inlet of the disc channels, leading to the energy loss at those places. The energy loss on the downstream wall of the N-R chamber wall is the main source of the local loss in the nozzle. That at the inlet of the disc channels is a part of the disc loss, which also includes the friction loss.

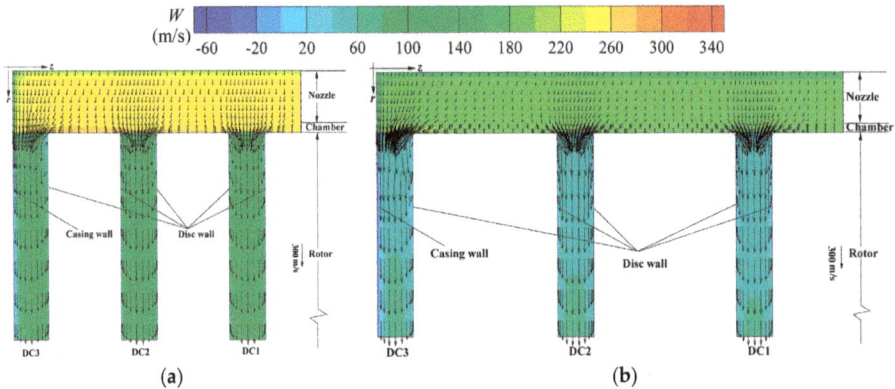

Figure 14. Contours of relative tangential velocity and vector distributions on radial cross-sections of 0° for the one-to-many Tesla turbine cases with different disc thickness: (**a**) Case 1-0.5; and (**b**) Case 2-0.5.

Compared with turbine Case 1-0.5, the flow velocity at the nozzle outlet for turbine Case 2-0.5 is lower, thus leading to a larger flow angle at the rotor inlet. The radial velocity at the rotor inlet is higher and the relative tangential velocity is lower. In addition, the vortex on the downstream wall of the N-R chamber and at the inlet of the disc channels for turbine Case 2-0.5 is much larger than that for turbine Case 1-0.5, therefore the energy loss caused by the variation of the flow area for Case 2-0.5 is much higher.

Combining Figures 12 and 14, the relationship of the mass flow rate percentages in each disc channel of the one-to-many turbine with disc thickness is analyzed. Compared with Case 1-0.5, the flow angle in DC1 for Case 2-0.5 is much larger, causing more air to flow through DC1 for Case 2-0.5. The mass flow rate of the air flowing into each disc channel of the one-to-many turbine has a significant difference from that of the one-to-one turbine. Figure 14 shows that the nozzle outlet area of the air flowing into DCs 1 and 2 is the same, and equals to the corresponding area of one disc spacing distance and one disc thickness. It is larger than that flowing into DC3, which equals to the corresponding area of one disc spacing distance and half of disc thickness. The area ratio of the working fluid flowing into DC1 to DC3 increases with disc thickness, leading to increasing mass flow rate percentage in DC1. Based on the two factors, the mass flow rate percentage in DC1 for Case 2-0.5 is higher than that for Case 1-0.5, which agrees with the results in Figure 6.

3.3.2. One-To-Many Turbine with Different Disc Spacing Distance

Mach number contours and streamlines on the mid-sections of DCs 1 and 3 for the one-to-many turbine cases with different disc spacing distance are presented in Figure 15. With increasing disc spacing distance, the Mach number increases at the nozzle outlet, which leads to lower flow angle at the rotor inlet. Thus, the turbine cases with larger disc spacing have longer path lines. Moreover, the sudden increase in Mach number at the rotor inlet decreases significantly with increasing disc spacing distance.

As discussed above, some air in DC1 passes through the N-R chamber and finally flows into DC3, which leads to more air in DC3 than that in DC1. For the turbine cases with disc thickness of 1 mm, with increasing disc spacing distance, the working fluid in DC1 flowing into the N-R chamber becomes much less due to less flow angle; with increasing disc spacing distance, the nozzle outlet area ratio of the air flowing into DC1 to that into DC3 decreases, causing less air in DC1. Under the combined actions of the two factors, the air in DC1 becomes less with increasing disc spacing distance.

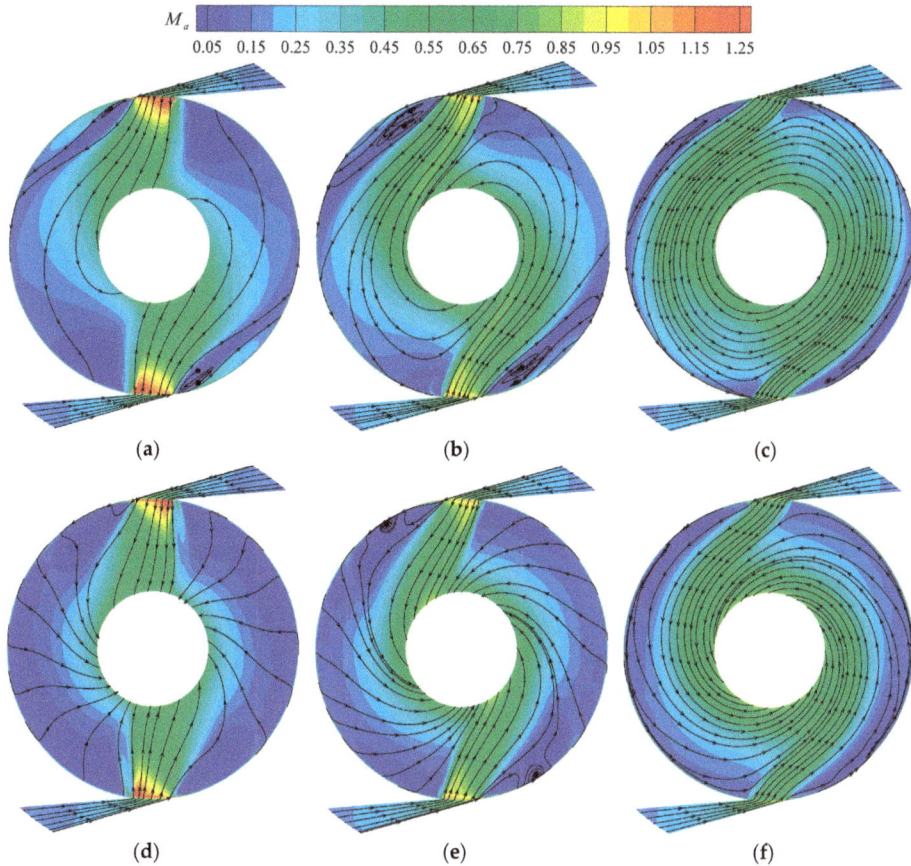

Figure 15. Mach number contours and streamlines on the mid-sections of DCs 1 and 3 for the one-to-many turbine cases with different disc spacing distance at 30,000 r/min: (**a**) DC1 for Case 1-0.3; (**b**) DC1 for Case 1-0.5; (**c**) DC1 for Case 1-1; (**d**) DC3 for Case 1-0.3; (**e**) DC3 for Case 1-0.5; and (**f**) DC3 for Case 1-1.

With increasing disc spacing distance, the local loss in the N-R chamber decreases, resulting from a decrease in flow area ratio of the N-R chamber to the disc channels. Based on the above analysis, the friction loss in the nozzle of the one-to-many turbine is quite little and has few contributions to the nozzle loss. Thus, the nozzle loss decreases with increasing disc spacing distance (Table 6).

For the one-to-many turbine, the energy loss in the disc channels also increases with disc spacing distance, which is the same as the one-to-one turbine. In addition, the local loss occurring at the inlet of the disc channels decreases with increasing disc spacing distance significantly. Thus, the disc loss of the one-to-many turbine decreases significantly with increasing disc spacing distance. Moreover, the leaving-velocity loss goes up with disc spacing distance, resulting from increasing Mach number at the turbine outlet (Figure 15).

Figure 16 shows the contours of pressure ratio on the mid-sections of DC1 for the one-to-many turbine cases with different disc spacing distance. The pressure ratio variation with increasing disc spacing distance is same as that with decreasing disc thickness, which is because both decreasing disc thickness and increasing disc spacing distance lead to a decrease in flow area ratio of the nozzle channel to the disc channels for the one-to-many turbine. In general, an increase in flow area ratio of the nozzle to disc channels leads to an increase in pressure at the nozzle outlet and a significant decrease in pressure at the rotor inlet facing the nozzle outlet. Therefore, the flow velocity at the nozzle outlet decreases and that at the rotor inlet increases much more.

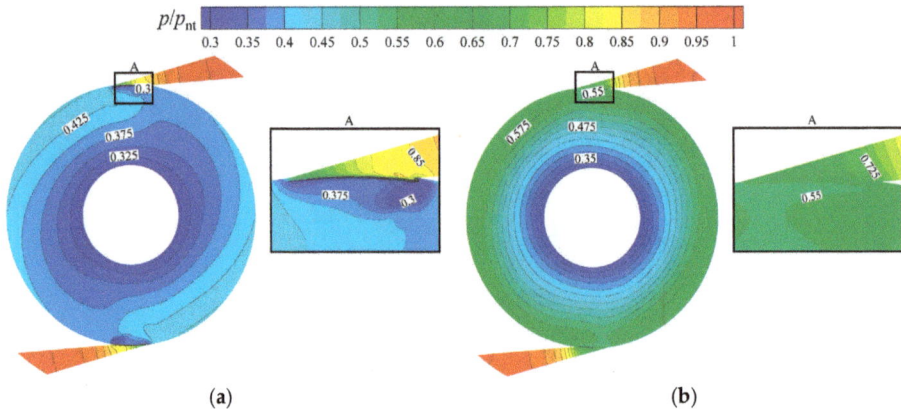

Figure 16. Contours of pressure ratio on the mid-sections of DC1 for one-to-many Tesla turbine cases with different disc spacing distance at 30,000 r/min: (**a**) Case 1-0.3; and (**b**) Case 1-1.

3.4. Influence of Disc Thickness and Spacing Distance on Energy Loss

As discussed above, the disc thickness has a little influence on the aerodynamic performance of the one-to-one Tesla turbine. In detail, the isentropic efficiency has a slight decrease with increasing disc thickness (less than 7% reduction for the turbine with 0.5 mm in disc spacing distance). That of the one-to-many turbine decreases significantly with disc thickness (decrease of about 45% for the turbine with 0.5 mm in disc spacing distance), which is due to higher leaving-velocity. For both the one-to-one turbine and the one-to-many turbine, the isentropic efficiency goes up first and then down with increasing disc spacing distance.

For the one-to-one turbine, the flow coefficient remains unchanged with the variation of disc thickness, and it increases with disc thickness for the one-to-many turbine. In addition, it increases with decreasing disc spacing distance for both two kinds of turbines. This coefficient of the one-to-one turbine is much lower than that of the one-to-many turbine with same disc spacing distance and disc thickness (The former is about half of the latter for turbine Case 1-0.5).

The nozzle loss of the multichannel Tesla turbine includes the friction loss and the local loss. That of the one-to-one turbine decreases with increasing disc spacing distance; that of the one-to-many turbine decreases with increasing disc spacing distance and increases slightly with increasing disc thickness. In general, the one-to-one turbine has higher nozzle loss than the one-to-many turbine (18.49% and 5.78%, respectively, for Case 1-0.5) due to much higher friction loss of the one-to-one turbine, although the local loss of the one-to-one is lower than that of the one-to-many turbine. Combining the discussion in the previous sections, it can be concluded that most of the nozzle loss is the friction loss for the one-to-one turbine, while, for the one-to-many turbine, the local loss in the nozzle is competitive with the friction loss in the nozzle. The friction loss decreases rapidly with a decrease in proportion ratio of the boundary layer in the nozzle channel. In detail, it drops with increasing disc spacing distance for the one-to-one turbine and increasing disc spacing distance or disc thickness for the one-to-many turbine, and the friction loss of the one-to-many turbine is quite low. The local loss in the nozzle decreases with decreasing flow area ratio of the N-R chamber to the disc channels, which is increasing disc spacing distance or decreasing disc thickness.

For the one-to-one turbine, the disc loss goes up first and then down with increasing disc spacing distance, and its variation range is much less (about 2%), while, for the one-to-many turbine, it decreases significantly with increasing disc spacing distance and decreases a little with increasing disc thickness (the maximum variation is 24.56%). For both the one-to-one turbine and the one-to-many turbine, the disc energy loss in the disc channels rises with disc spacing distance. In addition, the local loss at the inlet of the disc channels decreases with increasing disc spacing distance. According to the variation rule of the local loss with disc spacing distance, it can be concluded that the local loss of the one-to-one turbine is less than that of the one-to-many turbine.

The leaving-velocity loss of the one-to-one turbine rises with disc spacing distance, while that of the one-to-many turbine goes up with both disc thickness and spacing distance. Moreover, compared with the one-to-one turbine, the leaving-velocity loss of the one-to-many turbine with same disc thickness and spacing distance is much higher (33.1% and 13.03%, respectively, for Case 1-0.5).

4. Conclusions

As one of the best alternative options for small-scale conventional bladed turbines, Tesla turbines can be comparable with conventional turbines when they are well designed. Two kinds of multichannel Tesla turbines (one-to-one Tesla turbine and one-to-many Tesla turbine) with different disc spacing distance and disc thickness were studied numerically to analyze the influence of the two rotor geometrical parameters on their aerodynamic performance and flow characteristics, which offers a theoretical guidance in the design method. The main conclusions are summarized as follows.

(1) For both the one-to-one turbine and the one-to-many turbine, the isentropic efficiency goes up first and then down with increasing disc spacing distance, thus an optimal disc spacing distance exists to allow the multichannel Tesla turbine to obtain the highest efficiency. With increasing disc thickness, the isentropic efficiency of the one-to-many turbine decreases dramatically, while that of the one-to-one turbine decreases slightly. For turbine cases with disc spacing distance of 0.5 mm, as disc thickness increases from 1 mm to 2 mm, the isentropic efficiency of the one-to-one turbine was lowered by less than 7% (relative variation), while that of the one-to-many turbine drops by about 45% (relative variation). In addition, the isentropic efficiency of the one-to-one turbine is more sensitive to rotational speed than the one-to-many turbine.

(2) For the one-to-one turbine, with increasing disc thickness, the flow field is similar for turbine cases with same disc spacing distance; the flow velocity in the N-R chamber and the relative tangential velocity gradient close to the disc walls decrease slightly, causing the isentropic efficiency to decrease slightly. With increasing disc spacing distance, the flow velocity at the nozzle outlet decreases, and the flow angle at the rotor inlet increases, leading to an increase in mass flow rate percentage of DCs 1 and 2.

(3) For the one-to-many turbine, with increasing disc thickness, the flow velocity at the nozzle outlet decreases and its sudden increase at the rotor inlet facing the nozzle outlet is greater; the flow angle in the disc channels increases slightly and the flow velocity at the turbine outlet also increases. With increasing disc spacing distance, the flow velocity at the nozzle outlet goes up and its sudden increase becomes less; the flow angle in the disc channels decreases and the velocity at the rotor outlet increases.

Author Contributions: Conceptualization, W.Q. and Q.D.; Methodology, W.Q. and Q.D.; Investigation, W.Q. and Y.J.; Data Curation, W.Q. and Y.J.; Writing—Original Draft Preparation, W.Q.; Writing—Review and Editing, Q.D., Q.Y. and Z.F.; Visualization, W.Q. and Y.J.; and Supervision, Q.Y. and Z.F.

Funding: This research was supported financially by National Natural Science Foundation of China (grant number 51376143).

Conflicts of Interest: The authors declare no conflict of interest.

Nomenclature

b	disc spacing distance, mm
c	radial clearance of nozzle-rotor chamber, mm
c_p	specific heat at constant pressure, J/kg·K
C_m	flow coefficient
d	diameter, mm
h	static enthalpy, m^2/s^2
h_{tot}	mean total enthalpy, m^2/s^2
k	turbulent kinetic energy,
m	mass flow rate, kg/s
M	torque, N·m
M_a	Mach number
n	rotational speed of the rotor, r/min
N	number
p_{abs}	absolute pressure
p_{nt}/p_i	ratio of total pressure at the nozzle inlet to pressure at the turbine outlet
P	power, kW
Pr_t	turbulent Prandtl number
r	radial coordinate or radius, mm
R	specific gas constant of air,
S_E	energy source, kg/(m·s^3)
S_M	momentum source, kg/(m^2·s^2)
t	time, s
th	disc thickness, mm
T	temperature, K
T_{nt}	total temperature at the nozzle inlet, K
U_j	averaged velocity in three directions, m/s
v	average radial velocity, m/s
W	relative tangential velocity, m/s
$\overline{W}$	average relative tangential velocity, m/s
x_j	coordinates in three directions, m
z	axial coordinate, m
α	nozzle exit geometrical angle (relative to the tangential direction), °
γ	specific heat ratio
Γ_t	eddy diffusivity, kg/(m·s)
δ	relative variation of parameters
Δh_{isen}	isentropic enthalpy drop of the whole turbine, J/kg
η	isentropic efficiency
θ	circumferential coordinate, rad
μ_t	eddy viscosity or turbulent viscosity, kg/(m·s)
ρ	density, kg/m^3
$\rho\overline{u_i u_j}$	Reynolds stresses
$\rho\overline{u_j h}$	turbulence flux
$\rho\overline{u_i \phi}$	Reynolds flux
τ	molecular stress tensor, kg/(m·s^2)
ω	turbulent frequency, s^{-1}
Ω	rotational angular speed, rad/s

Subscripts

d	disc
dc	disc channel
i	inner
n	nozzle
o	outer

References

1. Gerendas, M.; Pfister, R. Development of a very small aero-engine. In Proceedings of the ASME Turbo Expo 2000, Munich, Germany, 8–11 May 2000.
2. Capata, R.; Sciubba, E. Experimental fitting of the re-scaled Balje maps for low-Reynolds radial turbomachinary. *Energies* **2015**, *8*, 7686–8000. [CrossRef]
3. Capata, R.; Saracchini, M. Experimental campaign tests on ultra micro gas turbines, fuel supply comparison and optimization. *Energies* **2018**, *11*, 799. [CrossRef]
4. Epstein, A.H. Millimeter-scale, MEMS gas turbine engines. *J. Eng. Gas Turbines Power* **2015**, *126*, 205–226. [CrossRef]
5. Fu, L.; Feng, Z.P.; Li, G.J. Experimental investigation on overall performance of a millimeter-scale radial turbine for micro gas turbine. *Energy* **2017**, *134*, 1–9. [CrossRef]
6. Fu, L.; Feng, Z.P.; Li, G.J. Investigation on design flow of a millimeter-scale radial turbine for micro gas turbine. *Microsyst. Technol.* **2018**, *24*, 2333–2347. [CrossRef]
7. Shan, X.C.; Zhang, Q.D.; Sun, Y.F.; Wang, Z.F. Design, fabrication and characterization of an air-driven micro turbine device. *J. Phys.* **2006**, *34*, 316–321. [CrossRef]
8. Isomura, K.; Murayama, M.; Teramoto, S.; Hikichi, K.; Endo, Y.; Togo, S.; Tanaka, S. Experiment verification of the feasibility of a 100 W class micro-scale gas turbine at an impeller diameter of 10 mm. *J. Micromech. Microeng.* **2006**, *16*, s254–s261. [CrossRef]
9. Kim, M.J.; Kim, J.H.; Kim, T.S. The effects of internal leakage on the performance of a micro gas turbine. *Appl. Energy* **2018**, *212*, 175–184. [CrossRef]
10. Deam, R.T.; Lemma, E.; Mace, B.; Collins, R. On scaling down turbines to millimeter size. *J. Eng. Gas Turbines Power* **2008**, *130*, 052301. [CrossRef]
11. Lampart, P.; Kosowski, K.; Piwowarski, M.; Jedrzejewski, L. Design analysis of Tesla micro-turbine operating on a low-boiling medium. *Pol. Marit. Res.* **2009**, *16*, 28–33. [CrossRef]
12. Tesla, N. Turbine. U.S. Patent 1,061,206, 6 May 1913.
13. Beans, E.W. Investigation into the performance characteristics of a friction turbine. *J. Spacecr.* **1966**, *3*, 131–134. [CrossRef]
14. Breiter, M.C.; Pohlhausen, K. Laminar Flow between Two Parallel Rotating Disks. Available online: https://apps.dtic.mil/dtic/tr/fulltext/u2/275562.pdf (accessed on 8 April 2016).
15. Rice, W. An analytical and experimental investigation of multiple-disk turbines. *J. Eng. Power* **1965**, *87*, 29–36. [CrossRef]
16. Adams, R.; Rice, W. Experimental investigation of the flow between corotating disks. *J. Appl. Mech.* **1970**, *37*, 844–849. [CrossRef]
17. Lemma, E.; Deam, R.T.; Toncich, D.; Collins, R. Characterisation of a small viscous flow turbine. *Exp. Therm. Fluid Sci.* **2008**, *33*, 96–105. [CrossRef]
18. Carey, V.P. Assessment of Tesla turbine performance for small scale Rankine combined heat and power systems. *J. Eng. Gas Turbines Power* **2010**, *132*, 122301. [CrossRef]
19. Sengupta, S.; Guha, A. A theory of Tesla disc turbines. *Proc. Inst. Mech. Eng. Part A J. Power Energy* **2012**, *226*, 650–663. [CrossRef]
20. Guha, A.; Sengupta, S. The fluid dynamics of the rotating flow in a Tesla disc turbine. *Eur. J. Mech. B/Fluid* **2013**, *37*, 112–123. [CrossRef]
21. Sengupta, S.; Guha, A. Analytical and computational solutions for three-dimensional flow-field and relative pathlines for the rotating flow in a Tesla disc turbine. *Comput. Fluids* **2013**, *88*, 344–353. [CrossRef]
22. Talluri, L.; Fiaschi, D.; Neri, G.; Ciappi, L. Design and optimization of a Tesla turbine for ORC applications. *Appl. Energy* **2018**, *226*, 300–319. [CrossRef]
23. Hoya, G.P.; Guha, A. The design of a test rig and study of the performance and efficiency of a Tesla disc turbine. *Proc. Inst. Mech. Eng. Part A J. Power Energy* **2009**, *223*, 451–465. [CrossRef]
24. Guha, A.; Smiley, B. Experiment and analysis for an improved design of the inlet and nozzle in Tesla disc turbines. *Proc. Inst. Mech. Eng. Part A J. Power Energy* **2010**, *224*, 261–277. [CrossRef]
25. Schosser, C.; Lecheler, S.; Pfitzner, M. A test rig for the investigation of the performance and flow field of Tesla friction turbines. In Proceedings of the ASME Turbo Expo 2014, Düsseldorf, Germany, 16–20 June 2014.

26. Lampart, P.; Jedrzejewski, L. Investigations of aerodynamics of Tesla bladeless microturbines. *J. Theor. Appl. Mech.* **2011**, *49*, 477–499.

27. Deng, Q.H.; Qi, W.J.; Feng, Z.P. Improvement of a theoretical analysis method for Tesla turbines. In Proceedings of the ASME Turbo Expo 2013, San Antonio, TX, USA, 3–7 June 2013.

28. Qi, W.J.; Deng, Q.H.; Feng, Z.P.; Yuan, Q. Influence of disc spacing distance on the aerodynamic performance and flow field of Tesla turbines. In Proceedings of the ASME Turbo Expo 2016, Seoul, Korea, 13–17 June 2016.

29. Qi, W.J.; Deng, Q.H.; Jiang, Y.; Feng, Z.P.; Yuan, Q. Aerodynamic performance and flow characteristics analysis of Tesla turbines with different nozzle and outlet geometries. *Proc. Inst. Mech. Eng. Part A J. Power Energy* **2018**, accepted. [CrossRef]

30. Hidema, T.; Okamoto, K.; Teramoto, S.; Nagashima, T. Numerical investigation of inlet effects on Tesla turbine performance. In Proceedings of the AJCPP, Miyazaki, Japan, 4–6 March 2010.

31. Sengupta, S.; Guha, A. Inflow-rotor interaction in Tesla disc turbines: Effects of discrete inflows, finite disc thickness, and radial clearance on the fluid dynamics and performance of the turbine. *Proc. Inst. Mech. Eng. Part A J. Power Energy* **2018**. accepted. [CrossRef]

32. Okamoto, K.; Goto, K.; Teramoto, S.; Yamaguchi, K. Experimental investigation of inflow condition effects on Tesla turbine performance. In Proceedings of the ISABE Conference 2017, Manchester, UK, 3–8 September 2017.

33. Carlson, D.R.; Widnall, S.E.; Peeters, M.F. A flow-visualization study of transition in plane Poiseuille flow. *J. Fluid Mech.* **1982**, *121*, 487–505. [CrossRef]

34. ANSYS. *CFX-Solver Theory Guide*; Release 12.1; ANSYS Inc.: Canonsburg, PA, USA, 2009.

energies

MDPI

Article

Effects of the Second-Stage of Rotor with Single Abnormal Blade Angle on Rotating Stall of a Two-Stage Variable Pitch Axial Fan

Lei Zhang, Liang Zhang, Qian Zhang *, Kuan Jiang, Yuan Tie and Songling Wang

School of Energy, Power and Mechanical Engineering, North China Electric Power University, Baoding 071003, China; zhang_lei@ncepu.edu.cn (L.Z.); zhangliang@ncepu.edu.cn (L.Z.); jiangkuan@ncepu.edu.cn (K.J.); tieyuan@ncepu.edu.cn (Y.T.); wsl@ncepu.edu.cn (S.W.)
* Correspondence: zhangqian@ncepu.edu.cn; Tel.: +86-0312-7522197

Received: 11 October 2018; Accepted: 23 November 2018; Published: 26 November 2018

Abstract: It is of great value to study the impact of abnormal blade installation angle on the inducement mechanism of rotating stall to achieve the active control of rotating stall in an axial fan. Based on throttle value function and SST k-ω turbulence model, numerical simulations of the unsteady flow process in stall condition of an axial flow fan with adjustable vanes were carried out, and the influence mechanism of abnormal stagger angle of a single blade in the second stage rotor on induced position and type of stall inception and evolution process of rotating stall were analyzed. The results show that compared with synchronous adjustment of blade angle, the blade with abnormal stagger angle will cause the increase of flow rate at the beginning of stall and make the fan fall into an unstable condition in advance. The existence of blade with abnormal angle does not cause the change of the induced position and type of stall inception and the inducement mechanism of rotating stall, which are the same as the axial fan with normal blade angle. Moreover, the single blade with abnormal deviation angle has important impacts on the 3D unsteady evolution process from stall inception to stall cell formation in two rotors.

Keywords: two-stage axial fan; numerical simulation; abnormal blade installation angle; rotating stall

1. Introduction

With the wide application of the two-stage variable pitch axial fan, its safety and reliability has become increasingly vital in engineering applications. Rotating stall is the instability phenomenon that occurs when the fan operates at flow rates below the design conditions, which can not only deteriorate the flow field, generate noise [1], and threaten the service life of rotor blades owing to the alternating stress, but even cause fan blades to break and endanger the operation of the fan seriously. Hence, it is necessary to study the inducement mechanism of rotating stall, which is significant for improving the performance and safety of the axial fan and realizing the active control of stall [2].

Initially, Emmons et al. [3] theoretically studied the mechanism of instability and believed the separation of airflows caused by the imbalance resulting from manufacturing and installation of blades and any other reasons lead to the occurrence of rotational stall. In fact, the system instability is the result of flow collapse, and the internal flow of impellers is the internal cause and mechanism of instability phenomenon, such as rotating stalls. At present, great progresses have been made in researches of the mechanism of stall experimentally [4]. The mechanism of rotating stall in a centrifugal fan has been experimentally studied by Hou [5], who found the intermittent phenomenon of energy exchange when the fan works under weak rotating stall and the stall on the pressure side of the G4-73 fan blades; he further pointed out that stall at the pressure side of blades requires two conditions, i.e., the actual inlet flow of fan being greater than that at designed point and the inlet prewhirl being caused by deflector adjustment, separately. However, experimental researches still have limitations. With the development of computer technology and computational methods, it was found that numerical simulations can be used to analyze the flow inside flow passages in more details, and are expected to break through the constraints of experimental researches. Zhang et al. [6] conducted a numerical study on the mechanism and dynamics of rotating stalls in a centrifugal fan, revealing that the relative positions in circumferential directions and the propagation velocity of stall cells are important factors affecting the total pressure fluctuation and the characteristic frequency in the fan.

Researches on the inducement mechanism, patterns, and impact factors of stall inceptions are the premise of active control of stall. Researches on stall inceptions are a hot spot at present. Many scholars have conducted a lot of experiments and numerical simulations to attain a deeper understanding of the patterns, mechanism, and impact factors of stall inceptions in turbo machinery [7–9]. Based on the M-G model, Greitzer and Moore predicted that stall inceptions should appear before rotating stall occurs, considering the stability of the compression system [10]. There are two stall inception patterns in turbo machinery that are generally accepted currently, namely the spike type stall inception and model type stall inception. Day [11] first discovered the spike wave experimentally which is characterized by short term in time and small scale in space, and which is frequently originated in the tip area and has a high rotation speed. In the meantime, its propagation speed and range rapidly increased. Through the measurement of pulsation of the axial velocity in a single-stage axial compressor, Mcdougall [12] first discovered the model type stall inception characterized by large-scale disturbance, smaller rotation speed, and lack of specific time and origin of space. In recent years, relevant researches at home and abroad have shown that the adjustment of structural and operating parameters can change complex flow fields in flow passages, affecting the forms of stall inception [13,14]. Bianchi S et al. [15] conducted an experimental study on stall inception patterns under different installation angles of rotor blades in a low-speed axial fan. It was found that owing to the interactions of inflow, the blade tip leakage flow and the back flow from end walls, the stall inception patterns belong to the spike-type stall at the designed angle and the modal wave type inception at an angle less than the designed one. Gaetani et al. [16] found experimentally that the intensity of separation vortexes in turbo machinery is related to the axial spacing, which affects the wake disturbance in stages and the mixing intensity of mainstream. It was found that leakage vortexes in the tip can produce significant unsteady fluctuations under certain flow conditions and affect the total performance of compressors by looking through the relevant literature [17]. Tong et al. [18] studied the relationship between the unsteadiness of leakage vortexes and rotating stall and discovered that the places that experience the unsteady activity of leakage vortexes are the source of the spike-type stall inception as it has been experimentally proved. Vo et al. [19] combined experimental data and multichannel numerical simulations to deeply investigate the influence of the tip clearance flow on the spike-type stall inception and the inducing conditions. The tip leakage flow directly affected by the size of the tip clearance is thought to be the unsteady flow in the top region caused by the pressure difference between the two sides of the blade, and the main factor that induces the spike-type stall inception. A large number of experimental and numerical studies have confirmed the close relationship between the tip clearance flows and stall inception from various perspectives [20–22]. Choi [23] conducted a numerical study on stall in the same fan at different speeds and found that the development of stall is affected by speed changing and stall inception patterns are closely related to the speed. Domestic researches on stall inception patterns and the influence factors mainly focus on the tip clearance, the axial spacing, the rotating speed [24] and so on now. It has been showed in the literature that an abnormal regulation of the stagger angle may have an important impact on the fan performance and the stability control. However, the influence on the rotating stall inception patterns and the evolution process of stall has often been taken not seriously.

During the actual operation of two-stage variable pitch axial fan, the nonsynchronous adjustment of one or several rotating blades under variable working conditions occurs sometimes, due to the block caused by the ash accumulation of the petiole, the installation errors after the blade inspection and other reasons. Researches on the abnormal deviation from the design blade angle mainly focus on the aerodynamic and noise performance at present [25]. Li et al. [26] studied the aerodynamic performance of axial fan when the abnormal deviation of the blade angle occurred. By analyzing the simulation results, they found that the performance of fan is significantly deteriorated and the noise is increased once the deviation of the stagger angle occurs, and the main source of noise is the tip and leading edge of the suction surface of blades. Through researches based on numerical simulations on characteristics of the internal flow and performance curves of a single blade in an OB-84 axial fan under normal and abnormal deviations, Ye et al. [27] found that with the increase of installation angle, the loss of the trailing edge on abnormal blades increases and the gradient on the pressure side decreases from the leading edge to the trailing edge in the presence of abnormal deviation of the blade angle; the abnormal installation angle has a significant influence on the radial total pressure distribution in intermediate blade regions, and the fan efficiency is significantly decreased. However, the impact of the abnormal deviation on stall inceptions has not been reported. The local flow field and interaction between two stages will be changed with a stagger angle anomaly of one or several rotating blades in the two-stage variable pitch axial fan, and the collapse of the local flow field may induce the fan to experience stall in advance.

In the research of this paper, the influence of blade with an abnormal stagger angle on stall inception patterns, starting positions and induced mechanism are studied through simulations in the presence of an abnormal deviation of a second-stage blade angle in a two-stage axial fan.

This paper is organized as follows. The numerical calculation method, including the physical model, computational method, mesh generation, and the grid independence verification, is described in Section 2. The static pressure distribution at the exit, the relative speed at monitoring points and the dynamic characteristics in flow passages under different deviations of the stagger angle of blade are discussed in Section 3. Finally, the conclusions of this work are summarized in Section 4.

2. Computational Method

2.1. Geometric Model

The axial fan model consists of four components: a bell mouth, two stage rotors, two stage guide vanes, and a diffuser, as drawn in Figure 1, which shows the geometric model of a two-stage variable pitch axial fan. The main structural parameters of the fan are listed in Table 1.

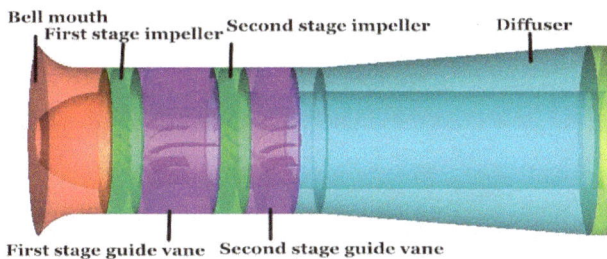

Figure 1. Geometry model of the axial fan.

Table 1. Main structural parameters of the fan.

Structural Parameters	Value
Rotation speed (r/min)	1490
Hub ratio	0.668
Number of rotor blades	2 × 24
Number of guide blades	2 × 23
Inlet diameter (m)	2.312
Outlet diameter (m)	2.305

The abnormal stagger angle of rotating blades in the second stage is shown in the Figure 2 and the value, recorded as $\Delta\beta_y < 0°$, is negative when the blade is deflected in the clockwise direction. On the contrary, the value is positive when the direction is anticlockwise.

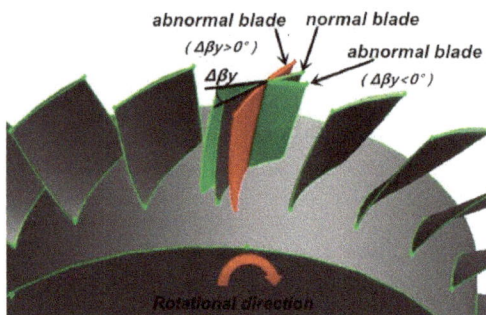

Figure 2. Diagram of the abnormal blade angle.

2.2. Mesh generation

The meshes were generated in the GAMBIT software (GAMBIT 2.4.6, ANSYS, Canonsburg, PA, USA). To satisfy the flow requirements in the impeller, the grids were generated using a T-Grid type and Tet/Hybrid elements. The leading and trailing edges and the tip clearance of blades are covered with grids intensively using a size stretching function. The specific parameters of the size function are as follows, the type is meshed, the source and attachment entities are the surfaces of all blades and the impeller volume, the growth rate is 1.2, and the maximum size is 20 mm. The types of elements employed in the mesh and their characteristic size in the different regions of the geometry are shown in Table 2. T-Grid type indicates that the fluid domain is divided into tetrahedron mesh, containing hexahedral mesh, pyramidal, and wedge mesh in the appropriate position. Cooper type means dividing the fluid domain according to the established source surface. A data exchange at the interfaces of different domains was accomplished using the Interface scheme. Figure 3 shows computational grid around the main blade edge and the local grid in rotor blade surface. The steady calculations use multiple reference frames (MRF) technique. The unsteady simulations use the moving mesh model. Blade surfaces were added on the boundary layer meshes. The boundary layer meshes are set to 15 rows with a growth factor of 1.2, and the first row is set to 0.02 mm. In this paper, the Y+ values are equal to 1.1 on the abnormal blades as well as on other blades.

In order to eliminate the influence of the number of elements of the mesh on results, this paper selected the grid number of elements of about 3.75 million, 5.87 million, 6.71 million, and 7.28 million to verify the grid independence with MRF technique and moving mesh technique, respectively. As shown in Figure 4a,b, when the grid numbers reach 6.71 million, the numerical results agree well with the experimental results obtained by the fan manufacturer. In the grid independence calculation, we obtained the number of elements by varying the number of internal points on the edge of the blades and the maximum size from 10 to 30 mm.

Table 2. The details of the mesh in the different regions.

Regions	Element	Type	Spacing
First stage guide vane	Tet/Hybrid	T-Grid	25
Second stage guide vane	Tet/Hybrid	T-Grid	25
Diffuser	Hex/Wedge	Cooper	40
Bell mouth	Tet/Hybrid	T-Grid	30
First stage rotor	Tet/Hybrid	T-Grid	20
Second stage rotor	Tet/Hybrid	T-Grid	20

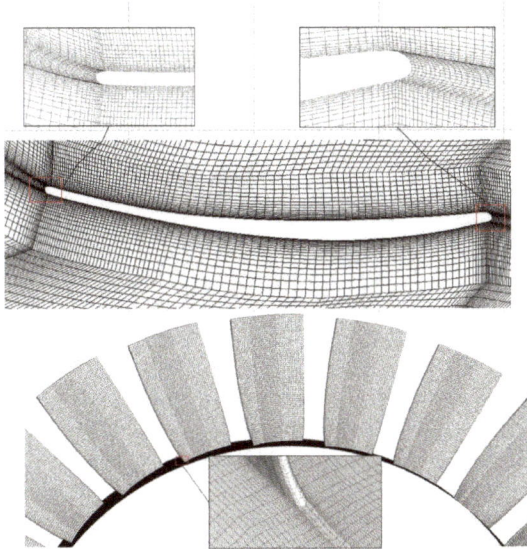

Figure 3. Computational grid around main blade edge and local grid in rotor blade surface.

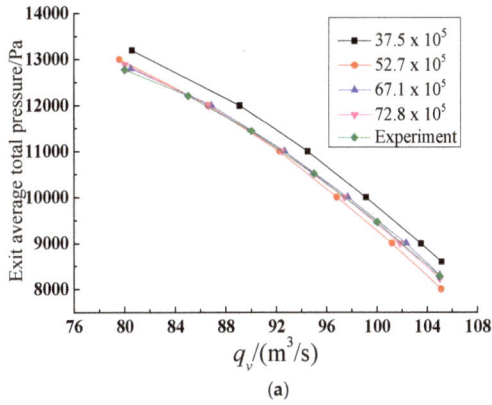

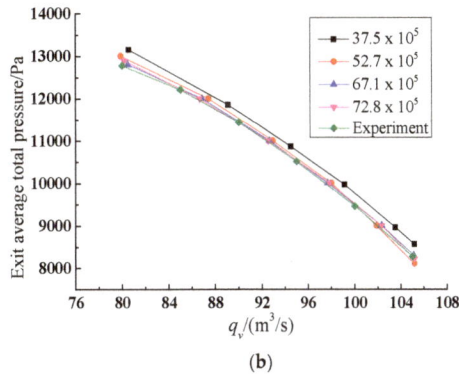

Figure 4. Grid independence calculation. (**a**) Grid independence calculation with multiple reference frames (MRF) technique; (**b**) Grid independence calculation with moving mesh technique.

2.3. Governing Equation and Boundary Conditions

A commercial CFD scheme of Fluent 15.0 is used to simulate the internal flow field of the axial fan. Rotating stall is an unsteady flow that changes with time, the unsteady Reynolds time-averaged Navier–Stokes equation is used to describe its flowing law. The temperature and density are constant, and the values are 288.16 K and 1.205 kg/m³. Due to the fact that the internal flow of axial flow fans is extremely complex and flow separation phenomenon are found, the SST k-ω turbulence model is selected to predict the flow separation accurately, such a choice is useful in the application to instability rotating stall and were verified in published literature [28]. The wall function is used near the wall, the finite volume method is used to discretize the governing equations, the SIMPLE algorithm is used to address the pressure-velocity coupling. In order to improve the accuracy, the convection term and the diffusion term are discretized by the second-order upwind scheme and the center scheme, and the unsteady calculation time adopts an implicit dual time step approach, and the time precision is second order.

In the unsteady calculations, the time step is 0.000839 s in searching for the flow rate at the onset of rotating stall. The time step has been adjusted to 0.0000839 s after the flow rate at the onset of rotating stall is found. Meanwhile we have run simulation with time step 0.000112 s, that is the delta angle = 1°, and the results of the static pressure of exit are only 0.1% different than with the delta angle = 7.5°.

The collector inlet and diffuser outlet are selected as the inlet and outlet of the calculation regions, separately, and the boundary conditions are set as inlet total pressure and outlet static pressure, separately. The static pressure is given in steady calculations, the self-defined throttle valve model [29] is loaded at the outlet in unsteady calculations to carry out iterative calculations for each step. The accuracy of the throttle valve model has been verified by applying it to a centrifugal fan in literature [6].

The throttle value function can be expressed as follows

$$Ps_{out}(t) = Pi_{in} + \frac{1}{2}\frac{k_0}{k_1}\rho U^2, \tag{1}$$

where Ps_{out} and Pi_{in} represent the outlet back pressure and environmental pressure, k_0 and k_1 are constant and throttle value, respectively, ρ represents the air density, and U is the axial velocity at the outlet. The throttle valve opening is set to 1 at the initial, the flow rate decreases and approaches to the occurrence of rotating stall gradually by decreasing the value of k_0 until to zero.

3. Results and Discussions

3.1. Analysis of Static Pressure Characteristics of Fan Outlet

Curves of the outlet static pressure with time can help shed light on the time when stall inception occurs, which can induce steep drop of the pressure. Figure 5 shows the variation of the static pressure at the outlet of the fan with time under conditions of design stagger angle and one of blades in the second-stage impeller with abnormal angles −6° and +6° respectively. k_s and k_s + 0.001 represent the throttle value (k_1) when the fan operate under the conditions of near and induced stall respectively.

The static pressure at the outlet of fan is basically stable at 13230 Pa at the throttle value of k_s + 0.001 in Figure 5a. When the throttle value is reduced to k_s, the outlet static pressure begins to decrease approximately at T = 12 (in this paper, the rotation period is represented by T), and is followed by periodic and stable fluctuations with small ranges again at approximately T = 18 after a steep drop. That is to say, stall inceptions occur at T = 12 when the flow disturbance increases, with a throttle value at k_s in the present of the design installation angle in the axial fan. After approximately six rotor cycles, stall inceptions evolve into complete stall cells, and the static pressure at the outlet of the fan shows periodic fluctuations.

Figure 5b shows the outlet static pressure under the condition of an abnormal deviation equal to 6°. The static pressure at outlet is stable at about 13049 Pa at the throttle value of k_s + 0.001. The fluid disturbance in the fan occurs at about $T = 8$ at the throttle value of k_s, and then the generation and deterioration of stall inceptions cause a trend of steep drop in outlet static pressure. After approximately nine rotor cycles, it exhibits stable periodic fluctuations at approximately $T = 17$, resulting in stable stall cells. The outlet static pressure under the condition of an abnormal deviation angle equal to $-6°$ is shown in Figure 5c. The static pressure is stable after a short period of time and is maintained at approximately $13047 Pa$ afterwards at the throttle value of $k_s + 0.001$. When the throttle value is set to k_s, the outlet static pressure begins to decrease at appoximately $T = 2.5$. The deterioration of the flow field causes the occurrence of stall in the impeller. The steep drop trend ends at approximately $T = 16.5$, and the value of pressure is stable around 10930 Pa. What is more, the static pressure exhibits stable periodic fluctuations afterwards.

When the throttle value at $k_s + 0.001$, the steady value of outlet static pressure is lower than that when there is no abnormal deviation, by comparing diagrams of Figure 5 with the abnormal deviation angle being $-6°$ or $+6°$. Moreover, it reaches a steady state at $+6°$ in a short time. When the throttle value is set to k_s and stall is about to occur, the occurrence of stall inceptions is earlier in time than that with a normal stagger angle. Comparing Figure 5a,b, the static pressure drops rapidly and then presents a stable periodic fluctuation with large amplitude when the abnormal stagger angle is $-6°$, and the outlet static pressure exhibits a stable periodic fluctuation after a short-time steep drop, with amplitude greater than the one under the normal stagger angle when the abnormal deviation angle is $-6°$.

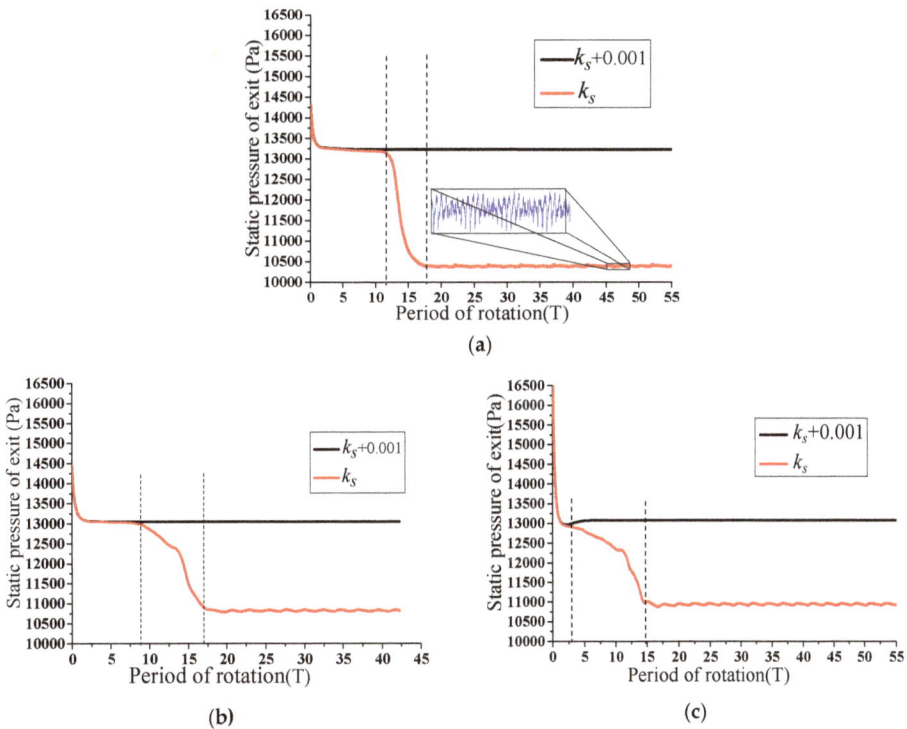

Figure 5. Variation of the outlet static pressure with time. (**a**) Normal blades; (**b**) an abnormal stagger angle of a blade equal to $-6°$; and (**c**) an abnormal stagger angle of a blade equal to $+6°$.

3.2. Analysis of Flow Rate at the Beginning of Stall and Stall Margin

The flow rate corresponding to the working conditions when stall inception occurs during simulations is called the flow rate at the beginning of stall, and is an important parameter to characterize the stable operating range of fan and calculate the stall margin. Figure 6 shows the flow rates at the beginning of stall under different stagger angles in the fan. The flow rates at the beginning of stall are 65.44m³/s and 67.07m³/s, in response to the single rotor blade with abnormal stagger angle of −6° and +6°, respectively, and the difference is little. The flow rate is only 59.5 m³/s under normal design angle. Compared with those at the design angle, the flow rates at the beginning of stall increase when the abnormal blade deviation is −6° or +6°. The data above shows that the stable operating range of fan is narrowed when the single blade with abnormal stagger angle in the second stage rotor is −6° or +6°. The stable operating range is narrower under an abnormal stagger angle of +6°, with a great impact on stall of the fan, compared with that under the abnormal stagger angle of −6°.

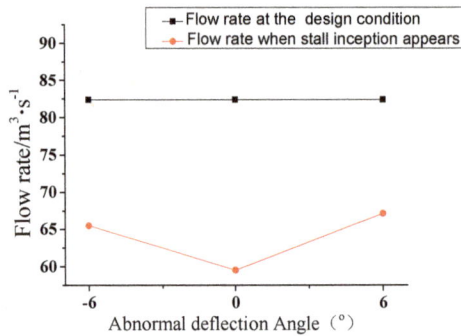

Figure 6. The flow rate when stall inception appears in.

The stall margin mainly represents the ratio of the flow rate range under the steady operation conditions to the design total flow, which can reflect the stable operating range of the fan clearly. The stall margin (Δm) under the abnormal deviation which can be calculated according to Equation (2), is calculated and analyzed in the following.

$$\Delta m = \frac{q_v{}^* - q_{vs}}{q_v{}^*} \times 100\%,\qquad(2)$$

where $q_v{}^*$ and q_{vs} represent the flow rate under the design conditions, which is 82.4 m³/s, and the flow rate of the starting stall, respectively. The calculation results are shown in Table 3 below, where 0° is the design angle of the blade.

Table 3. Stall margins.

Abnormal Angles	Stall Margins (%)
−6°	20.58
0°	27.79
+6°	18.60

It can be seen from Table 3 that the presence of the abnormal deviation of blades in the second stage, namely −6° and +6°, leads to the drop of the stall margin in the fan. Whereas the angles above, namely −6° and +6, have impacts on the stall margin in the fan, 7.21% and 9.19%, respectively, compared with that of the normal blades. And the difference of the value is only ~2%.

The curves above about the starting flow rate of stall and the outlet static pressure with time can be used to judge that stall occurs at a biggish flow rate, owing to the existence of the abnormal deviation of the blade in the second stage impeller, with values of −6° and +6, separately. They can also be used to analyze the moment when stall inception is generated, developed and evolve into complete stall cells, but cannot account for the reason why the changes occur completely. Therefore, the induced position and pattern of stall inception will be discussed below based on the relative velocities monitored at the monitoring points in the first and second rotors.

3.3. Analysis of Induced Position and Pattern of Stall Inception

The induced position of stall can be judged by the change of the relative speed with time in the first and second rotors, using the monitoring points. The numerical probes are placed at 50% and 90% of the radial height of the two-stage rotors and are circumferentially spaced with three impeller flow paths being between them. The numerical probes in the first rotor and the second rotor are labeled as m1-1, m1-2, m1-3, h1-1, h1-2, h1-3, and m2-1, m2-2, m2-3, h2-1, h2-2, h2-3, separately. The arrangement of the numerical probes in the first stage rotor is shown in Figure 7, and the probes in the second rotor are distributed in the corresponding passages according to placement in the first rotor.

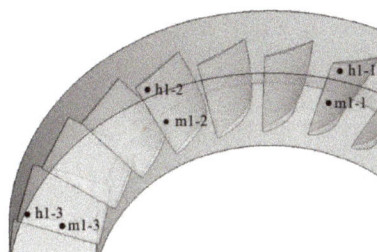

Figure 7. Distribution of the numerical probes in the first stage rotor.

Figure 8 shows curves of the relative velocities obtained by numerical probes in the first and second stage rotors when the throttle value is at k_s, wherein the curves of Figure 8a,b show the phenomenon under the design angle. The relative velocity values of h1-2 and h2-2 are shifted upwards by 30 units and the relative velocity values of h1-3 and h2-3 are shifted upwards by 60 units for comparison. In Figure 8a,b, it can be seen that, after ~13 rotor cycles, the relative velocity of each monitoring point begins to show fluctuations with large amplitude, stall inception induced in the impeller, and stall inception develop into stall cell after 3–4T with the relative velocity exhibiting periodic oscillation and the cycle is regarded as the rotating cycle of the stall cell. The relative velocity in the second stage rotor shows a significant jump at around 12T when stall inception is formed and the appearance time is about one rotation period earlier than that of the first stage rotor. Afterwards, the overall rule of the relative velocities in the process from the occurrence of the stall inception to the complete stall cell is approximately the same as that of the first stage rotor. It can be seen from the analysis above that stall inception first occurs in the second stage rotor, and then gradually affects the flow field in the first stage rotor, causing the flow field to deteriorate.

It can also be drawn from Figure 8a,b that the difference of the moments when stall inception occurs at the monitoring points h1-1 and h1-3 is ~0.334T at the range of 12 to 14T, and the value is 0.458T when stable stall cells are formed. The interval angle between the monitoring points h1-1 and h1-3 is 90° and the spread speed can be calculated as

$$\omega_s = \frac{P_m}{P_s}\omega_r = \frac{90}{0.334 \times 360}\omega_r = 0.750\omega_r, \tag{3}$$

$$\omega_T = \frac{P_m}{P_T}\omega_r = \frac{90}{0.458 \times 360}\omega_r = 0.549\omega_r, \tag{4}$$

The cycle of stall cell is $1.833T$ and the number of stall cell can be calculated as

$$N_c = \frac{P_T}{T_T \cdot P_m} = \frac{0.458}{1.833 \times \frac{90}{360}} \approx 1, \tag{5}$$

where ω_r stands for the speed of the rotor, P_m, P_s, P_T stand for the phase difference between two monitoring points, stall inceptions and stall cells, respectively. T_T is the period of the stall cell.

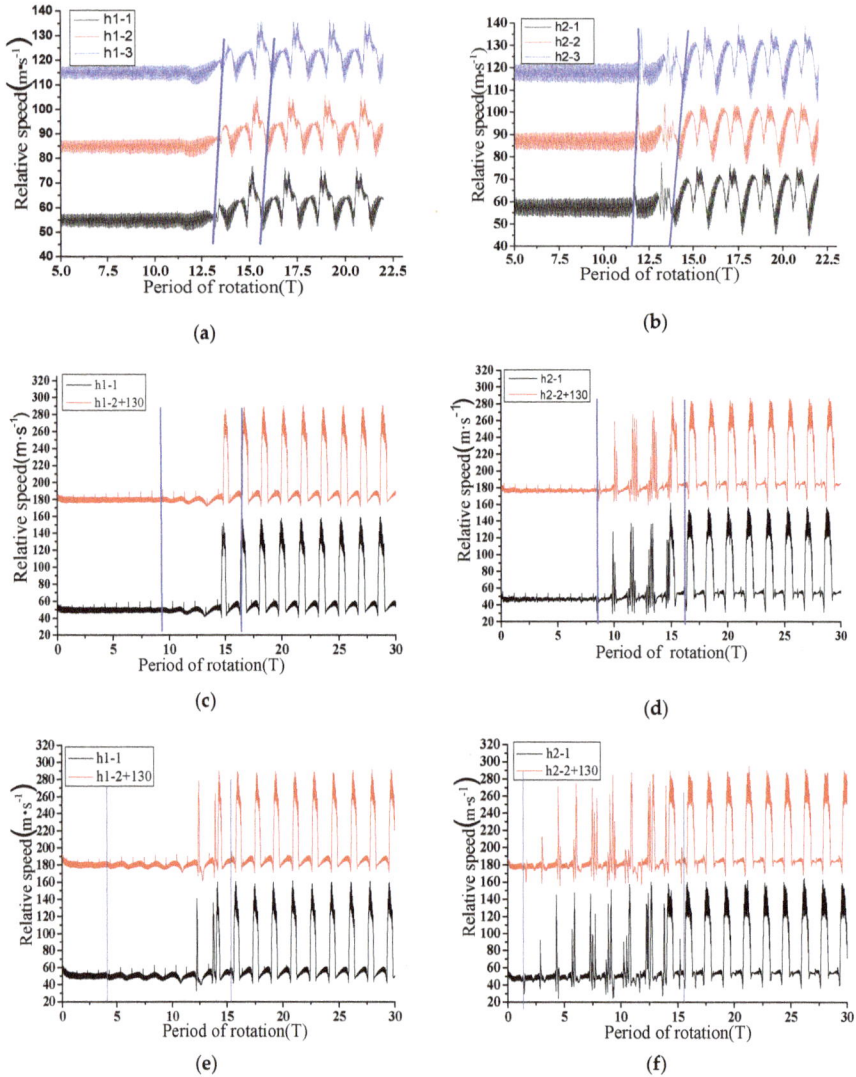

(a)

(b)

(c)

(d)

(e)

(f)

Figure 8. Curves of the relative speed at the monitoring points. (**a**) The first stage rotor under design angle. (**b**) The second stage rotor under design angle. (**c**) The first stage rotor when the abnormal deviation angle equal to $-6°$. (**d**) The second stage rotor when the abnormal deviation angle equal to $-6°$. (**e**) The first stage rotor when the abnormal deviation angle equal to $+6°$. (**f**) The second stage rotor when the abnormal deviation angle equal to $+6°$.

The values of spread speeds of stall inceptions and stable stall cells in the second stage rotor are calculated to be $0.75\omega_r$ and $0.549\omega_r$, respectively, and the number of stall cell is one, using the same method. It can be seen that the stall inception in the second-stage rotor occurs initially, the spread speed of stall inceptions in both rotors are the same, and the spread speeds of stall cells are also consistent after the development of stall cell is completed. Moreover, the stall inceptions appearing in the two stage impellers show identical characteristics: the unsteady fluctuations with small scale in impellers, the high speed of stall inceptions, the evolution into the stable stall cells within three rotor cycles, and the significant decrease of the spread speed of stall cells. Comparing the above characteristics with the researches on stall inceptions at home and abroad [30], the stall inception patterns in the two stage rotors belong all to the spike model under design blade angle condition.

The fluctuation ranges of the speed of monitoring points are large owing to the existence of blades with the abnormal stagger angle; therefore, only the relative speed of two adjacent monitoring points is analyzed. The relative velocity values of h1-2 and h2-2 are shifted by 130 units for comparison. The curves of the relative velocity of monitoring points are showed in the Figure 8c,d, with an abnormal deviation angle of $-6°$. It can be seen that stall inceptions occur in the second stage rotor at ~$8T$, and develop into stall cells after nine rotor cycles. Stall inceptions occur in the first stage rotor owing to the influence of the stall inception in the second sage rotor and turn into stable stall cells at $17T$ eventually. The curves of relative velocities of monitoring points under the abnormal stagger deviation angle of $+6°$ are showed in the Figure 8e,f. It can be seen that stall inception first occur in the second rotor, at ~$2T$, and develop into complete stall cell after about 13 rotor cycles. The relative velocity of monitoring points in the first stage rotor fluctuates at $4.5T$ and appears periodic at ~$15T$. The analysis above shows that the occurrence of stall inception in the second stage rotor affects the flow field in the first stage rotor, causing flow instability at $4.5T$, which continues to deteriorate. Finally, stall inceptions in the first and second stage impellers evolve into complete stall cells at approximately the same time ($15T$).

The Equations (4)–(6) are used to calculate the circumferential propagation velocity of stall inception and stall cell under the abnormal stagger angle and the results are listed in the Table 4. R1 and R2 represent the first stage rotor and the second stage rotor separately.

Table 4. Circumferential propagation velocity of stall inceptions and stall cells.

Abnormal Stagger Angles	$-6°$		$0°$		$+6°$	
	R1	R2	R1	R2	R1	R2
Spread speed of stall inception (ω_r)	0.638	0.718	0.750	0.750	0.672	0.753
Spread speed of stall cell (ω_r)	0.61	0.584	0.549	0.549	0.584	0.658

The circumferential propagation velocity of the first and second stage impellers is ~70% of the rotor speed with a blade of an abnormal stagger angle of $+6°$ or $-6°$, compared with that of the design blade angle. The propagation velocity decreases after developing into stable stall cell, and the propagation velocities in two rotors are different. There are two reasons for the asynchronization of speed: considering the cases of abnormal blades, the rotation of stall cell is not strictly periodic, and there may be some errors in data acquisition and calculation; otherwise, the existence of abnormal blades may also affect the propagation speed of stall inceptions and stall cells to a certain extent, leading to the difference of the propagation speed in different impellers.

3.4. Analysis of Dynamic in Rotating Impellers before and after Stall Induced

The evolution process of stall inception in the two rotors is basically the same under the design blade angle condition, so this paper analyzes the relative velocity and the turbulent kinetic energy contours of the middle span-wise cross section in the second stage rotor at typical moments, as shown in Figure 9.

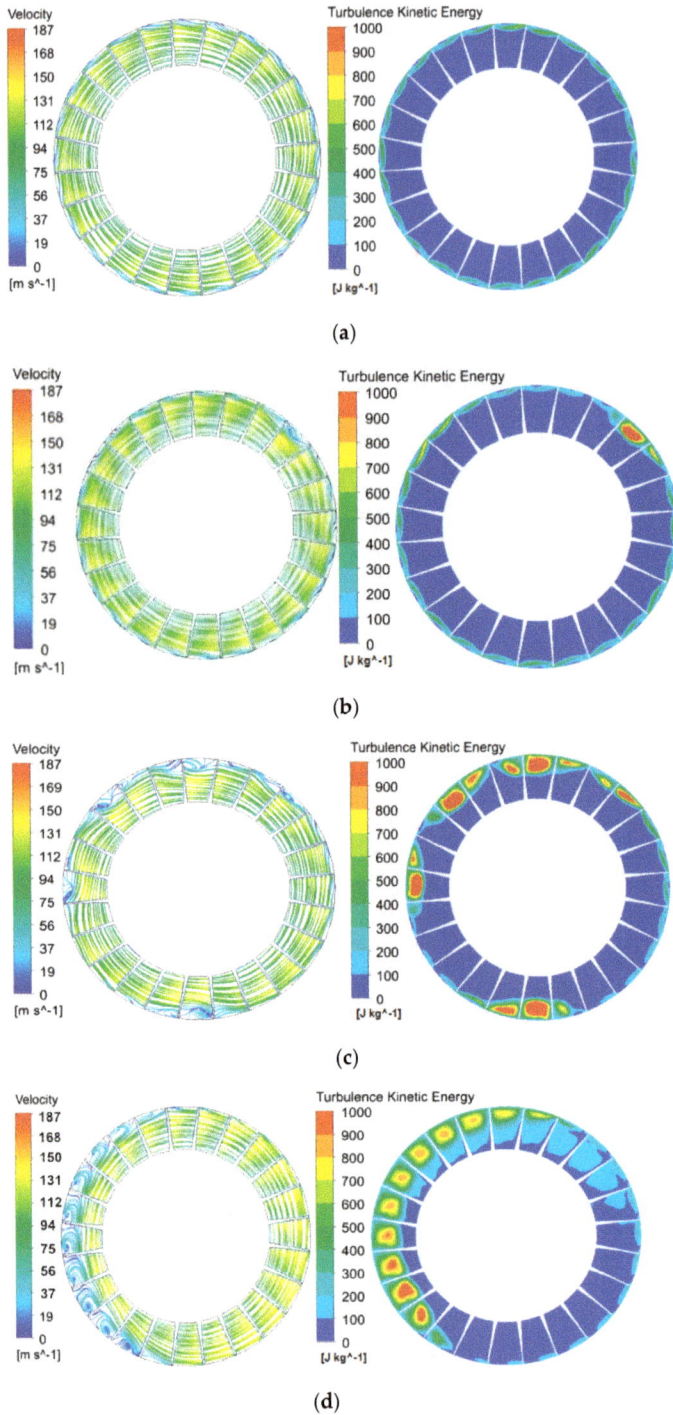

Figure 9. Streamline and the turbulent kinetic energy contour of the second rotor midsection. (**a**) Near stall inception at 11*T*, (**b**) stall inception stat at 12*T*, (**c**) development of stall inception at 15*T*, and (**d**) complete stall cell at 18*T*.

Figure 9a shows that the flow field in each passage is relatively uniform near the induced stall inception. However, there is a small radial velocity at the tip of the blade; the small fluctuation of the flow rate is caused by the enhancement of the tip leakage vortex. The kinetic energy is evenly distributed along the circumference, the turbulent kinetic energy is obviously increased near the tip of the blade, covering almost the entire tip clearance area. This is due to the positive angle that gradually increases with the decrease of flow rate, and the vortex area that increases, causing the improvement of turbulent pulsation. The low kinetic energy zone in the middle of the flow passages which is elliptical and has an increasing trend, develops toward the top of the blade.

As shown in Figure 9b, stall inception appears. The flow field disturbance is further enhanced, and large reflow and flow separation in the tip region are observed in three flow passages as the streamline diagram illustrates. The largest separation vortex occupies about two flow passages, the recirculation zone in the middle flow passages is large, and the recirculation zone in several flow passages adjacent along the counterclockwise direction is small, with an increasing trend; on the contrary, the area of the recirculation zone in the flow passages along the clockwise direction is also small with a tendency to decrease. The recirculation process in flow passages involves air flowing from the suction side of the vane to the pressure surface, and then flowing back to the suction side after the fluid is mixed.

The turbulent energy distribution corresponding to the flow field has a similar law. The region of turbulent kinetic energy occupies a large area near the tip of the blade to the 70% blade height, and turbulent energy increases. The area of turbulent kinetic energy in adjacent flow passages along the counterclockwise direction has an increasing tendency, with a decreasing tendency in the adjacent flow paths along the clockwise direction, owing to that the propagation speed of stall inception is smaller than the rotor speed. From the perspective of the relative coordinate system, stall inceptions rotate along the counterclockwise direction, resulting in the deterioration of the flow in the adjacent passages along the counterclockwise direction and the flow inside the adjacent passages along the clockwise direction. With the evolution of stall inception, the flow field disturbance is further enhanced. The separation vortex appears in the four regions of the flow diagram, the largest of which occupies approximately five flow passages, and the recirculation zone is larger in the middle flow passages. Corresponding to the turbulent kinetic energy contour, there is a similar distribution law, that is, the area of high kinetic energy in the middle flow passages is large, occupying the area from the tip to 60% of the blade height. There is an increasing trend in the adjacent flow passages along the clockwise direction, and a decreasing trend along the clockwise direction, as shown in Figure 9c.

The velocity and distribution of turbulent energy at a certain moment after stall inception evolve into stall cell are shown in Figure 9d, when stall cell rotate in the circumferential direction at a fixed speed. It can be seen that stall cell occupy about 11 passages along the circumferential direction, and the blocked area in five passages in the middle of stall cell is larger, occupying the area from the tip to 50% of the blade height. Compared with Figure 9a, the stall cell have a larger influence on the flow filed, and evolve from stall inceptions to stall cells in a short period of time, which is consistent with the spike-type stall inception. It can be seen from the Figure 8d that a zone of high kinetic energy appears in the region corresponding to the stall cell. Compared with Figure 9c, the maximum value of turbulent kinetic energy in passages is almost constant, whereas the area of the turbulent kinetic energy is significantly increased. As stall cell spreads along the circumferential direction, the region of turbulent kinetic energy in passages undergoes a dynamic process that first increases and then decreases in the radial direction.

The turbulent kinetic energy indicates the kinetic energy of the turbulent pulsation, it can reflect the magnitude of the normal value of the Reynolds stress and the intensity of the turbulent pulsation. Analyzing the contours of turbulent energy distribution under the design blade angle condition, the variation of regions where the kinetic energy is high in impellers can be used to clearly determine the generation and development process of stall inceptions. Therefore, taking on the change of kinetic energy in the first and second rotors as an example when the abnormal blade deviation angle is −6°, the evolution process of stall inception is analyzed.

Figure 10 shows the distribution of the turbulent energy in the middle section of the second stage rotor at four typical moments when the abnormal deviation angle is −6°. It can be seen that the kinetic energy in three passages corresponding to regions of stall inceptions is higher. The high kinetic energy fully occupies the tip area of the flow paths in the counter-clockwise direction of the abnormal deflection blade; the area of high kinetic energy in clockwise passages is particularly small. As the flow field evolves, as shown in Figure 10b, the region of high kinetic energy occupies about five impeller flow passages, with regions of the highest turbulent energy occupying at approximately from 50% to 90% of blade height. The development Figure 10b,c also shows that flow fields are affected by stall cell and zones of the high kinetic energy develop in radial direction from the tip of the blade to 50% blade height and in the direction along the circumference, more passages being influenced. The stall inception is further developed, eventually forming a stable stall cell occupying by about nine passages, as shown in Figure 10d.

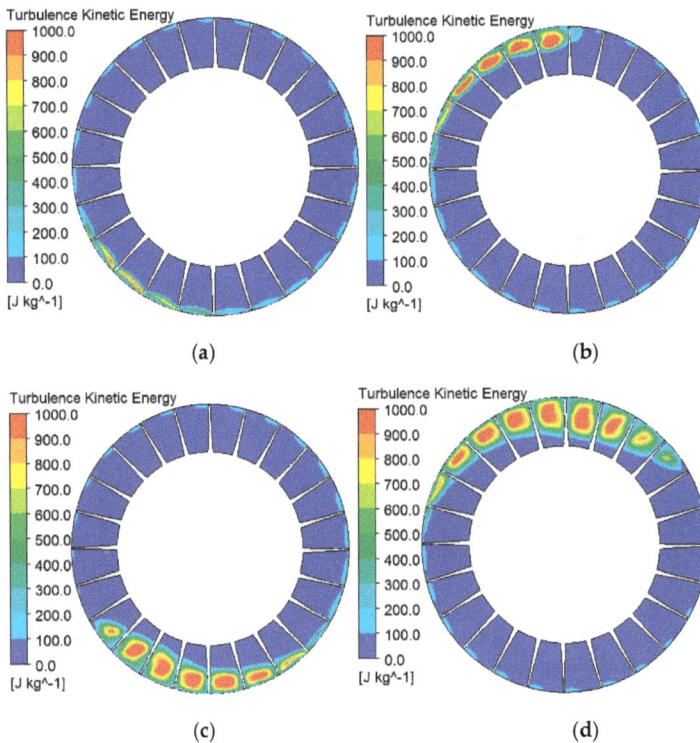

Figure 10. Turbulent kinetic energy contour of the first rotor midsection when the abnormal angle equal to −6°. (**a**) Near stall inception at 11*T*, (**b**) stall inception start at 13*T*, (**c**) development of stall inception at 15*T*, and (**d**) complete stall cell at 17*T*.

The distribution of the turbulent energy in the middle section of the second stage rotor at four typical moments is shown in Figure 10. Figure 11a shows the stage of stall inception generation, where the zone of high kinetic energy occupies about two impeller passages. Subsequently, four immature small stall cells are generated almost simultaneously in passages, and then develop stably and occupy two passages respectively, as shown in Figure 11b. Zones of sorghum kinetic energy develop from the tip to the 70% blade height, and three small stall cells are combined into a large one owing to the impact of the generation of stall in the first-stage rotating impeller, as shown in Figure 11c. It can be clearly seen from Figure 11d that the flow filed is further deteriorated, four small stall cells developing into a stable one and occupying about 11 passages. Regions of higher turbulent energy occupy the area from the 40% to the 90% of blade height, making passages easy to be blocked.

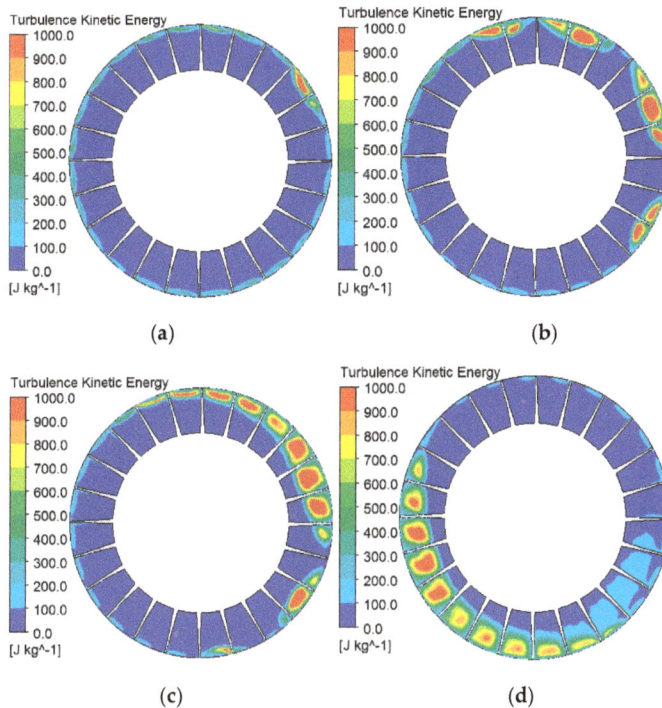

Figure 11. Turbulent kinetic energy contours of the second rotor midsection when the abnormal angle is −6°. (**a**) Near stall inception at 8*T*. (**b**) Stall inception start at 11*T*. (**c**) Development of stall inception at 15*T*. (**d**) Complete stall cell at 17*T*.

3.5. Analysis of the Streamline at 95% Radial Blade Height

The streamline diagram can clearly show the flowing law of the tip leakage flow to reveal the induced mechanism of rotating stall of the impeller. Figure 12 is a chart showing streamline diagrams at the 95% radial blade height in the second-stage rotor under three conditions. It can be seen that the streamline diagrams near the tip are not completely the same in the three cases, but the flowing laws are basically the same: a leakage vortex is generated near the leading edge of blade, and appears near suction surface, opposite to the main flow direction. A low velocity region is generated in the passage and changes the fluid direction to flow from the suction side of the blade to the pressure surface after mixing with the mainstream fluid. The reflow is therefore generated in the low velocity zone, causing deterioration of the flow field and the occurrence of rotating stall. In the case of abnormal and normal deviation angle of the blade, the same flowing law near the tip of the blade when the stall inception appears indicates that the inducement mechanism of stall is the same under the three conditions.

(a)

(b)

(c)

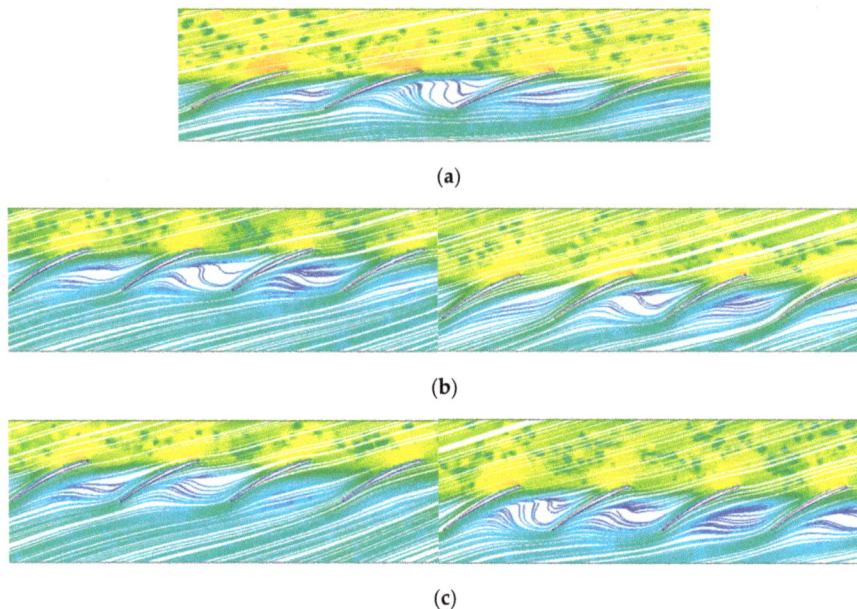

Figure 12. Streamline at the 95% radial blade height under three conditions: (**a**) No abnormal stagger angle; (**b**) abnormal stagger angle is $-6°$; and (**c**) abnormal stagger angle is $+6°$.

4. Conclusions

By simulating and analyzing the evolution process of stall inception in the two-stage variable pitch axial fan under normal and abnormal stagger angles, the following conclusions can be drawn.

(1) The existence of the abnormal deflection blade causes the fan to fall into rotating stall at a large flow rate, and the fan stall margin is reduced, that is, the stable operating range of the fan is reduced. A single blade with abnormal deviation angle induces the fan to get into stall in advance.

(2) Compared with the designed stagger angle condition, stall inception still first appears around the leading edge of the second-stage rotor blades, and the stall inception spreads at a high speed, about 70% of the rotor speed. After developing into a complete stall cell, the spread speed is significantly reduced, and shows characteristics of the spike-type stall inception. An abnormal deviation blade in the second stage rotor has little effect on induced position and type of stall inception.

(3) By observing and analyzing the flow diagram and the turbulent kinetic energy contours inside impellers, it was found that the number of stall cell and the evolutionary time change during the evolution process of stall inception. The existence of an abnormal deviation blade has great influence on the evolution process from stall inception to stall cell.

(4) By analyzing the streamlines at 95% radial blade height, it can be seen that the flowing laws in passages are basically the same, and inducement mechanisms of rotating stall are consistent under three conditions. An abnormal deviation blade in the second stage rotor does not change the inducement mechanism of rotating stall.

Author Contributions: L.Z. (Lei Zhang) and L.Z. (Liang Zhang) proposed the ideas of the research and analyzed and explained the data obtained by numerical simulation. Q.Z. carried out part of numerical simulation work, determined the writing method, and revised the paper. K.J. wrote the manuscript and made charts based on the numerical simulation data. Y.T. collected the relevant references and the data of the numerical simulation, and revised the paper. S.W. participated in part of the numerical simulation work in this study.

Funding: This research was supported by National Natural Science Foundation of China (Grant No. 11602085), Natural Science Foundation of Hebei Province, China (Grant No. E2016502098), and Fundamental Research Funds for the Central Universities, China (Grant No. 2018MS107).

Conflicts of Interest: The authors declare no conflicts of interest.

References

1. Zhang, L.; Yan, C.; He, R.Y.; Zhang, Q. Numerical Study on the Acoustic Characteristics of a Axial Fan under Rotating stall Condition. *Energies* **2017**, *10*, 1945. [CrossRef]
2. Niazi, S.; Stein, A.; Sankar, L. Numerical studies of stall and surge alleviation in a high-speed transonic fan rotor. *Mech. Eng.* **2000**, *225*, 38–39.
3. Emmons, H.W.; Pearson, C.E.; Grant, H.P. Compressor surge and stall propagation. *Trans. ASME* **1955**, *77*, 455–469.
4. Tang, Y.; Liu, Y.; Lu, L. Solidity Effect on Corner Separation and Its Control in a High-Speed Low Aspect Ratio Compressor Cascade. *Int. J. Mech. Sci.* **2018**, *142–143*, 304–321. [CrossRef]
5. Hou, J.H. *Research on Condition Monitoring and Fault Diagnosis of Fan Based on Multi-Parameter*; North China Electric Power University: Baoding, China, 2003.
6. Zhang, L.; Wang, S.L.; Zhang, Q.; Wu, Z.R. Dynamic Characteristics of Rotating Stall for Centrifugal Fans. *Proce. CSEE* **2012**, *32*, 95–102.
7. Kazutoyo, Y.; Hiroaki, K.; Ken-ichiro, I.; Furukawa, M.; Gunjishima, S. An Explanation for Flow Features of Spike-Type Stall Inception in an Axial Compressor Rotor. *J. Turbomach.* **2013**, *135*, 1–8.
8. Salunkhe, P.B.; Pradeep, A.M. Stall Inception Mechanism in an Axial Flow Fan under Clean and Distorted Inflows. *J. Fluids Eng.* **2010**, *132*, 1–11. [CrossRef]
9. Sheard, A.G.; Corsini, A.; Bianchi, S. Stall Warning in a Low-Speed Axial Fan by Visualization of Sound Signals. *J. Eng. Gas Turbines Power* **2011**, *133*, 1–9. [CrossRef]
10. Moore, F.K.; Greitzer, E.M. A theory of post-stall transients in axial compression systems. I: Development of equations. *J. Eng. Gas Turbines Power* **1986**, *108*, 68–76. [CrossRef]
11. Day, I.J. Stall inception in axial flow compressors. *J. Turbomach.* **1993**, *115*, 1–9. [CrossRef]
12. McDougall, N.M.; Cumpsty, N.A.; Hynes, T.P. Stall inception in axial compressors. *J. Turbomach.* **1990**, *112*, 116–125. [CrossRef]
13. Anish, S.; Sitaram, N.; Kim, H.D. A Numerical Study of the Unsteady Interaction Effects on Diffuser Performance in a Centrifugal Compressor. *J. Turbomach.* **2014**, *136*, 1–9. [CrossRef]
14. Biela, C.; Brandstetter, C.; Schiffer, H.P.; Heinichen, F. Unsteady wall pressre measurement in a one-and-a-half stage axial transonic compressor during stall inception. *Proc. Inst. Mech. Eng. Part A J. Power Energy* **2013**, *227*, 643–653. [CrossRef]
15. Bianchi, S.; Corsini, A.; Mazzucca, L.; et al. Stall Inception, Evolution and Control in a Low Speed Axial Fan with Variable Pitch in Motion. In Proceedings of the ASME 2011 Turbo Expo: Turbine Technical Conference and Exposition, Vancouver, BC, Canada, 6–10 June 2011; pp. 445–456.
16. Gaetani, P.; Persico, G.; Osnaghi, C. Effects of Axial Gap on the Vane-Rotor Interaction in a Low Aspect Ratio Turbine Stage. *J. Propuls. Power* **2015**, *26*, 325–344. [CrossRef]
17. Lakshminarayana, B.; Horlock, J.H. Review: Secondary flows and losses in cascades and axial-flow turbomachines. *Int. J. Mech. Sci.* **1963**, *5*, 287–307. [CrossRef]
18. Tong, Z.T. *Experimental Study on the Unsteady Correlation of Tip Leakage Vortex, Stall Inception and Micro Tip Injection*; Graduate School of the Chinese Academy of Sciences: Shenzhen, China, 2006.
19. Vo, D.H.; Tan, C.S.; et al. Criteria for Spike Initiated Rotating Stall. *J. Turbomach.* **2008**, *130*, 11–23. [CrossRef]
20. Torresi, M.; Camporeale, S.M.; Pascazio, G. Detailed CFD Analysis of the Steady Flow in a wells Turbine Under Incipient and Deep Stall Conditions. *J. Fluids Eng.* **2009**, *131*, 1–9. [CrossRef]
21. Tomita, I.; Ibaraki, S.; Furukawa, M. The Effect of Tip Leakage Vortex for Operating Range Enhancement of Centrifugal Compressor. *J. Turbomach.* **2013**, *135*, 1–8. [CrossRef]
22. Cameron, J.D.; Bennington, M.A.; Ross, M.H.; Morris, S.C. The Influence of Tip Clearance Momentum Flux on Stall Inception in a High-Speed Axial Compressor. *J. Turbomach.* **2013**, *135*, 1–11. [CrossRef]
23. Choi, M.; Vahdati, M.; Imregun, M. Effects of Fan Speed on Rotating Stall Inception and Recovery. *J. Turbomach.* **2011**, *133*, 1–8. [CrossRef]

24. Pavesi, G.; Cavazzini, G.; Ardizzon, G. Numerical Analysis of the Transient Behaviour of a Variable Speed Pump-Turbine during a Pumping Power Reduction Scenario. *Energies* **2016**, *9*, 534. [CrossRef]

25. Ye, X.M.; Li, C.X.; Yin, P. Effect of Abnormal Blade Reverse Deviation on Performance of the Axial Fan. *J. Chin. Soc. Power Eng.* **2013**, *9*, 702–710.

26. Li, C.X.; Lin, Q.; Ding, X.L.; Ye, X.M. Performance, aeroacoustics and feature extraction of an axial flow fan with abnormal blade angle. *Energy* **2016**, *103*, 322–339. [CrossRef]

27. Ye, X.M.; Ding, X.L.; Zhang, J.K.; Li, C. Numerical simulation of pressure pulsation and transient flow field in an axial flow fan. *Energy* **2017**, *129*, 185–200. [CrossRef]

28. Zhang, L.; Jiang, K.; Wang, S.L.; Zhang, Q. Effect of the First-stage of Rotor With Single Abnormal Blade Angle on Rotating Stall of a Two-stage Variable Pitch Axial fan. *Proc. CSEE* **2017**, *37*, 1721–1730.

29. Gourdain, N.; Burguburu, S.; Leboeuf, F.; Michon, G.J. Simulation of rotating stall in a whole stage of an axial compressor. *Comput. Fluids* **2010**, *39*, 1644–1655. [CrossRef]

30. Tan, C.S.; Day, I.; Morris, S.; Wadia, A. Spike-Type Compressor Stall Inception, Detection, and Control. *Annu. Rev. Fluid Mech.* **2010**, *42*, 275–300. [CrossRef]

energies

MDPI

Article

Designing Incidence-Angle-Targeted Anti-Cavitation Foil Profiles Using a Combination Optimization Strategy

Di Zhu [1,2], Ruofu Xiao [1,2,*], Ran Tao [2,3] and Fujun Wang [1,2]

[1]   College of Water Resources and Civil Engineering, China Agricultural University, Beijing 100083, China; zhu_di@cau.edu.cn (D.Z.); wangfj@cau.edu.cn (F.W.)
[2]   Beijing Engineering Research Center of Safety and Energy Saving Technology for Water Supply Network System, China Agricultural University, Beijing 100083, China; randytao@mail.tsinghua.edu.cn
[3]   Department of Energy and Power Engineering, Tsinghua University, Beijing 100083, China
*   Correspondence: xrf@cau.edu.cn

Received: 15 October 2018; Accepted: 5 November 2018; Published: 9 November 2018

Abstract: In hydraulic machinery, the surface of the blade can get damaged by the cavitation of the leading-edge. In order to improve the cavitation performance, the anti-cavitation optimization design of blade leading-edge is conducted. A heuristic-parallel locally-terminated improved hill-climbing algorithm, which is named as the global dynamic-criterion (GDC) algorithm was proposed in this study. The leading-edge shape of NACA 0009-mod foil profile was optimized by combining the GDC algorithm, CFD prediction, Diffusion-angle Integral (DI) design method and orthogonal test. Three different optimal foil geometries were obtained for specific incidence angles that 0, 3, and 6 degrees. According to the flow field analyses, it was found that the geometric variation of the optimized foil fits the incoming flow better at the respective optimal incidence angles due to a slighter leading-edge flow separation. The pressure drops become gentler so that the cavitation performance get improved. Results show that the GDC algorithm quickly and successfully fits the target condition by parallel running with the ability against falling into local-best tarps. The $-C_{pmin}$ of the optimal foils was improved especially by +11.4% and +14.5% at 3 and 6 degrees comparing with the original foil. This study provided a reference for the anti-cavitation design of hydraulic machinery blades.

Keywords: leading edge; global optimization; cavitation inception; orthogonal test; CFD simulation

---

1. Introduction

Leading-edge (LE) cavitation is the most common type of cavitation in hydraulic machinery [1,2]. The occurrence of LE cavitation is strongly related to the pressure changes generated by fluid flow around the geometry [3,4]. The direct effect of geometric shape on LE cavitation will be considered in the design of blades or other flow-around bodies. The cavity or traveling bubbles of LE cavitation often covers the surface [5] which may cause direct material damage due to the collapse of cavitation bubbles [6]. Therefore, it is very important to investigate and improve the LE cavitation (including delaying its occurrence, reducing its size, etc.) for the design of hydraulic machinery [7–9]. Compared with the design of the flow passage, the blade will exert a direct force on the fluid medium while working and generate a pressure difference between the two sides of blade surface [10–12]. The blade cavitation, especially the LE cavitation characteristics, has become a key factor restricting the blade design under specific conditions. In the past, many researchers have studied the occurrence and characteristics of LE cavitation [13–15]. Generally, cavitation occurs when pressure drops below the saturation pressure. Because of the flow separation and local pressure drop at the blade LE, cavitation often occurs at the LE [16]. Experimental studies found that the development of the cavity had a negative effect on the flow separation due to inter-phase surface tension [17]. The interaction effect causes a special relationship between the separation position and cavitation inception position in the laminar flow case around a smooth body [18]. This relationship will be different in turbulent flow case, rough surface case and other different boundary layer cases [19]. Generally, the flow separation and pressure drop caused by the geometric characteristics of blade LE will regularly affect the occurrence and development of cavitation at a certain incidence angle [20].

Foil was commonly used as the simplified model of hydraulic machinery blades in engineering [21]. Meanline-symmetrical foils can further reduce the complexity of geometric parameters and improve the pertinence of the research [22]. The smooth symmetrical NACA0009 foil profile can be used to study the laminar flow under a certain Reynolds number. NACA0009 and its modified profiles are popular in the former studies of hydrodynamics and hydraulic machinery including the LE cavitation, tip-clearance cavitation, wake and hydraulic damping, excitation and response [23,24]. Also, numerous researches have been carried out on the LE cavitation of hydraulic machineries [25–30].

Based on the study of LE cavitation on foils and turbo blades, there have been abundant researches on the mechanism, occurrence, development and variation. However, there is still a problem that has not been solved perfectly that how to obtain a hydraulic geometry for improving the LE cavitation. Optimization is an effective way to solve this problem by properly defining the optimized target and choosing the reasonable algorithm. In this study, the foil shape with better LE cavitation characteristics was set as the optimization target. Feasible algorithms including genetic algorithm, ant colony algorithm, hill climbing algorithm, etc. can be used. Each of them has its own advantages and disadvantages and requires improvements or adjustments in different cases [31–35]. Huang et al. [36] used the genetic algorithm to drive the computational fluid dynamics simulations, guiding the configuration of a suction jet and a blowing jet on the airfoil's upper surface. Liu et al. [37] selected an adaptive simulated annealing algorithm to solve the energy performance calculation model. The weighted average efficiency of the impeller after the three-condition optimization has increased by 1.46% than that of original design. Liu et al. [38] used a multi-objective optimization design system to develop an ultrahigh-head runner with good overall performance. Compared to the initial runner, the preferred runner's efficiency under turbine mode is increased by about 0.7% and the pump efficiency by about 0.6%, while the runner's cavitation is greatly promoted. Liu et al. [39] proposed the hydraulic design method of controllable blade angle for rotodynamic multiphase pump with impeller and diffuser. The orthogonal optimization method was employed to optimize the geometry parameters. The distributions of gas volume fraction and the pressure became more uniform after optimization, and improved the transporting performance of the multiphase pump. In this study, the global dynamic-criterion (GDC) algorithm which can run in parallel, stop from dropping into local-best trap was chosen. Finally, reducing the number of effective tests and reducing time cost is the top priority of engineering optimization. Therefore, the strategy combining the GDC algorithm, CFD prediction, Diffusion-angle Integral (DI) method [40] for foil thickness geometry design and orthogonal test was used for optimization. The new optimization algorithm proposed in this paper can complete the optimization process quickly and simply with high optimization efficiency. Based on the strategy above, the optimal hydraulic shape and improved cavitation performance of NACA0009 foil can be obtained at different specific incidence angles. This research shows an example for designing the anti-cavitation turbo blade of hydraulic machinery including axial-flow pump, mixed-flow pump, centrifugal pump and other bladed pumps.

2. The Studied Hydrofoil Object

The NACA0009-mod profile was used as the objective hydrofoil [24]. As shown in Figure 1, it is a symmetrical foil, m is the meanline direction and t is the thickness. The total meanline length is $m_{total} = L$. The thickness t distribution along m direction can be expressed as:

$$\begin{cases} \frac{t}{L} = a_0\left(\frac{m}{L}\right)^{\frac{1}{2}} + a_1\left(\frac{m}{L}\right) + a_2\left(\frac{m}{L}\right)^2 + a_3\left(\frac{m}{L}\right)^3 & 0 \le \frac{m}{L} \le 0.5 \\ \frac{t}{L} = b_0 + b_1\left(1 - \frac{m}{L}\right) + b_2\left(1 - \frac{m}{L}\right)^2 + b_3\left(1 - \frac{m}{L}\right)^3 & 0.5 \le \frac{m}{L} \le 1 \end{cases} \tag{1}$$

where

$$\begin{cases} a_0 = +0.1737 \\ a_1 = -0.2422 \\ a_2 = +0.3046 \\ a_3 = -0.2657 \end{cases} \quad \begin{cases} b_0 = +0.0004 \\ b_1 = +0.1737 \\ b_2 = -0.1898 \\ b_3 = +0.0387 \end{cases} . \tag{2}$$

The hydrofoil was built based on Equations (1) and (2), and another trailing edge modification. The TE modification was cut on the trailing edge to $l = 0.9091L$. The TE was processed by circular arc, and the radius is $r = 0.0168L$. The final hydrofoil model is shown in Figure 1. The incidence angle was defined as α. The 2D flow domain ($1.5l \times 7.5l$) for computational fluid dynamics (CFD) simulations was built as shown in Figure 2. The commercial software ANSYS CFX was used for numerical simulation.

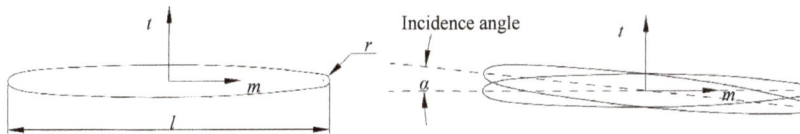

Figure 1. Hydrofoil model and its parameters.

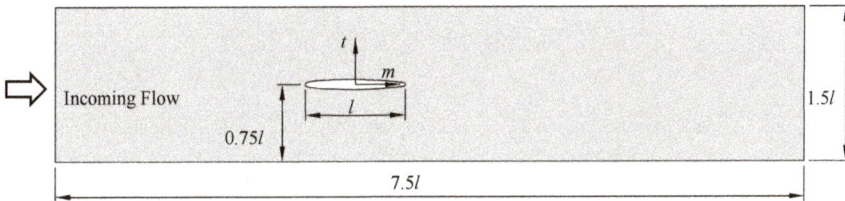

Figure 2. Flow domain for computational fluid dynamics (CFD) simulation.

3. Methods

3.1. Numerical Model of Turbulent Flow

In this study, the Reynolds-averaged Navier–Stokes (RANS) equations were used to solve the turbulent flow. Considering the turbulence isotropic assumption and rotating/curvature insensitivity of eddy viscosity models, the v^2–f model, which can be more suitable especially for the viscous flow separation from a curved hydrofoil surface, was used. The v^2–f model was based on the turbulence kinetic energy k equation, dissipation rate ε equation, velocity variance scale $\overline{v^2}$ equation, and the elliptic relaxation function f. It can be expressed as [41–43]:

$$\frac{\partial(\rho k)}{\partial t} + \frac{\partial(\rho u_i k)}{\partial x_i} = P - \rho\varepsilon + \frac{\partial}{\partial x_j}\left[\left(\mu + \frac{\mu_t}{\sigma_k}\right)\frac{\partial k}{\partial x_j}\right] + S_k \tag{3}$$

$$\frac{\partial(\rho k)}{\partial t} + \frac{\partial(\rho u_i k)}{\partial x_i} = \frac{C'_{\varepsilon 1}P - C_{\varepsilon 2}\rho\varepsilon}{T} + \frac{\partial}{\partial x_j}\left[\left(\mu + \frac{\mu_t}{\sigma_\varepsilon}\right)\frac{\partial\varepsilon}{\partial x_j}\right] + S_\varepsilon \tag{4}$$

$$\frac{\partial\left(\rho\overline{v^2}\right)}{\partial t} + \frac{\partial\left(\rho\overline{v^2}u_i\right)}{\partial x_i} = \rho k f - 6\rho\overline{v^2}\frac{\varepsilon}{k} + \frac{\partial}{\partial x_j}\left[\left(\mu + \frac{\mu_t}{\sigma_k}\right)\frac{\partial\overline{v^2}}{\partial x_j}\right] + S_{\overline{v^2}} \tag{5}$$

$$f - L^2\frac{\partial^2 f}{\partial x_j^2} = (C_1 - 1)\frac{2k - 3\overline{v^2}}{3kT} + C_2\frac{P}{\rho k} + \frac{5\overline{v^2}}{kT} + S_f \tag{6}$$

where C_1, C_2, $C_{\varepsilon1}$, $C'_{\varepsilon1}$, $C_{\varepsilon2}$, C_η, C_μ, and C_L are model constants; σ_k and σ_ε are the turbulent Prandtl numbers; and S_k, S_ε, $S_{\overline{v^2}}$, and S_f are source terms. Term T is the turbulent time scale and L is the turbulent length scale:

$$T' = \max\left[\frac{k}{\varepsilon}, 6\sqrt{\frac{\nu}{\varepsilon}}\right] \tag{7}$$

$$T = \min\left[T', \frac{\alpha}{\sqrt{3}}\frac{k}{\overline{v^2}C_\mu\sqrt{2S^2}}\right] \tag{8}$$

$$L' = \min\left[\frac{k^{3/2}}{\varepsilon}, \frac{1}{\sqrt{3}}\frac{k^{3/2}}{\overline{v^2}C_\mu\sqrt{2S^2}}\right] \tag{9}$$

$$L = C_L\max\left[L', C_\eta\left(\frac{v^3}{\varepsilon}\right)^{1/4}\right] \tag{10}$$

where α is a model constant, ν is the kinematic viscosity, and S is the strain rate tensor. Thus, the turbulent viscosity μ_t can be expressed as

$$\mu_t = \rho C_\mu\overline{v^2}T. \tag{11}$$

All the default values of the model constants are listed in Table 1. Based on $C_{\varepsilon1}$, the value of $C'_{\varepsilon1}$ can be calculated by

$$C'_{\varepsilon1} = C_{\varepsilon1}\left(1 + 0.045\sqrt{\frac{k}{\overline{v^2}}}\right). \tag{12}$$

Table 1. Default value of the model constants in the v^2–f turbulence model.

Constant	α	C_1	C_2	$C_{\varepsilon1}$	$C_{\varepsilon2}$	C_η	C_μ	C_L	σ_k	σ_ε
Value	0.6	1.4	0.3	1.4	1.9	70	0.22	0.23	1.0	1.3

3.2. CFD Setup

Based on the setting of the turbulent flow, the flow domain was discretized using unstructured mesh elements as shown in Figure 3. The mesh scheme was determined by an independence check and had 64,868 nodes in total. Considering the usage of the v^2–f turbulence model, the near-wall mesh in the boundary layer was checked and refined with three prism layers (first layer height 1×10^{-4} mm, growth rate 1.2). The y^+ value was in the range of 0.08–6.86, which fits the requirement of a direct near-wall solution. In the simulation, the Reynolds number Re was set to 5×10^5 with the incidence angle between 0 and 6 degrees. The boundary conditions were set as follows: Firstly, the velocity inlet boundary was set at the inflow, the velocity was perpendicular to the inlet boundary, and the pressure followed the Neumann condition. Secondly, the pressure boundary was set at the outflow with average 0 Pa, and the velocity followed the Neumann condition. Thirdly, the no-slip condition was applied on the wall boundaries including the upper domain boundary, lower domain boundary, and the foil surface. Moreover, to simplify the 3D case to 2D, symmetry boundaries were given perpendicular to the 2D domain plane. The fluid medium was set as water at 20 °C with density 1×10^3 kg/m^3 and dynamic viscosity 1.01×10^{-3} Pa·s. The steady-state simulation was conducted with a maximum iteration number of 1000, and the convergence criterion was set to 1×10^{-5}. The transient-state simulation was conducted based on the steady-state simulation. For a better convergence performance and timely flow regime resolution, the total time and time step were determined based on the Reynolds number and Courant number. Finally, the total time was set as 2 s with a constant time step of 2×10^{-5} s which can be suitable in this case.

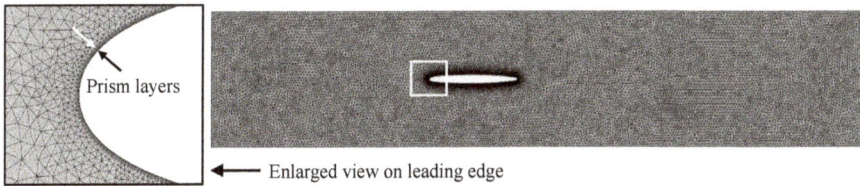

Figure 3. Schematic map of mesh elements with an enlarged view on LE.

3.3. Brief Introduction to the Diffusion Angle Integral Method

The Diffusion-angle Integral (DI) method [40] can improve the cavitation performance by changing the geometry of the foil LE. First, the DI method requires the geometric deconstruction of the foil, as shown in Figure 4. The shape of the arc and diffusion section of the foil changes rapidly. So, the DI method was mainly used to design the geometry of the foil LE. In this study, the length of the design section was $0.15l$ along the length of the meanline, which is divided into circular and integral sections. Through the DI method, the number of design parameters can be simplified to three, which greatly improves the design efficiency. The DI method is mainly divided into five steps as follows:

1. Providing the long/short axis ratio $R_{ab} = a_{LE}/b_{LE}$;
2. Based on R_{ab}, scaling the ellipse arc into an arc;
3. Providing the diffusion angle γ_s and calculating the scaled LE arc r_{LE};
4. Providing the thickness integral coefficient B (the change rate of Part 2 in Figure 4) and integrating the thickness diffusion part;
5. Based on R_{ab}, scaling the designed arc back to an ellipse arc.

Based on the steps above, the LE ellipse arc can be calculated under coordinate scaling:

$$a_{LE} = R_{ab} \frac{t_{O'} - 2m_{O'} \tan \gamma_s}{2(\sin \gamma_s \tan \gamma_s - \tan \gamma_s + \cos \gamma_s)} \tag{13}$$

$$b_{LE} = \frac{a_{LE}}{R_{ab}} \tag{14}$$

where $t_{O'}$ is the thickness at O' and $m_{O'}$ is the m position at O'. The increase in t at the thickness diffusion section can be calculated by

$$\Delta t = C_s \int_{m_A}^{m_{O'}} \tan \gamma(m) \, dm \tag{15}$$

where m_A is the m position at point A, C_s is the scale factor, and $\gamma(m)$ is the thickness integral expression which is defined as

$$\gamma(m) = \left(\frac{m_{O'} - m}{m_{O'} - m_A} \right)^B. \tag{16}$$

The variation law at thickness diffusion section can be controlled by the coefficient B. Applying DI method can be simple by following the steps above but the detailed mathematical deduction process of the DI method is complex. Hence, only the brief introduction is put here and the detail of DI method can be found in Ref. [40].

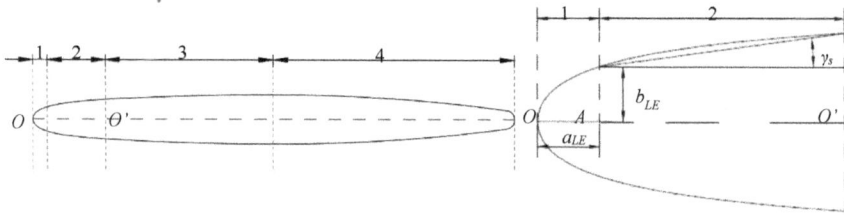

Figure 4. Geometry deconstruction and parameters of the Diffusion angle Integral method: 1—Elliptical arc section; 2—Thickness diffusion section; 3—Thickness transition section; 4—Thickness reduction section.

3.4. Orthogonal Testing

The orthogonal test method can be used instead of large scale comprehensive test with fewer test times. In this paper, 25 tests (3 factors and 5 levels) were used in the design of orthogonal test. The optimum values of 3 parameters in the DI method under different incidence angles were analyzed. The factor levels of R_{ab}, γ_s, B are shown in Table 2.

Table 2. Factor levels.

Level	Factor		
	R_{ab}	γ_s [degree]	B
1	1	1	1
2	1.5	2	2
3	2	3	3
4	2.5	4	4
5	3	5	5

In the process of flow, the dimensionless pressure coefficient Cp is used to characterize the surface pressure of the foil:

$$C_p = \frac{p - p_\infty}{\frac{1}{2}\rho V_\infty^2} \tag{17}$$

where p is the foil surface pressure, p_∞ is the reference position average pressure (calculation domain inlet), and V_∞ is the reference position average speed (calculation domain inlet). Typically, the number of cavitation number C_σ is defined as:

$$C_\sigma = \frac{p_\infty - p_v}{\frac{1}{2}\rho V_\infty^2}, \tag{18}$$

when cavitation begins, we have $p_{min} = p_v$. Therefore, the minimum pressure coefficient C_{pmin} on the foil surface is negatively correlated with the inception cavitation number $C_{\sigma i}$:

$$-C_{pmin} = C_{\sigma i}. \tag{19}$$

The minimum pressure coefficient of the foil surface C_{pmin} was chosen as the evaluation index. As shown in Table 3, the minimum pressure coefficients $C_{p\alpha}$ of foils with different incidence angles of 0, 3, and 6 degrees were obtained by numerical simulation. Then, the range analyses were conducted based on the results of orthogonal test as shown in Table 4. $K_1 \sim K_5$ and $k_1 \sim k_5$ are the polar difference values and averaged polar difference values, respectively. R is the range value based on orthogonal test.

Table 3. Orthogonal test table with factors, levels and test values.

Order Number	R_{ab}	γ_s [deg]	B	C_{p0}	C_{p3}	C_{p6}
1	1	1	1	−1.4987	−2.889	−4.4240
2	1	2	2	−1.4316	−2.8391	−4.4486
3	1	3	3	−1.3561	−2.8080	−4.5451
4	1	4	4	−1.2756	−2.7677	−4.5856
5	1	5	5	−1.1910	−2.7169	−4.5187
6	1.5	1	2	−1.0299	−2.1063	−3.5227
7	1.5	2	3	−0.9875	−2.0877	−3.5436
8	1.5	3	4	−0.9407	−2.0723	−3.5969
9	1.5	4	5	−0.8888	−2.0526	−3.5870
10	1.5	5	1	−0.8322	−2.0367	−3.6896
11	2	1	3	−0.7890	−1.7245	−3.1430
12	2	2	4	−0.7610	−1.7185	−3.1881
13	2	3	5	−0.7299	−1.7103	−3.2384
14	2	4	1	−0.6940	−1.7091	−3.2879
15	2	5	2	−0.6526	−1.7020	−3.3610
16	2.5	1	4	−0.6439	−1.5140	−3.0710
17	2.5	2	5	−0.6262	−1.5180	−3.1082
18	2.5	3	1	−0.6048	−1.5189	−3.1723
19	2.5	4	2	−0.5783	−1.5251	−3.2313
20	2.5	5	3	−0.5467	−1.5277	−3.3146
21	3	1	5	−0.5469	−1.4048	−3.1355
22	3	2	1	−0.5357	−1.4083	−3.1153
23	3	3	2	−0.5216	−1.4109	−3.2224
24	3	4	3	−0.5027	−1.4163	−3.2881
25	3	5	4	−0.4768	−1.4230	−3.3829

Table 4. Range analysis results.

ParamETERS	C_{p0}			C_{p3}			C_{p6}		
	R_{ab}	γ_s [deg]	B	R_{ab}	γ_s [deg]	B	R_{ab}	γ_s [deg]	B
K_1	−9.0567	−6.3002	−6.0576	−14.144	−9.8917	−9.8353	−20.095	−15.252	−15.524
K_2	−6.5729	−6.1712	−6.1082	−10.546	−9.8318	−9.8667	−15.855	−15.314	−15.609
K_3	−5.3795	−6.0521	−6.0855	−8.8359	−9.8091	−9.8531	−14.194	−15.587	−15.642
K_4	−4.7500	−5.8966	−6.0104	−7.9727	−9.7685	−9.7886	−13.778	−15.725	−15.619
K_5	−4.3877	−5.7268	−5.8853	−7.5234	−9.7216	−9.6791	−13.880	−15.924	−15.409
k_1	−1.8114	−1.2601	−1.2115	−2.8289	−1.9783	−1.9671	−4.0190	−3.0503	−3.1048
k_2	−1.3146	−1.2342	−1.2216	−2.1093	−1.9664	−1.9733	−3.1711	−3.0629	−3.1217
k_3	−1.0759	−1.2104	−1.2171	−1.7672	−1.9618	−1.9706	−2.8388	−3.1175	−3.1284
k_4	−0.9500	−1.1793	−1.2021	−1.5945	−1.9537	−1.9577	−2.7556	−3.1450	−3.1238
k_5	−0.8775	−1.1454	−1.1771	−1.5047	−1.9443	−1.9358	−2.7760	−3.1848	−3.0817
R	0.9338	0.1147	0.0446	1.3242	0.0340	0.0375	1.2634	0.1345	0.0467

It can be seen from Tables 2 and 3 that R_{ab} has the greatest influence on the minimum pressure coefficient C_p of the foil surface. When the incidence angle is 0 degree, the optimal parameter combination is $R_{ab} = 3$, $\gamma_s = 5$, and $B = 4$. When the incidence angle is 3 degrees, the optimal parameter combination is $R_{ab} = 3$, $\gamma_s = 2$, and $B = 1$. When the incidence angle is 6 degrees, the optimal parameter combination is $R_{ab} = 2.5$, $\gamma_s = 1$, and $B = 4$. The optimal parameter combination given by the orthogonal testing is the basis for the next step in finding the optimal design.

4. Optimization

4.1. Global Dynamic Criterion Algorithm

To further optimize the design, an improved hill climbing algorithm with heuristic parallel characteristics was proposed in this paper. The algorithm was used to search for the distribution of multiple local optimal values in a certain area. It can optimize the design parameters of foil LE.

The optimal value of the design parameters of foil LE is searched quickly and efficiently. The minimum pressure coefficient of foil surface can be improved. The algorithm includes the following steps:

(1) The algorithm is based on hill climbing algorithm. Firstly, 10 parameter combinations denoted as F_n ($n = 1, 2, 3 \ldots, 10$) are randomly generated near the optimal values obtained by the orthogonal experiments of 3 parameters R_{ab}, γ_s, B in the DI method. The corresponding foil is numerically simulated for each set of parameters F_n. Three different cases of incidence angle 0, 3 and 6 degrees are calculated respectively. The minimum pressure coefficient $C_{p\alpha}$ of the foil LE shape under each incidence angle is obtained. Therefore, the minimum pressure coefficient obtained from the LE parameters of each group is defined as C_{pmin}:

$$C_{pmin} = I \times C_{p0} + j \times C_{p3} + k \times C_{p6} \qquad (i + j + k = 1). \qquad (20)$$

When focus on cavitation performance at 0 degree of incidence angle, here are $i = 0.8$, $j = 0.1$, $k = 0.1$. For 3 degrees, here are $i = 0.1$, $j = 0.8$, $k = 0.1$. For 6 degrees, here are $i = 0.1$, $j = 0.1$, $k = 0.8$.

(2) We then set the initial decision condition C_{pt}. We decide whether the C_{pmin} obtained from each set of parameters F_n satisfies the criteria. If $C_{pmin} \geq C_{pt}$, we keep this set of parameters and search in the neighborhood of each parameter. The search range should not exceed 5% of the total parameter range. If $C_{pmin} < C_{pt}$, we discard this set of parameters, and re-generate a group for the next round of the search. The optimal value of each search is T_n ($n = 1, 2, 3, \ldots, 10$).

(3) When searching in a small neighborhood, the new set of parameters F'_n will be used for the foil LE design. A new two-dimensional foil is thus obtained. Three different cases of incidence angle—0, 3, and 6 degrees—are calculated. The minimum pressure coefficient $C_{p\alpha}$ of the foil LE shape under each incidence angle is obtained. Taking the case of 0 degree as an example, C'_{pmin} is obtained using Equation (16). For each group of F'_n, the optimal T_1 value obtained from the first search is the criterion for the second search. If $C'_{pmin} \geq T_1$, we keep this set of parameters and search in the neighborhood of the new set of parameters F'_n. The search range should not exceed 5% of the total parameter range. If $C'_{pmin} < T_1$, we discard this set of parameters, and re-generate a group for the next round of the search.

(4) For the third search round, it is necessary to continue to change the criteria T_3. It is changed to the weighted value of the optimal value T_1 of the first search round and the optimal value T_2 of the second search round. T_3 is defined as:

$$T_3 = p \times T_1 + q \times T_2 \qquad (p + q = 1) \qquad (21)$$

The purpose of changing the criteria is to continuously improve the goal of optimization. Here, $p = 0.4$ and $q = 0.6$.

(5) In order to avoid falling into the local optimal solution, we set a small probability to terminate the current search. The termination probability Y and iteration number x have a certain functional relationship. The function has the following characteristics. With increasing x, the termination probability Y increases. When iteration number $x \rightarrow +\infty$, termination probability $Y \rightarrow 1$. According to the above characteristics, the function can be written as

$$\begin{cases} Y(x) = 0, x < 3 \\ Y(x) = \left[-e^{\left(\frac{b-x}{a}\right)} + c \right]\%, x \geq 3 \end{cases} \qquad (22)$$

where a, b, and c are constants. The function can adjust the specific values of a, b, and c according to the maximum iteration number, so as to adjust the distribution of the termination probability. Here, the values in this study are as follows: $a = 2.8$, $b = 10$, and $c = 1$.

(6) The maximum iteration number in this article is set to 50 to have a balance between searching time and improvement effect. Repeat the above steps until reach the maximum iteration number, and finally get the global optimal solution. The optimum design parameters of foil LE at 0 degree, 3 degrees and 6 degrees of incidence angles are obtained.

4.2. Comparison of Foil Geometry

After optimization, the optimal geometries were get respectively on 0 degree, 3 degrees and 6 degrees. As shown in Figure 5, differences can be found in the range of 0–0.15 m/l. The optimal geometry for 0 degree has the smallest thickness values within 0–0.15 m/l. The optimal geometry for 6 degrees has the largest thickness values within 0–0.15 m/l. The differences of geometry showed the adaptability of incoming flow striking and local separation caused by incidence angle. The mechanism was analyzed in detail in the next section.

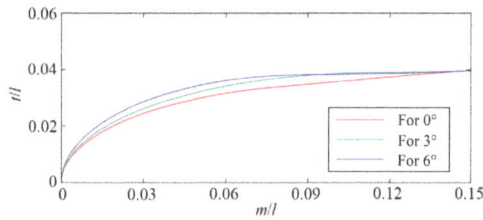

Figure 5. Schematic diagram of the optimal foil LE shape for 0, 3, and 6 degrees.

5. Results and Discussion

5.1. Comparison of $-C_{pmin}$ and Lift/Drag Ratio

The $-C_{pmin}$ of the original foil and the optimal three foils were compared, as shown in Figure 6. It can be seen from the Figure that the $-C_{pmin}$ of the optimal foils and the original foil at the 0 degree is similar with only difference of 1.2%. However, with the increase of the incidence angle, the optimal foils are lower than the original foil at 3 degrees and 6 degrees, which decreases by 11.4% and 14.5% respectively. It shows that the cavitation performance of the optimal foil at 3°~6° is improved. Comparing the three optimal foils, it can be seen that the $-C_{pmin}$ of the optimal for 0 degree' foil are lower in the range of 0~3 degrees, but higher at 6 degrees' incidence angle. However, the $-C_{pmin}$ of the optimal for 6 degrees' foil are lower at 6 degrees' incidence angle, but higher in the range of 0~3 degrees. Generally, the application of the DI method and GDC algorithm on optimizing the foil geometry can effectively improve the cavitation performance at large incidence angle. The foil with better cavitation performance can be obtained by focusing on a certain incidence angle, but the performance of both large and small incidence angles is difficult to be considered at the same time. Figure 7 compared the lift/drag ratio F_L/F_D of the original foil and the optimal foils. From the comparison of F_L/F_D, it can be seen that the F_L/F_D of the optimal foils are slightly lower than that of the original foil, but the overall difference is very small. The minimum difference is 0.1% and the maximum difference is 5%. The results showed that the performance of foil is not affected after optimization.

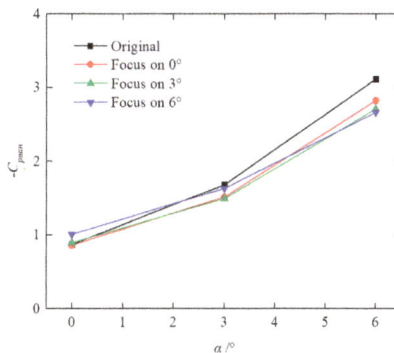

Figure 6. Comparison of $-C_{pmin}$ values.

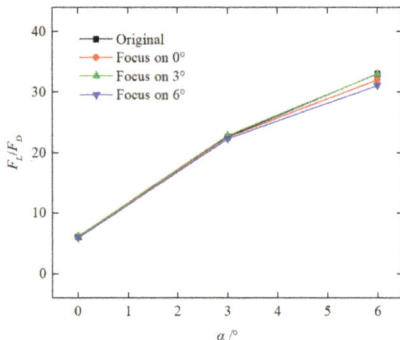

Figure 7. Comparison of lift/drag ratio F_L/F_D.

5.2. Pressure Distribution on the Foil Surface

Figure 8 shows a comparison of the surface pressure distribution curves of the three optimal foils under different incidence angles. It can be seen from the figure that the position of the lowest pressure point obtained by different LE shapes at the same incidence angle is relatively small. The lowest pressure coefficient has a certain difference. In Figure 8a, where $\alpha = 0$ degree, the optimal foil for 0 degree has the gentlest pressure drop on LE, while the optimal foil for 6 degrees has the most sudden pressure drop on LE. This shows that the optimal foil for 0 degree fits the incidence angle of 0 degree the best. The opposite relationship can be found in Figure 8c. However, the situation becomes complex in Figure 8b. At $\alpha = 3$ degrees, the optimal foil for 3 degrees fitted the incidence angle of 3 degrees the best. The optimal foil for 0 degree also performed well at the incidence angle of 3 degrees. The optimal foil for 6 degrees was the worst and had the minimum $-C_{pmin}$. Generally, the optimal design for a specific incidence angle has a gentler pressure drop on the LE. The cavitation scale would consequently be smaller after the design optimization.

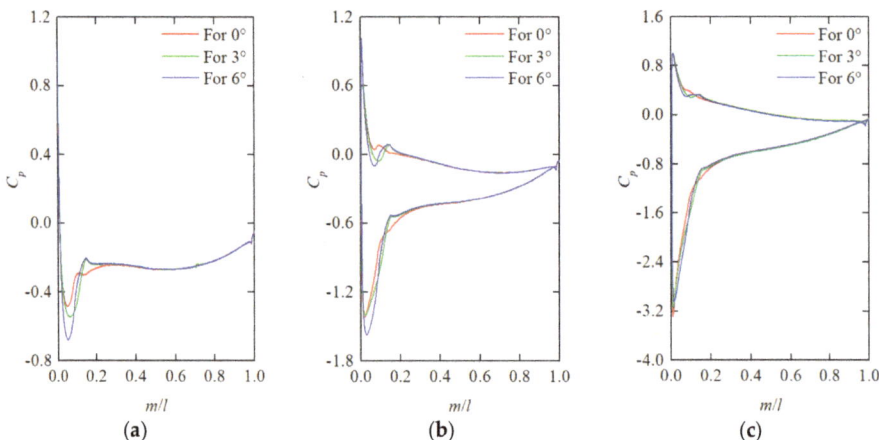

Figure 8. Surface pressure distribution curves of the three optimal foils under different incidence angles: (a) $\alpha = 0°$; (b) $\alpha = 3°$; (c) $\alpha = 6°$.

5.3. Flow Field around the Foil

The pressure and velocity vectors at the LE of the three optimal foils were compared and analyzed, as shown in Figure 9 and Figure 10. It can be seen from Figure 9 that when the incidence angle is 0 degree, the low-pressure area of the LE of the foil optimal for 0 degree is smaller. With the change of the LE geometry, the low-pressure zones of the foils optimal for 3 degrees and for 6 degrees gradually increase. After increasing the incidence angle, the pressure distributions of the three optimal foils' LE are similar. By analyzing the velocity vector, it is found that the three foils each have the smallest flow separation at their respective optimal incidence angle. This is because the gradient of geometric change is slower under the corresponding incidence angle. The local separation of the optimal foil LE is improved obviously. The pressure drop near the LE slows down. The shape of the foil LE can be adapted to the direction of incoming flow, making the flow more suitable for the foil. It is thus shown that the cavitation performances of the optimal foils at various incidence angles can be significantly improved by using the DI method and the GDC algorithm.

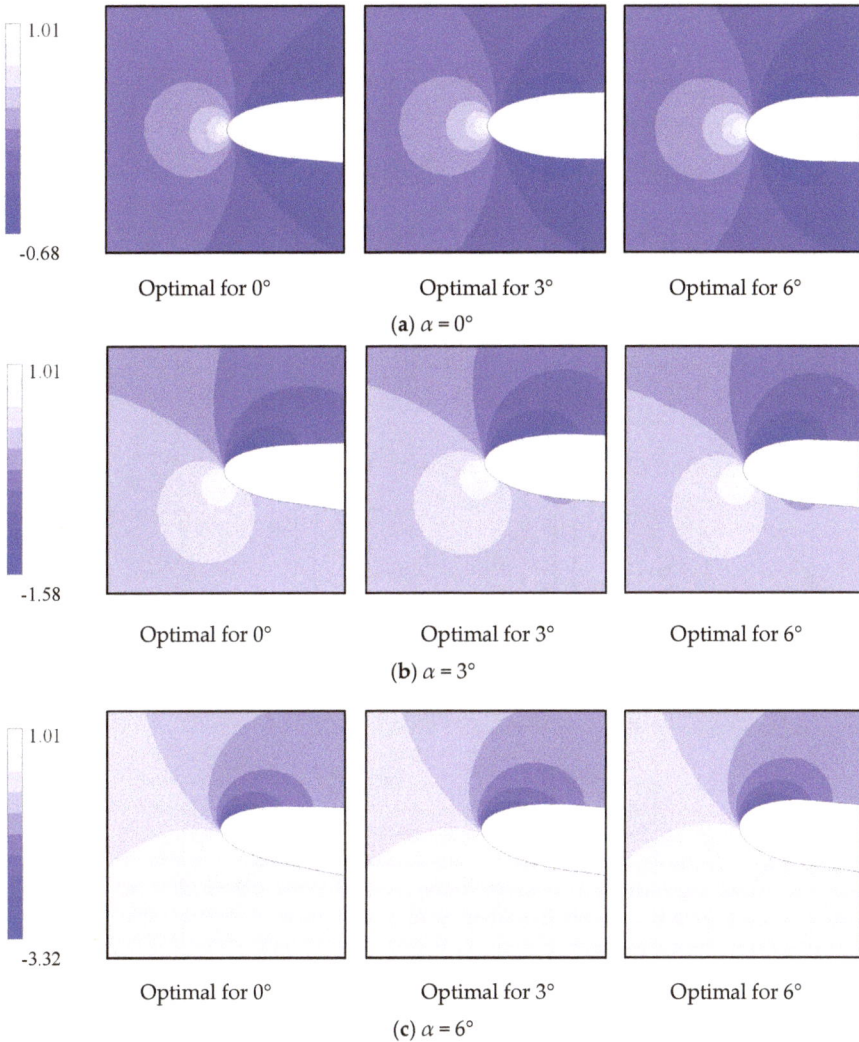

Optimal for 0°   Optimal for 3°   Optimal for 6°

(a) $\alpha = 0°$

Optimal for 0°   Optimal for 3°   Optimal for 6°

(b) $\alpha = 3°$

Optimal for 0°   Optimal for 3°   Optimal for 6°

(c) $\alpha = 6°$

Figure 9. Contour of the dimensionless pressure coefficient C_p.

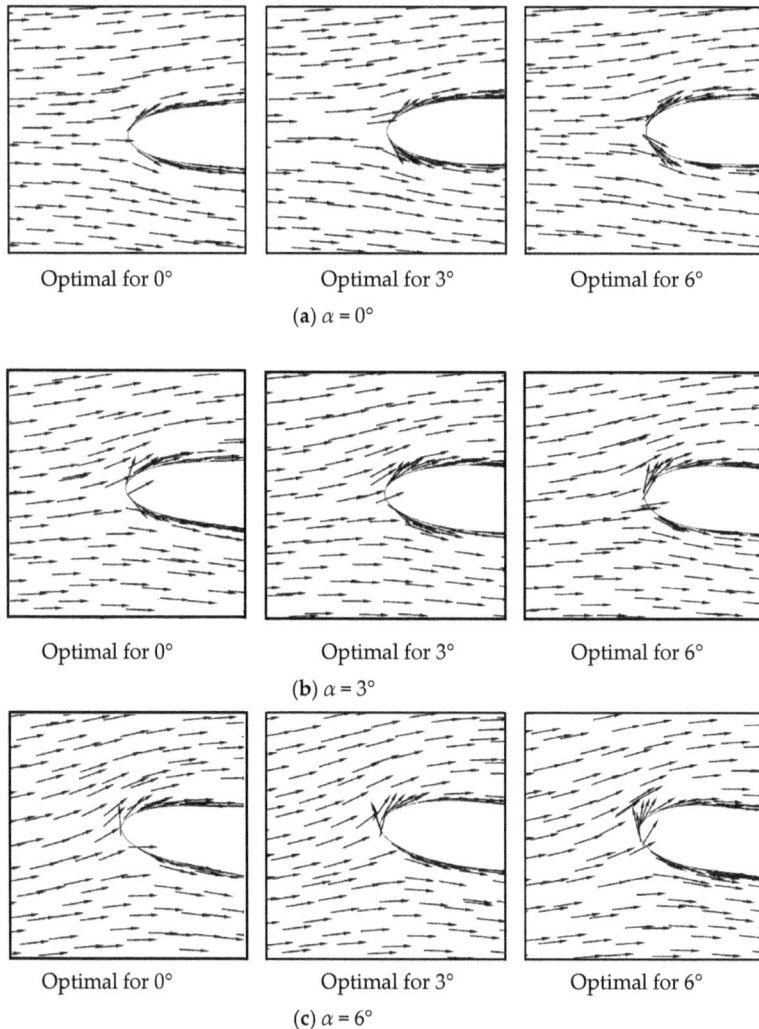

Optimal for 0°   Optimal for 3°   Optimal for 6°

(**a**) $\alpha = 0°$

Optimal for 0°   Optimal for 3°   Optimal for 6°

(**b**) $\alpha = 3°$

Optimal for 0°   Optimal for 3°   Optimal for 6°

(**c**) $\alpha = 6°$

Figure 10. Velocity vectors around the foils.

6. Conclusions

According to the study above, the following three conclusions can be drawn:

(a) The global dynamic-criterion algorithm based on the improved hill-climbing algorithm was introduced. It can initially filter the input parameters/conditions, run in a parallel mode, and set a small probability for falling into the local optimum trap. In this optimization of the cavitation performance of foils or impeller blades, the relationship between cavitation behavior and foil/blade profile is strongly nonlinear. Thus, the global dynamic criterion algorithm provided a reasonable and convenient solution for this cavitation optimization problem.

(b) Several typical methods and some new methods were combined for the optimization. The typical methods include the CFD simulation for turbulent flow and the orthogonal test. The new methods include the Diffusion-angle Integral method and the global dynamic-criterion algorithm. They worked together to search the optimal geometry for improving the cavitation performance for 0, 3 and 6 degrees' incidence angles.

(c) The front 15% geometry was re-designed with slight changing on geometry and impressive enhancement on cavitation performance. The $-C_{pmin}$ of the optimal foils was improved by 1.2%, 11.4% and 14.5% at 0 degree, 3 degrees and 6 degrees comparing with the original foil. In the design range, different parameter values of Diffusion-angle Integral method caused different geometries. The geometries fitted the incoming flow better with smaller scale LE separation and gentler pressure drop. The cavitation performance was enhanced at different incidence angles.

Above all, the geometry re-design around leading-edge can efficiently improve the cavitation performance of hydrofoil. This study provided a successful work for applying anti-cavitation design to the impellers of hydraulic turbomachinery. This study will be helpful for readers who need the improvement of cavitation performance for foils and impellers.

Author Contributions: Methodology, D.Z. and R.T.; Investigation, D.Z.; Writing-Original Draft Preparation, D.Z.; Writing-Review & Editing, R.X.; Supervision, F.W.; Funding Acquisition, R.T., R.X. and F.W.

Funding: This research was funded by National Natural Science Foundation of China, grant number 51836010; National Natural Science Foundation of China, grant number 51879265; China Postdoctoral Science Foundation, grant number 2018M640126.

Conflicts of Interest: The authors declare no conflict of interest.

References

1. Hammitt, F.G. *Cavitation and Multiphases Flow Phenomena*, 1st ed.; McGraw-Hill: New York, NY, USA, 1980; ISBN 9780070259072.
2. Luo, X.; Ji, B.; Tsujimoto, Y. A review of cavitation in hydraulic machinery. *J. Hydrodyn.* **2016**, *28*, 335–358. [CrossRef]
3. Pan, Z.Y.; Yuan, S.Q. *Fundamentals of Cavitation in Pumps*, 1st ed.; Jiangsu University Press: Zhenjiang, China, 2013.
4. Bishop, R.J.; Totten, G.E. Effect of pump inlet conditions on hydraulic pump cavitation: A review. *ASTM Spec. Tech.* **2001**, *339*, 318–332.
5. Lauterborn, W.; Bolle, H. Experimental investigation of cavitation-bubble collapse in the neighbourhood of a solid boundary. *J. Fluid Mech.* **1975**, *72*, 391–399. [CrossRef]
6. Wang, G.; Liu, S.; Shintani, M. Study on Cavitation Damage Characteristics around a Hollow-Jet Valve. *JSME Int. J.* **1999**, *42*, 649–657. [CrossRef]
7. Visser, F.C.; Backx, J.J.M.; Geerts, J. Pump impeller lifetime improvement through visual study of leading-edge cavitation. In Proceedings of the International Pump Users Symposium, Houston, TX, USA, 3–5 March 1998.
8. Tao, R.; Xiao, R.; Zhu, D. Predicting the inception cavitation of a reversible pump-turbine in pump mode. In Proceedings of the 9th International Symposium on Cavitation, Lausanne, Switzerland, 6–10 December 2015.
9. Ruan, H.; Luo, X.; Liao, W. Effects of low pressure edge thickness on cavitation performance and strength for pump-turbine. *Trans. Chin. Soc. Agric. Eng.* **2015**, *31*, 32–39.
10. Adhikari, R.; Vaz, J.; Wood, D. Cavitation Inception in Crossflow Hydro Turbines. *Energies* **2016**, *9*, 237. [CrossRef]
11. Anaka, T.; Tsukamoto, H. Transient behavior of a cavitating centrifugal pump at rapid change in operating conditions-Part 2: Transient phenomena at pump startup/shutdown. *J. Fluids Eng.* **1999**, *121*, 850–856.
12. Balasubramanian, R.; Sabini, E.; Bradshaw, S. Influence of impeller leading edge profiles on cavitation and suction performance. In Proceedings of the 27th International Pump Users Symposium, Houston, TX, USA, 12–15 September 2011.
13. Obeid, S.; Jha, R.; Ahmadi, G. RANS Simulations of Aerodynamic Performance of NACA 0015 Flapped Airfoil. *Fluids* **2017**, *2*, 2. [CrossRef]

14. Wang, G.; Ostoja-Starzewski, M. Large eddy simulation of a sheet/cloud cavitation on a NACA0015 hydrofoil. *Appl. Math. Model.* **2007**, *31*, 417–447. [CrossRef]

15. Neill, G.D.; Reuben, R.L.; Sandford, P.M. Detection of incipient cavitation in pumps using acoustic emission. *J. Process. Mech. Eng.* **1997**, *211*, 267–277. [CrossRef]

16. Fukaya, M.; Okamura, T.; Tamura, Y. Prediction of cavitation performance of axial flow pump by using numerical cavitating flow simulation with bubble flow model. In Proceedings of the 5th International Symposium on Cavitation, Osaka, Japan, 1–4 November 2003.

17. Barre, S.; Rolland, J.; Boitel, G. Experiments and modeling of cavitating flows in venturi: Attached sheet cavitation. *Eur. J. Mech. B/Fluids* **2009**, *28*, 444–464. [CrossRef]

18. Acosta, A.J.; Tsujimoto, Y.; Yoshida, Y. Effects of leading edge sweep on the cavitating characteristics of inducer pumps. *Int. J. Rotating Mach.* **2007**, *7*, 397–404. [CrossRef]

19. Numachi, F.; Oba, R.; Chida, I. Effect of surface roughness on cavitation performance of hydrofoils-report 1. *J. Basic Eng.* **1965**, *87*, 495–502. [CrossRef]

20. Yao, Z.; Xiao, R.; Wang, F. Numerical investigation of cavitation improvement for a francis turbine. In Proceedings of the 9th International Symposium on Cavitation, Lausanne, Switzerland, 1–5 December 2015.

21. Kinnas, S.A. Non-Linear Corrections to the Linear Theory for the Prediction of the Cavitating Flow around Hydrofoils. Ph.D. Thesis, Massachusetts Institute of Technology, Boston, MA, USA, 1985.

22. Tao, R.; Xiao, R.; Farhat, M. Effect of leading edge roughness on cavitation inception and development on a thin hydrofoil. *J. Drain. Irrig. Mach. Eng.* **2017**, *35*, 921–926.

23. Bouziad, Y.A. Physical modelling of leading edge cavitation: Computational methodologies and application to hydraulic machinery. *EPFL* **2005**, *3353*. [CrossRef]

24. Ausoni, P. Turbulent vortex shedding from a blunt trailing edge hydrofoil. *EPFL* **2009**. [CrossRef]

25. Coutier-Delgosha, O.; Reboud, J.L.; Delannoy, Y. Numerical simulation of the unsteady behavior of cavitating flows. *Int. J. Numer. Methods Fluids* **2003**, *42*, 527–548.

26. Kunz, R.F.; Boger, D.A.; Stinebring, D.R. A preconditioned Navier–Stokes method for twophase flows with application to cavitation prediction. *Comput. Fluids* **2000**, *29*, 849–875. [CrossRef]

27. Yang, W.; Xiao, R.; Wang, F. Influence of splitter blades on the cavitation performance of a double suction centrifugal pump. *Adv. Mech. Eng.* **2014**, *6*. [CrossRef]

28. Gopalan, S.; Katz, J. Flow structure and modeling issues in the closure region of attached cavitation. *Phys. Fluids* **2000**, *12*, 895–911. [CrossRef]

29. Luo, X.; Zhang, Y.; Peng, J. Impeller inlet geometry effect on performance improvement for centrifugal pumps. *J. Mech. Sci. Technol.* **2008**, *22*, 1971–1976. [CrossRef]

30. Tan, L.; Cao, S.; Wang, Y. Numerical Simulation of Cavitation in a Centrifugal Pump at Low Flow Rate. *Chin. Phys. Lett.* **2012**, *29*, 014702. [CrossRef]

31. Ting, C.K. On the mean convergence time of multi-parent genetic algorithms without selection. *Adv. Artif. Life* **2005**, 403–412.

32. Akbari, R.; Ziarati, K. A multilevel evolutionary algorithm for optimizing numerical functions. *Int. J. Ind. Eng. Comput.* **2011**, *2*, 419–430. [CrossRef]

33. Whitley, D. A genetic algorithm tutorial. *Stat. Comput.* **1994**, *4*, 65–85. [CrossRef]

34. Colorni, A.; Dorigo, M.; Maniezzo, V. Ant system for job-shop scheduling. *Oper. Res. Stat. Comput. Sci.* **1994**, *34*, 39–53.

35. Cormen, T.H.; Leiserson, C.E.; Rivest, R.L. *Introduction to Algorithms*, 2nd ed.; McGraw-Hill: New York, NY, USA, 2001.

36. Huang, L.; Huang, P.G.; Lebeau, R.P. Optimization of aifoil flow control using a genetic algorithm with diversity control. *J. Aircr.* **2007**, *44*, 1337–1349. [CrossRef]

37. Liu, H.; Wang, K.; Yuan, S. Multicondition Optimization and Experimental Measurements of a Double-Blade Centrifugal Pump Impeller. *J. Fluids Eng.* **2013**, *135*, 111031. [CrossRef] [PubMed]

38. Liu, L.; Zhu, B.; Bai, L. Parametric design of an ultrahigh-head pump-turbine runner based on multiobjective optimization. *Energies* **2017**, *10*, 1169. [CrossRef]

39. Liu, M.; Tan, L.; Cao, S. Design method of controllable blade angle and orthogonal optimization of pressure rise for a multiphase pump. *Energies* **2018**, *11*, 1048. [CrossRef]

40. Tao, R.; Xiao, R.; Wang, F. Improving the cavitation inception performance of a reversible pump-turbine in pump mode by blade profile redesign: Design concept, method and applications. *Renew. Energy* **2019**, *133*, 325–342. [CrossRef]

41. Behnia, M.; Parneix, S.; Shabany, Y. Numerical study of turbulent heat transfer in confined and unconfined impinging jets. *Int. J. Heat Fluid Flow* **1999**, *20*, 1–9. [CrossRef]

42. Durbin, P.A. Separated Flow Computations with the k–ε–v2 Model. *AIAA J.* **1995**, *33*, 659–664. [CrossRef]

43. Parneix, S.; Durbin, P.A.; Behnia, M. Computation of 3-D Turbulent Boundary Layers Using the V2F Model. *Flow Turbul. Combust.* **1998**, *60*, 19–46. [CrossRef]

![energies logo] **energies**

MDPI

Article

Numerical Study of the Axial Gap and Hot Streak Effects on Thermal and Flow Characteristics in Two-Stage High Pressure Gas Turbine

Myung Gon Choi and Jaiyoung Ryu *

School of Mechanical Engineering, Chung-Ang University, 84, Heukseok-ro, Dongjak-gu, Seoul 06974, Korea; mgon1122@cau.ac.kr
* Correspondence: Jairyu@cau.ac.kr; Tel.: +82-2-820-5279

Received: 12 September 2018; Accepted: 1 October 2018; Published: 4 October 2018

Abstract: Combined cycle power plants (CCPPs) are becoming more important as the global demand for electrical power increases. The power and efficiency of CCPPs are directly affected by the performance and thermal efficiency of the gas turbines. This study is the first unsteady numerical study that comprehensively considers axial gap (AG) in the first-stage stator and first-stage rotor (R1) and hot streaks in the combustor outlet throughout an entire two-stage turbine, as these factors affect the aerodynamic performance of the turbine. To resolve the three-dimensional unsteady-state compressible flow, an unsteady Reynolds-averaged Navier–Stokes (RANS) equation was used to calculate a $k - \omega$ SST γ turbulence model. The AG distance d was set as 80% (case 1) and 120% (case 3) for the design value case 2 (13 mm or $d/Cs_1 = 0.307$) in a GE-E^3 gas turbine model. Changes in the AG affect the overall flow field characteristics and efficiency. If AG decreases, the time-averaged maximum temperature and pressure of R1 exhibit differences of approximately 3 K and 400 Pa, respectively. In addition, the low-temperature zone around the hub and tip regions of R1 and second-stage rotor (R2) on the suction side becomes smaller owing to a secondary flow and the area-averaged surface temperature increases. The area-averaged heat flux of the blade surface increases by a maximum of 10.6% at the second-stage stator and 2.8% at R2 as the AG decreases. The total-to-total efficiencies of the overall turbine increase by 0.306% and 0.295% when the AG decreases.

Keywords: gas turbine; axial gap; hot streak; heat transfer

---

1. Introduction

The global electrical power demand is expected to increase by 46% from 2015 to 2040, and this will lead to a large increase in electricity generation. Natural gas will be used for electricity generation instead of coal to minimize environmental problems caused by CO_2 emissions. Electricity generation using natural gas is expected to increase from 5.22 trillion kWh in 2015 to 9.6 trillion kWh in 2040, an increase of 83.9%. Therefore, the importance of combined cycle power plants (CCPPs) that use natural gas is expected to considerably increase in the future. CCPPs consist of gas turbines that use natural gas and steam turbines that use steam, which is emitted from heat recovery steam generators. Increase in gas turbine power cause a similar increase of power in steam turbine and improves the overall efficiency of CCPP. The gas turbines' aerodynamic performance and thermal efficiency have a direct effect on the cost of generated power of CCPPs.

One method of increasing gas turbine efficiency is to increase the turbine inlet temperature (TIT). However, a TIT distribution that is higher than the melting point of the material of the turbine blades results in a high thermal load on the turbine blades. Without a suitable cooling system, this leads to high-temperature corrosion and becomes a major factor in reducing the life of a turbine. To increase the TIT, it is necessary to more closely analyze the heat transfer characteristics of the surface of the turbine blade through an analysis of the transformed temperature distribution of the turbine inlet. The TIT distribution is directly affected by the fluid flow characteristics and temperature distribution of the combustor outlets. The temperature distribution of the turbine inlet is called a hot streak (HS), and it creates a complex heat transfer environment in the fluid flow passage and blade surface of the turbine [1]. HS has a different effect on the turbine blade surface compared with uniform temperature distribution; therefore, it is important to consider HS to analyze the overall performance and efficiency. As such, various numerical analyses and experimental studies have already been performed to describe the effect of HS on the heat transfer phenomena of a turbine blade. Butler et al. observed that high-temperature gas is concentrated on the pressure side of the rotor owing to an increased incidence angle and discovered points where the heat transfer effect is weakened [2]. Povey and Qureshi performed an experimental study on the temperature distribution in a combustor outlet and developed enhanced OTDF (EOTDF), which has a temperature distribution ratio of 1.65 [3]. EOTDF was used to create HS and they studied the influence of uniform temperature distributions at the turbine inlet as well as changes in the position of HS in the radial direction on the stator and rotor [1,4]. Bai-Tao An et al. studied the effect of uniform temperature distributions and HS inlet conditions on aerodynamic parameters such as total temperature, static pressure, and velocity [5]. Feng et al. found that the second-stage stator exhibits a high efficiency and low thermal load when its clocking position and HS were aligned [6]. Smith found that the time-averaged heat load when the HS is aligned with the stator had a large effect on the stator and a small effect on the rotor compared with the case when the HS is aligned with the stator passage [7]. However, these studies did not consider the effect of axial gap (AG) on the thermal and flow characteristics of the blade surface.

To increase gas turbine efficiency, it is important to analyze not only HS but also factors that affect aerodynamic performance. One factor that affects aerodynamic performance is the AG, i.e., a length of the straight line from the stator to the rotor. AG is a factor that directly affects the design and operation of turbomachines. It not only determines the overall size, length, and weight of a turbomachine but it also affects the unsteady flow in the rotor, noise, and aerodynamic performance of the turbine blade [8,9]. Furthermore, if the AG is too short, problems such as reduced fatigue life due to high inlet temperature occur. Therefore, experimental and numerical studies have been performed on the AG as an element that affects the overall turbine performance and efficiency. The AG affects the heat transfer coefficient (HTC) of the blade midspan in large-scale axial-flow turbines as well as the flow at the hub [10,11]. Funazaki performed an experimental and numerical analysis on changes in the flow angle in the stator outlet according to the AG in the first stage of the turbine [12]. Syed performed a numerical study on the composite effect of tip clearance and the AG of a stator blade in a multistage compressor [9]. It was found that changes in the first-stage stator and rotor AG affected turbine performance, but changes in the distance between the first-stage rotor and second-stage stator did not have a considerable effect on performance improvement. Previous studies found that as the AG becomes shorter, it affects the rotor torque and improves aerodynamic performance. However, they did not consider the thermal and flow characteristics in which the turbine blade surface is affected by the nonuniform temperature distribution of the turbine inlet according to the AG.

Accurate predictions of thermal and flow characteristics in a high-pressure gas turbine at the turbine blade and passage in more than one stage can have a considerable effect on turbine design. However, numerical investigations on multistage gas turbines are still expensive, and not many studies have been conducted thus far [13]. Adel performed a numerical analysis on a two-stage gas turbine with steady and unsteady states. The first stage was not affected by the second stage, but the second stage was strongly affected by the first stage. It was found that the upstream flow caused distortion in the downstream flow along the circumferential direction, and the flow interacts with the secondary flow and tip leakage flow of the blade [14]. Therefore, it is necessary to accurately understand the effect of the first stage flow on the second-stage and predict the thermal and flow characteristics within the passage and the heat transfer distribution of the blade surface.

Previous numerical studies for the AG effects applied uniform inlet temperature distributions to examine the aerodynamic performance of a turbine. Nonuniform inlet temperatures have been applied to predict the heat transfer distribution of the blade surface and the thermal and flow characteristics, but the AG was not considered. Furthermore, numerical studies that considered AG or HS have been performed on turbines with 1.5 or fewer stages. The gas turbine efficiency is affected by various design elements such as the AG and temperature distribution and varies in each stage. Therefore, this study performed a numerical analysis to investigate the effect of the AG on the thermal and flow characteristics of the blade surface and passage when an HS is applied to a two-stage turbine.

2. Numerical Method

2.1. Numerical Model and Grid

The gas turbine configuration used in this study is a GE-E^3 gas turbine model. The actual gas turbine consists of 46 stators and 76 rotors in the first stage and 48 stators and 70 rotors in the second stage [15]. If one stator and two rotors have the same pitch angle, calculation errors during the numerical analysis can be minimized. As such, the number of blades was adjusted using a domain scaling method to create two rotors that correspond to the pitch of one stator in each stage [16]. In a state where the number of first stage rotors and the solidity of each blade are fixed, each chord length of the first stage stator (S1), second-stage stator (S2), and the rotor (R2) were magnified by 46/38, 48/38, and 70/76, respectively. Therefore, the number of blades used in the analysis was 38 for S1, 76 for the first stage rotor (R1), 38 for S2, and 76 for R2. Figure 1a shows the adjusted blade configuration for each stage and the computational domain used in this study, and Table 1 lists the information on each blade. The tip clearance of the rotor was 1% of the rotor height in R1 and 0.6% of the rotor height in R2.

In the numerical analysis, Ansys Turbogrid was used to create a hexahedral grid as shown in Figure 1b. To accurately predict the thermal and flow characteristics within the boundary layers, y^+ was set to less than 1 at all walls and less than 0.5 at the blade surface. A grid independence test was performed to determine the appropriate number of grids to be used in the study. Table 2 lists the number of meshes (to achieve the equal domain pitch angle, the R1, R2 meshes in the list is doubled) used in the test and the area-averaged heat flux of the blade. In the grid independence test, grids were created for each blade by multiplying 1.34 times based on Mesh-1. In Mesh-1 and 2 and Mesh-2 and 3, the relative errors of the area-averaged heat fluxes were both under 0.01% for S1, and they were 0.75% and 0.41% for R1, respectively; thus, the Mesh-2 grid was used for S1 and R1. In Mesh-2 and 3 and Mesh-3 and 4, the relative errors of area-averaged heat flux where 1.83% and 0.41% for S2, respectively, and 1.79% and 0.8% for R2, respectively, so the Mesh-3 grid was used for S2 and R2.

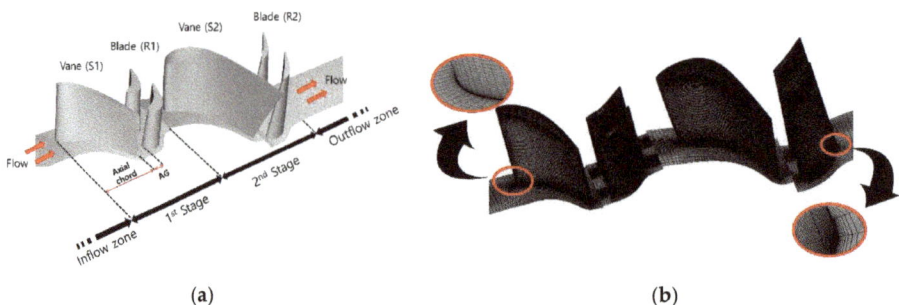

(a)   (b)

Figure 1. Two-stage stator and rotor geometry details (based on GE-E^3 (General Electric-energy efficient engine geometry [15]) turbine) (**a**) computational domain with domain scaling, and (**b**) computational grid of domain.

Table 1. Information on GE-E^3 turbine and domain.

	GE-E^3				Domain Scaling			
	S1	R1	S2	R2	S1	R1	S2	R2
Airfoil number	46	76	48	70	38	76	38	76
Chord length [mm]	70.412	41.821	78.370	40.734	85.211	41.821	98.991	37.504
TE pitch [mm]	47.042	28.452	45.337	31.066	56.646	28.452	57.268	28.613
Pitch chord ratio	0.668	0.680	0.579	0.668	0.668	0.680	0.579	0.668
Tip clearance [mm]	-	0.426	-	0.4191	-	0.426	-	0.4191

Table 2. Area-averaged heat flux at the surface of the four-row airfoil for applying the appropriate domain mesh (Mesh-2 is applied for 1st stage stator (S1) and rotor (R1), and Mesh-3 is applied for 2nd stage stator (S2) and rotor (R2)).

	Domain Node Number ($\times 10^6$)				Area Averaged Heat Flux [W/m^{2}K]				
	S1	R1	S2	R2	S1	Relative Error	R1 (Relative Error)	S2 (Relative Error)	R2 (Relative Error)
Mesh-1	0.99	2.48	0.97	2.34	277,226	Relative errors are less than 0.01%	382,851 (0.75%)	236,005 (1.83%)	185,058 (1.79%)
Mesh-2	**1.32**	**3.38**	1.33	3.12	277,022		385,727 (**0.41%**)	240,406 (0.96%)	188,448 (1.32%)
Mesh-3	1.82	4.40	**1.85**	**4.18**	276,993		387,291	242,747 (**0.52%**)	190,926 (**0.8%**)
Mesh-4	-	-	2.49	5.50	-		-	244,028	192,473

2.2. Numerical Details and Boundary Conditions

A continuity equation, momentum equation, and energy equation were used to analyze the compressible fluid flow of a three-dimensional unsteady state. These equations can be expressed as Equations (1)–(3), respectively.

$$\frac{\partial \rho}{\partial t} + \frac{\partial}{\partial x_i}(\rho u_i) = 0 \tag{1}$$

$$\frac{\partial}{\partial t}(\rho u_i) + \frac{\partial}{\partial x_i}(\rho u_i u_j) = -\frac{\partial P}{\partial x_i} + \frac{\partial}{\partial x_j}\left[\mu\left(\frac{\partial u_i}{\partial x_j} + \frac{\partial u_j}{\partial x_i} - \frac{2}{3}\delta_{ij}\frac{\partial u_k}{\partial x_k}\right)\right] + \frac{\partial}{\partial x_j}\left(-\rho\{u_i' u_j'\}\right) \tag{2}$$

$$\frac{\partial}{\partial t}(\rho E) + \frac{\partial}{\partial x_j}(u_j(\rho E + P)) = \frac{\partial}{\partial x_j}\left[\left(k_{eff}\right)\frac{\partial T}{\partial x_j}\right] + \frac{\partial}{\partial x_j}\left[u_i\mu_{eff}\left(\frac{\partial u_i}{\partial x_j} + \frac{\partial u_j}{\partial x_i} - \frac{2}{3}\delta_{ij}\frac{\partial u_k}{\partial x_k}\right)\right] \tag{3}$$

Here, ρ is the fluid's density. u is the fluid's velocity. P is the fluid's pressure. μ is the fluid's viscosity. In Equation (3), E is the specific internal energy. k_{eff} is the effective thermal conductivity. C_P is the specific heat capacity. μ_{eff} is the effective dynamic viscosity. To accurately predict the flow separation phenomena, a k − ω SST γ turbulence model was used. To solve the governing equation, a commercial computational fluid dynamics software ANSYS CFX was used.

The boundary conditions used in this simulation are indicated following the GE-E^3 turbine test performance report [15], air was used as the working fluid, and a total pressure of 344,740 Pa was used at the inlet and a static pressure of 50,000 Pa was used at the outlet. The rotor's rotation speed was 3600 RPM. At the inlet, the Mach number was 0.11 (45.1 m/s). The Reynolds number was 210,000 based on S1's axial chord length (Cs$_1$ = 42.4 mm). The inlet turbulence intensity was set as 5%. A no-slip condition was used on the wall surfaces. To calculate the HTC, simulations for both isothermal and adiabatic wall conditions were conducted. Under the isothermal wall condition, the turbine blade temperature was 389.95 K.

To analyze the internal thermal and flow characteristics according to the inlet temperature field, a nonuniform HS inlet temperature condition with a maximum temperature of 838 K at the center was used, and a uniform inlet temperature condition with a temperature of 728 K was used, as shown in Figure 2. In addition, to examine the effects according to the AG when an HS was applied, three cases were analyzed in which the distance of AG, d, of S1 and R1 was set as the design value (case 2: 13 mm or d/Cs_1 = 0.307), 80% of the design value (case 1: 10.4 mm or d/Cs_1 = 0.245), and 120% of the design value (case 3: 15.6 mm or d/Cs_1 = 0.368).

For numerical simulations, a 96-core workstation (4 Intel(R) Xeon(R) CPU E7-8890 v4 @ 2.20 GHz, RAM 512 GB) and a 44-core workstation (2 Intel(R) Xeon(R) CPU E5-2699 v4 @ 2.20 GHz, RAM 256 GB) were used, and the computation time required for a case was about 60 hours when using the 36-core.

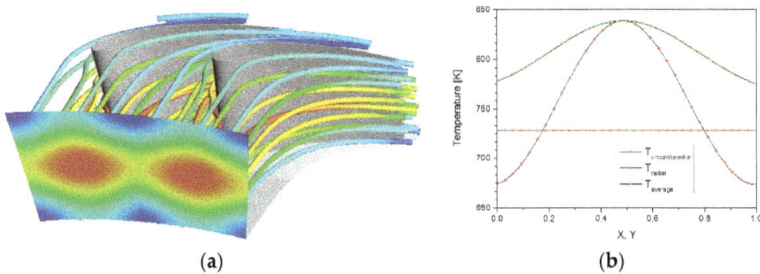

(a)	(b)

Figure 2. Temperature profile at the turbine inlet: (**a**) hot streak (HS) temperature contour with streamlines, and (**b**) circumferential, radial, and average temperature distributions.

2.3. Unsteady State

A steady state analysis was performed by setting the rotor–stator interface as a frozen rotor, and the results were used as the initial conditions for the unsteady state analysis. The transient rotor–stator model was used for the rotor–stator interface for the unsteady simulations. The pressure values of R1 at the midspan were compared among 16, 32, and 50 time step unsteady simulations to determine the step of each cycle in which one R1 completely passes through a pitch of S1. Figure 3 shows the pressure distribution plot at the 4.7° and 9.45° positions when the 1 pitch angle of S1 is 9.45°. The relative error rate was calculated by substituting the pressure value of each step at the R1 midspan in Equations (4) and (5) below.

$$Relative\ error = |step32 - step16|/step32 \tag{4}$$

$$Relative\ error = |step50 - step32|/step50 \tag{5}$$

The maximum relative error rates at the 16 and 32 time-step simulations were 3.22% at R1 position 4.7° and 6.81% at 9.45°. The mean relative error rates were 1.27% at 4.7° and 0.43% at 9.45°. The maximum relative error rates at 32 and 50 time-step simulations were 1.27% at 4.7° and 4.28% at 9.45°. The mean relative error rates were 0.25% at 4.7° and 0.25% at 9.45°. The maximum and mean relative errors at 32 and 50 time-step simulations were significantly reduced compared with those at 16 and 32 time-step simulations. Thus, 32 time steps for one pitch were used in each cycle to perform the unsteady state flow analysis.

The temperature and pressure at one point near each blade pressure side wall were monitored to confirm periodically constant convergence state. The initial 20 pitches were excluded as the initial transient from the total 30 pitches' unsteady state simulation, and the remaining 10 pitches were used in the analysis of the results. Figure 4 shows the temperature and pressure measured for eight pitches, excluding the initial transient. The pressure and temperature for each pitch cycle (=32 time steps) in the unsteady state analysis appear to be periodic.

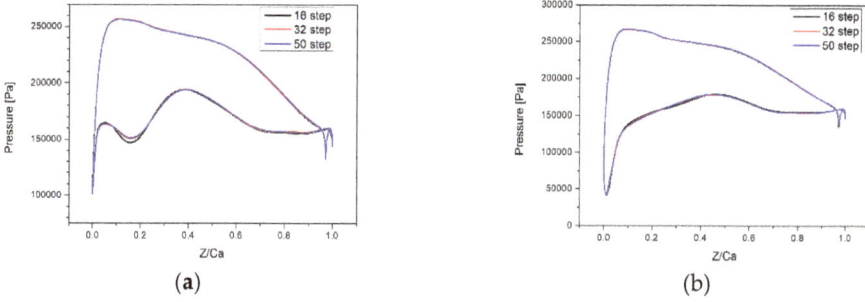

Figure 3. Comparing pressure differences between 16, 32, and 50 time step simulations for each angle on the R1 midspan to find the appropriate step in the unsteady state flow analysis: (**a**) θ = 4.7°, and (**b**) θ = 9.45°.

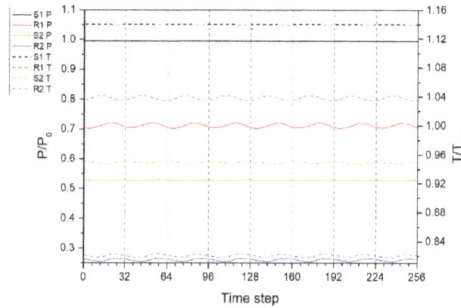

Figure 4. Periodic variation in temperature and pressure at four monitoring points (32 steps = 1 pitch, P_0 = 344,740 Pa, T_{ave} = 728 K).

2.4. Validation of the Turbulence Model

Deciding the appropriate turbulence model is important for more precise numerical analysis. Direct numerical simulations (DNS) and large eddy simulations (LES) provide a flow database for detailed turbulence statistics, but they require a high computational cost [17–21]. To replace such high-cost models, a Reynolds-average Navier-Stokes (RANS) model ($k - \varepsilon$, $k - \omega$, SST, SST γ etc. ...) is used in turbomachinery simulation and especially, SST γ and SST $\gamma - \theta$ transition models can predict more accurate transitional flows [22,23]. To determine the turbulence model for this study, a steady state analysis was performed. The validation for the model was performed using the relative pressure and HTC in comparison with Hylton's experiment values [24]. The configuration and boundary conditions can be seen in the experiment section of Hylton et al. [24]. The 50% span experiment values of the C3X cascade No. 4311 experimental stator were used for comparison. As for the computation domain used in validation, the inflow zone, which is between the inlet and the stator leading edge (LE), was set to be same as the C3X vane axial chord (AC), and the outflow zone, which is between the trailing edge (TE) and the outlet, was set at twice the AC. A grid independence test was performed and 4,015,728 hexahedron meshes were applied. The boundary conditions used in the simulation were set such that the uniform inlet condition's total pressure was 244,763 Pa, the Mach number was 0.17, the Reynolds number based on the C3X vane's AC was 3.9×10^5, the temperature was 802 K, and the turbulence intensity was 6.5%. The outlet conditions were set such that the static pressure was 131,800 Pa, and the outlet Mach number was 0.91. To calculate the HTC, simulations with adiabatic and isothermal wall (temperature 537 K) conditions were conducted. The HTC in this study was calculated using Equation (6), as given below:

$$h = q/(T_w - T_{aw}) \tag{6}$$

In Equation (6), q is the heat flux in the isothermal wall simulation, T_{aw} is the wall temperature value in the adiabatic wall simulation, T_w is the blade temperature value for the isothermal case, and h is the HTC. In Figure 5, the relative pressure and the HTC value of 50% span of the stator found in the experiment paper [24] are compared to the values found from using the $k - \omega$, SST, $k - \omega$ SST γ, and $k - \omega$ SST $\gamma - \theta$ turbulence models used in this study. For the relative pressure distribution in Figure 5a, the pressure at the C3X vane midspan normalized by the inlet total pressure P_0 were used. The comparison revealed a strong agreement between the simulations and experimental data. A relatively larger difference in the area of the suction side (SS) is attributed to the strong unsteadiness of the flow. Figure 5b shows the HTC distribution. In the $k - \omega$, SST, and $k - \omega$ SST $\gamma - \theta$ turbulence models, the SS transition region was different from the experiment values. The $k - \omega$ SST γ turbulence model shows a similar tendency with regard to the experiment values overall, including the SS transition regions. Therefore the $k - \omega$ SST γ turbulence model was used in this study with the onset Reynolds number of 150 [23].

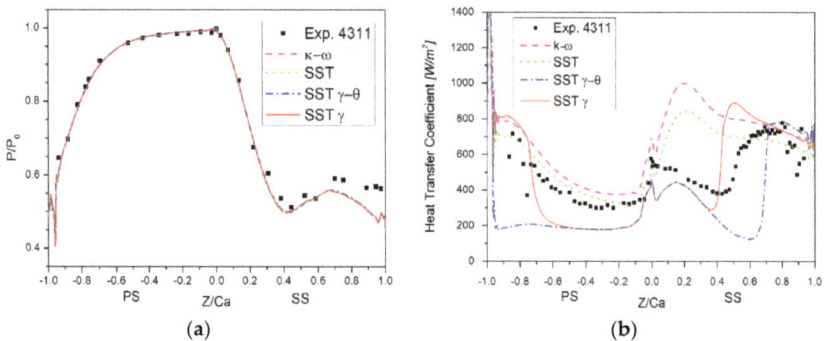

Figure 5. Comparison of experimental data with different turbulence models at 50% span of C3X vane (experimental geometry [24]): (**a**) relative pressure distribution (P$_0$ = 244,763 Pa), and (**b**) heat transfer coefficient (HTC) distribution.

3. Results and Discussions

3.1. Effect of Inlet Temperature Field

To examine the effect that the turbine inlet temperature conditions have on the blade, two steady state analyses were performed under uniform and nonuniform temperature conditions. In the uniform conditions, the turbine inlet temperature was maintained at 728 K. In the nonuniform conditions, the temperature distribution was nonuniformed with an average inlet temperature of 728 K, as shown in Figure 2. Figure 6 shows the surface temperature distribution of the S1 surface's according to each inlet condition on adiabatic walls. In the uniform inlet temperature conditions shown in Figure 6a, the maximum temperature was approximately 729 K, and the temperature distribution showed a trend of decreasing at the PS in the streamwise direction. In Figure 6b, it can be seen that the high temperature gas was centered on the midspan because of the temperature distribution caused by the HS. It is also clear that the high temperature gas formed in the radially inward direction due to the secondary flow transport effect that occurs at the tip and hub near the trailing edge of the SS. This led to fluid-mixing at the stator endwall and the temperature became lower around the hub and tip of the S1 TE. When an HS was applied, the S1 maximum surface temperature was 839 K, which shows a difference of over 110 K compared to the uniform inlet conditions.

Figure 7a shows a comparison of the temperature distributions along the span direction at AC 50% in the PS of S1 and R1. When a uniform temperature distribution was applied, there were few temperature changes along the span direction in S1 and R1. When an HS was applied, higher temperatures are observed at the midspan than around the endwall in S1 and R1. Furthermore, the temperature between 20% and 70% of the span of R1 was higher than the temperature of S1 when a uniform temperature was applied. The maximum temperature differences according to the inlet conditions were 110 K and 75 K for S1 and R1, respectively. Both Figure 7b,c shows the temperature contours at the first midspan according to each inlet temperature condition. Overall higher temperatures were observed in Figure 7c as compared to Figure 7b. Compared to the uniform inlet temperature distribution, the HS inlet temperature gradient, which formed differently along the radial and circumferential directions, had a direct effect on the overall temperature distribution of the blade. Figures 6 and 7 show that it is important to consider the HS inlet conditions in the numerical analysis to understand the thermal and flow characteristics.

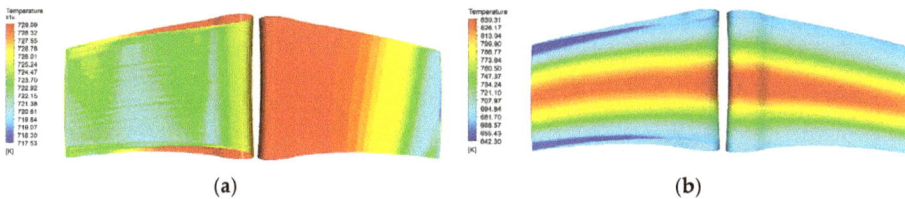

Figure 6. S1 surface temperature contour caused by inlet conditions (Right is pressure side (PS)): (a) uniform inlet temperature of 728 K, and (b) HS with an average temperature of 728 K.

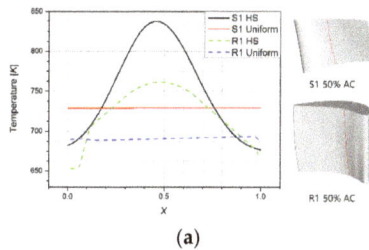

Figure 7. Temperature distribution and contour on the first stage with different inlet condition: (a) temperature distribution on S1, R1 PS axial chord (AC) 50%, (b) uniform inlet temperature of 728 K at the midspan, and (c) HS with an average temperature of 728 K at the midspan.

3.2. Flow and Thermal Characteristics at R1, R2

Flow in a turbine passage affects the surrounding blade surfaces. Figure 8a–c shows the time-averaged streamline of the SS for the surface of R1. Figure 8d–f shows the time-averaged streamline of the SS for the surface of R2. Compared to the PS where the S1 downstream flow acts directly, the flow characteristics of the surface according to the AG were greater at SS. In the figure, the part indicated with the solid line is the recirculation zone. If Figure 8a,c is compared, it can be seen that the AG was reduced (from c to a), and recirculation zone became larger and farther from the LE. This is because in case 3 the mass flow at the tip was 6.32 g/s and in case 1 it was 6.42 g/s. Thus, the mass flow that passes through the tip clearance increased by 1.61%. Owing to the decreased AG, the total pressure at the front of the R1 LE was 219,828 Pa for case 3 and 222,395 for case 1, which is an increase of 1.16%. In the R2 surface's time-averaged streamline, the flow effect at R2 was distributed such that the effect of the secondary flow and recirculation zone was not observed, unlike R1, which was strongly affected by the inlet. As the AG reduced, some streamlines formed downward from the tip (Figure 8d), as opposed to the most streamline formed around the LE in Figure 8f. The change in the AG of S1 and R1 is a factor that affects the rotor surface flow and the creation of secondary flow at the turbine passage after R1 owing to changes in the mass flow at the tip and R1 LE's total pressure.

These phenomena are also caused by a difference in velocity distribution in the main flow of the S1 downstream as the HS passes S1 until it reaches R1. Figure 9 shows the time-averaged velocity distribution at the R1 inlet for each case. The part indicated by the dotted line is where the S1 downstream flow exhibits high-speed flow in the vicinity of the R1 LE. As the AG decreases, the area with higher velocity forms from the midspan to the hub and exerts its influence.

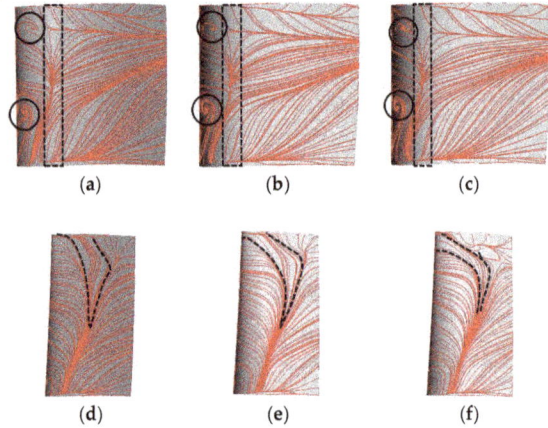

Figure 8. Surface streamlines on R1, R2 suction side (SS): (a) R1 case 1, (b) R1 case 2, (c) R1 case 3, (d) R2 case 1, (e) R2 case 2, and (f) R2 case 3.

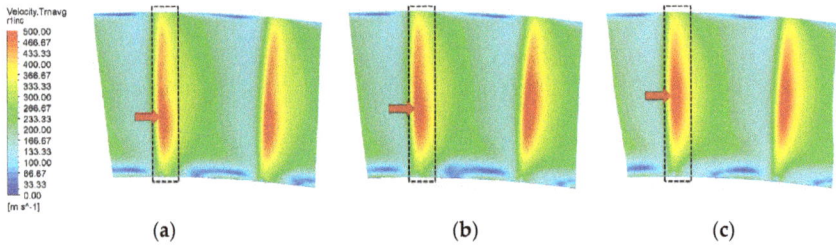

Figure 9. Time-averaged velocity profile at R1 inlet in three cases: (a) case 1, (b) case 2, and (c) case 3.

Figure 10 shows the time-averaged temperature contours of the PS and SS of the surface of R1. At the PS, where the HS turbine inlet condition is primarily concentrated, a high temperature area was observed; however, the effect of the axial gap was not large. At the SS, the high temperature area became larger at the TE in Figure 10a compared to Figure 10c owing to the secondary flow and tip leakage flow around the tip and hub as shown in Figure 8. Figure 11 shows the temperature distribution along the span at AC of 50% and 80% of the SS. Observing the effect that the AG had on the SS surface temperature, it can be seen that the effect of the tip leakage vortex at the tip was larger than the effect of the secondary flow at the hub. Near the tip, case 3 (in which the AG was large) was lower than case 1 by a maximum of 27 K at AC of 50% and a maximum of 43 K at AC of 80%.

The time-averaged temperature and pressure distribution results for each R1 were compared to analyze the thermal and flow characteristics according to the AG in a two-stage gas turbine under HS inlet conditions. Figure 12 shows the area-averaged and maximum temperature and pressure of the surface of R1. The time-averaged values in the area-averaged temperature distribution shown in Figure 12a were 689.15 K for case 1 and 687.69 K for case 3, which shows a difference of approximately 1.5 K. The maximum temperature distributions of case 1 and case 3 differed by over 3 K. The area-averaged pressure distributions in Figure 12b were 181,601 Pa for case 1, which has a short AG, and 179,779 Pa for case 3. This yields a difference of approximately 2000 Pa. In the maximum pressure distributions, there was a difference of approximately 400 Pa between case 1 and case 3, and a relatively small difference compared to the area-averaged pressure distribution was observed. Overall, a trend can be seen in which the AG decreased and the area-averaged and maximum pressure and temperature increased.

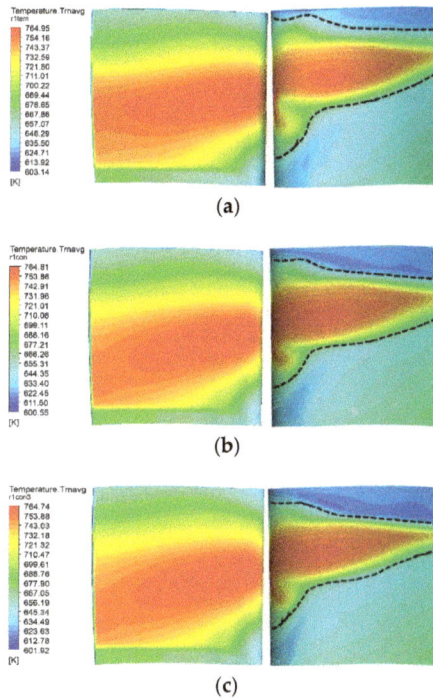

(a)

(b)

(c)

Figure 10. Time-averaged temperature contours on R1, PS, and SS (Left is PS) (**a**) case 1, (**b**) case 2, and (**c**) case 3.

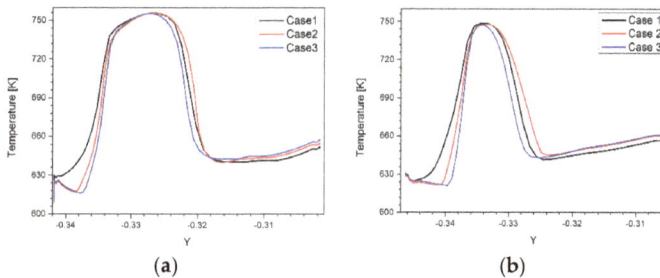

(a)

(b)

Figure 11. R1 SS span direction temperature distribution: (**a**) AC 50%, and (**b**) AC 80%.

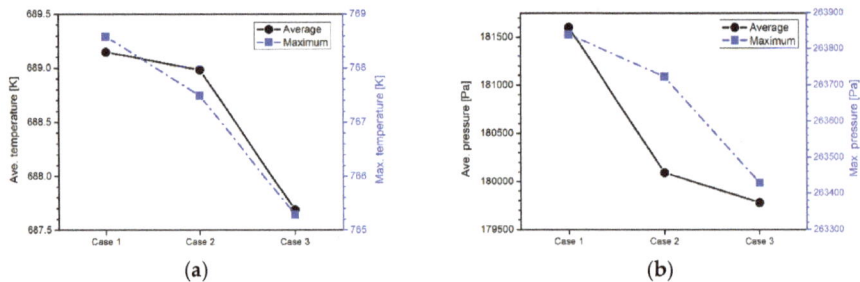

Figure 12. Time-averaged temperature and pressure distributions on R1 surface in three cases: (**a**) the distribution of area-averaged and maximum temperature on R1 surface, and (**b**) the distribution of area-averaged and maximum pressure on R1 surface.

3.3. Effect of Heat Flux on the Vane and the Blade Surface

The heat flux characteristics of the turbine blade can be understood by accurately quantifying the heat flux regions of each blade according to the AG, and a cooling technology can be developed accordingly. Figure 13 shows the time-average heat flux contour of the R1 and R2 surfaces. In the R1 PS shown in Figure 13a, in which the AG becomes close, it can be seen that the low heat flux region from the tip to the 40% span was reduced compared to Figure 13c. Furthermore, the low heat flux region that extended to an AC of 80% was reduced to an AC of 50% in Figure 13a. On the SS, there was a high heat flux around the LE in Figure 13a. When it moved to the TE side, and a contour formed in the radially inward direction and heat flux became lower. However, in Figure 13c, which has a long AG, a lower heat flux was formed at the LE than in Figure 13a, and a low heat flux region was centered on the TE. At R2's PS, the AG became closer, and the heat flux increased on the LE near the turbine hub. It was centered on the hub to mid span. The low heat flux region at 30% span formed at 40% span and the heat flux increased. On the SS, the AG became closer and heat flux increased at 20% span.

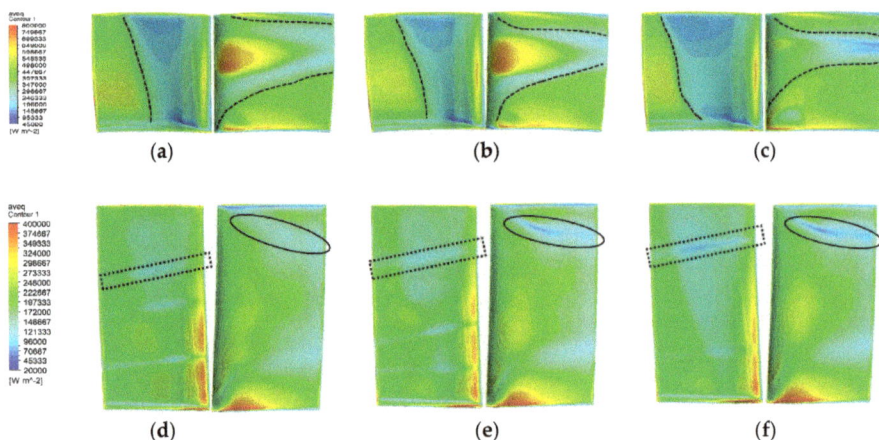

Figure 13. Time-averaged R1, R2 surface heat flux contour (Left is PS): (**a**) R1 case 1, (**b**) R1 case 2, (**c**) R1 case 3, (**d**) R2 case 1, (**e**) R2 case 2, and (**f**) R2 case 3.

Figure 14 shows the time-averaged distribution of the area-averaged heat flux for each blade (including the tip). At S1, the heat flux based on the AG showed a difference of less than 0.1%, thereby confirming that the S1-R1 AG did not have a significant effect on S1. At S2 and R2, the heat flux increased by 10.6% and 2.8%, respectively, as the AG decreased from case 3 to case 1. At R1, the heat flux was slightly lower in case 1 than in case 2. This is because the area-averaged heat flux at case 2's tip was higher than that of case 1. At the second-stage of S2 and R2, as the AG became closer, the heat flux became larger and the effect on the surface became larger.

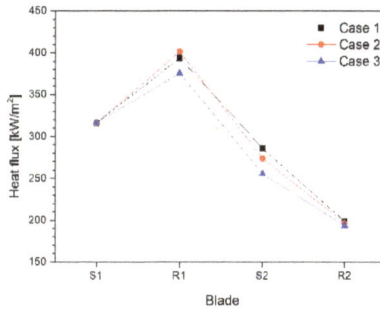

Figure 14. Area-averaged heat flux on the surface of S1, R1, S2, and R2.

3.4. Total-to-Total Efficiency

To quantitatively evaluate the aerodynamic performance of the two-stage turbine with regard to the AG, the total-to-total efficiency was calculated. The efficiency was calculated using Equation (7):

$$\eta = \frac{T\omega}{\dot{m}c_pT_1\left\{1 - (P_{02}/P_{01})^{\frac{\kappa-1}{\kappa}}\right\}} \tag{7}$$

Here, η is the total-to-total efficiency, T is the torque, ω is the angular velocity, $\dot{m}$ is the mass flow rate, c_p is the specific heat capacity of the turbine inlet, T_1 is the mean temperature of the turbine inlet, P_{02} is the mass-averaged total pressure of the turbine outlet, P_{01} is the total pressure of the turbine inlet, and the κ is the ratio of specific heat. Both T and P_{02} directly affect the efficiency whereas other parameter values remain constant throughout the whole cases. The X axis in Figure 15 indicates that the AG distance was divided by the axial chord length of S1, and the Y axis indicates the efficiency, torque, and pressure outlet (P_{02}) normalized by the value of case 2. The normalized values of efficiency, torque, and pressure outlet in cases 1 and 3 are 1.00295 and 0.99694, 1.00218 and 0.99735, and 1.00171 and 0.99911, respectively. The efficiency according to the decrease in the AG was increased. As the AG decreases, the turbine efficiencies increased by 0.306% and 0.295% for case 2 and case 1, respectively. In case 1, which had the shortest AG, the efficiency was 0.601% higher than case 3; however, R1's surface maximum temperature was the highest, as shown in Figure 12. Hence, it leads to an increase in thermal load, which has an effect on the blade's fatigue life.

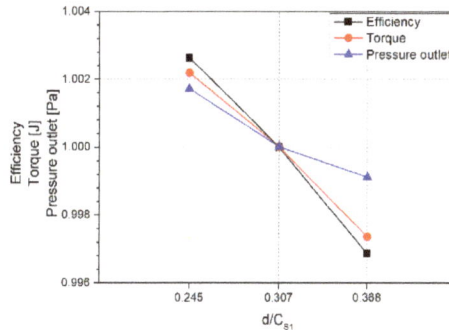

Figure 15. Variation in the total-to-total efficiency, torque, and pressure outlet with changes in the axial gap for the three cases.

4. Conclusions

Herein, we presented a numerical study on the unsteady state to understand how turbine blades and their passages are affected by changes in the AG between the stator and rotor of a GE-E^3 two-stage gas turbine that has a HS inlet temperature distribution. A $k - \omega$ SST γ turbulence model was used to examine the flow fields, heat flux, temperature, and other parameters, at the walls and the passage. In this process, the results were examined for 80% (case 1) and 120% (case 3) of the designed value for the AG of case 2 (13 mm or $d/Cs_1 = 0.307$).

The results showed that the blade surface's maximum temperature increased by over 110 K for the HS inlet conditions compared to uniform conditions. Therefore, an HS must be applied to clearly understand the turbine's thermal characteristics. When the HS inlet conditions were applied, the results according to the AG showed that when the AG was reduced from case 3 to case 1, the surface suction side streamline of first stage rotor (R1) and second-stage rotor (R2) become closer to the endwall because of the secondary flow. Thus, it was formed in the radially outward direction at R1's trailing edge. The area-averaged and maximum temperatures in R1 surface increased by 1.5 K and 3 K, respectively, and the area-averaged and maximum pressure increased by 2000 Pa and 400 Pa, respectively. The low temperature region near the tip and hub decreased. In the second-stage stator and R2, the area-averaged heat fluxes increased by 10.6% and 2.8%, respectively. As the AG decreased, the turbine overall efficiencies increased by 0.306% and 0.295%, respectively; however, this increases the blade surface's thermal load and reduces the turbine blade fatigue life. If an appropriate cooling technology is developed, it will lead to the development of a higher efficiency design using a shorter AG and reduce long-term operating costs.

The unsteady RANS model used in this study has a low computational cost; however, it has the drawback of being unable to precisely predict vortices and flow separation phenomena that occur around the tips and walls. DES is more accurate; however, it has a high computational cost and previous studies using DES have mostly focused on the local flow phenomena rather than full blade simulations [25–27]. With significant advances in computational resources, it is expected that DES or LES will soon be used to accurately predict the complex unsteady flow physics near the tips and walls within full blade simulations.

Author Contributions: Conceptualization, methodology, investigation: M.G.C. and J.R.; validation, formal analysis, writing—original draft preparation, visualization: M.G.C.; writing—review and editing, supervision, project administration, funding acquisition: J.R.

Funding: This research was funded by Basic Science Research Program through the National Research Foundation of Korea (NRF-2017R1C1B2012068). Also, this research was funded by the Chung-Ang University Research Grants in 2017.

Conflicts of Interest: The authors declare no conflict of interest.

Nomenclature

C_{S1}	Axial chord length of S1 [mm]
d	Axial chord length [mm]
q	Heat flux [W/m^2]
T_w	Wall temperature [K]
T_{aw}	Adiabatic wall temperature [K]
P	Pressure [Pa]
P_0	Inlet total pressure [Pa]
T	Temperature [K]
T_0	Inlet total temperature [K]
X, Y, Z	Cartesian coordinates

Abbreviations

HP	High pressure
HS	Hot streak
S1, R1, S2, R2	first stage stator, rotor, second-stage stator, rotor
HTC, h	Heat transfer coefficient
AG	Axial gap
LE	Leading edge
TE	Trailing edge
PS	Pressure side
SS	Suction side
AC, Ca	Axial chord
Subscripts	
ave	Averaged
max	Maximum value
min	Minimum value

References

1. Wang, Z.; Liu, Z.; Feng, Z. Influence of mainstream turbulence intensity on heat transfer characteristics of a high pressure turbine stage with inlet hot streak. *J. Turbomach.* **2016**, *138*, 041005. [CrossRef]
2. Butler, T.L.; Sharma, O.P.; Dring, R.P. Redistribution of an inlet temperature distortion in an axial flow turbine stage. *J. Propuls. Power* **1989**, *5*, 64–71. [CrossRef]
3. Povey, T.; Qureshi, I. A hot-streak (combustor) simulator suited to aerodynamic performance measurements. *Proc. Inst. Mech. Eng. Part G J. Aerosp. Eng.* **2008**, *222*, 705–720. [CrossRef]
4. Gundy-Burlet, K.L.; Dorney, D.J. Effects of radial location on the migration of hot streaks in a turbine. *J. Propuls. Power* **2000**, *16*, 377–387. [CrossRef]
5. An, B.T.; Liu, J.J.; Jiang, H.D. Numerical investigation on unsteady effects of hot streak on flow and heat transfer in a turbine stage. In Proceedings of the ASME Turbo Expo 2008: Power for Land, Sea, and Air, Berlin, Germany, 9–13 June 2008.
6. Feng, Z.; Liu, Z.; Shi, Y.; Wang, Z. Effects of hot streak and airfoil clocking on heat transfer and aerodynamic characteristics in gas turbine. *J. Turbomach.* **2016**, *138*, 021002. [CrossRef]
7. Smith, C.I.; Chang, D.; Tavoularis, S. Effect of inlet temperature nonuniformity on high-pressure turbine performance. In Proceedings of the ASME Turbo Expo 2010: Power for Land, Sea, and Air, Glasgow, UK, 14–18 June 2010.
8. Yamada, K.; Funazaki, K.; Kikuchi, M.; Sato, H. Influences of axial gap between blade rows on secondary flow and aerodynamic performance in a turbine stage. In Proceedings of the ASME Turbo Expo 2009: Power for Land, Sea, and Air, Orlando, FL, USA, 8–12 June 2009.
9. Danish, S.N.; Qureshi, S.R.; Imran, M.M.; Khan, S.U.; Sarfraz, M.M.; El-Leathy, A.; Al-Ansary, H.; Wei, M. Effect of tip clearance and rotor–stator axial gap on the efficiency of a multistage compressor. *Appl. Therm. Eng.* **2016**, *99*, 988–995. [CrossRef]
10. Dring, R.P.; Joslyn, H.D.; Hardin, L.W.; Wagner, J.H. Turbine rotor-stator interaction. *J. Eng. Power* **1982**, *104*, 729–742. [CrossRef]
11. Gaetani, P.; Persico, G.; Dossena, V.; Osnaghi, C. Investigation of the flow field in a HP turbine stage for two stator-rotor axial gaps: Part I—3D time averaged flow field. In Proceedings of the ASME Turbo Expo 2006: Power for Land, Sea and Air, Barcelona, Spain, 8–11 May 2006.
12. Funazaki, K.I.; Yamada, K.; Kikuchi, M.; Sato, H. Experimental studies on aerodynamic performance and unsteady flow behaviors of a single turbine stage with variable rotor-stator axial gap: Comparisons with time-accurate numerical simulation. In Proceedings of the ASME Turbo Expo 2007: Power for Land, Sea, and Air, Montreal, QC, Canada, 14–17 May 2007.
13. Sasao, Y.; Kato, H.; Yamamoto, S.; Satsuki, H.; Ooyama, H.; Ishizaka, K. Numerical and experimental investigations of unsteady 3-D flow through two-stage cascades in steam turbine model. In Proceedings of the International Conference on Power Engineering, Kobe, Japan, 16–20 November 2009.

14. Ghenaiet, A.; Touil, K. Characterization of component interactions in two-stage axial turbine. *Chin. J. Aeronaut.* **2016**, *29*, 893–913. [CrossRef]

15. Timko, L.P. *Energy Efficient Engine High Pressure Turbine Component Test Performance Report*; Technical Report; National Aeronautics and Space Administration (NASA): Cincinnati, OH, USA, 1 January 1984.

16. Arnone, A.; Benvenuti, E. Three-dimensional Navier-Stokes analysis of a two-stage gas turbine. In Proceedings of the ASME 1994 International Gas Turbine and Aeroengine Congress and Exposition, The Hague, The Neterlands, 13–16 June 1994.

17. Ryu, J.; Livescu, D. Turbulence structure behind the shock in canonical shock–vortical turbulence interaction. *J. Fluid Mech.* **2014**, *756*, R1. [CrossRef]

18. Ryu, J.; Lele, S.K.; Viswanathan, K. Study of supersonic wave components in high-speed turbulent jets using an LES database. *J. Sound Vib.* **2014**, *333*, 6900–6923. [CrossRef]

19. Celik, I.; Yavuz, I.; Smirnov, A. Large eddy simulations of in-cylinder turbulence for internal combustion engines: A review. *Int. J. Engine Res.* **2001**, *2*, 119–148. [CrossRef]

20. Meloni, R.; Naso, V. An insight into the effect of advanced injection strategies on pollutant emissions of a heavy-duty diesel engine. *Energies* **2013**, *6*, 4331–4351. [CrossRef]

21. Haworth, D.C.; Jansen, K. Large-eddy simulation on unstructured deforming meshes: Towards reciprocating IC engines. *Comput. Fluids* **2000**, *29*, 493–524. [CrossRef]

22. Hao, Z.R.; Gu, C.W.; Ren, X.D. The application of discontinuous Galerkin methods in conjugate heat transfer simulations of gas turbines. *Energies* **2014**, *7*, 7857–7877. [CrossRef]

23. Kim, J.; Park, J.G.; Kang, Y.S.; Cho, L.; Cho, J. A study on the numerical analysis methodology for thermal and flow characteristics of high pressure turbine in aircraft gas turbine engine. *Korean Soc. Fluid Mach.* **2013**, *17*, 46–51. [CrossRef]

24. Hylton, L.; Mihelc, M.; Turner, E.; Nealy, D.; York, R. *Analytical and Experimental Evaluation of the Heat Transfer Distribution over the Surfaces of Turbine Vanes*; Technical Report; National Aeronautics and Space Administration (NASA): Indianapolis, IN, USA, 1 May 1983.

25. Martini, P.; Schulz, A.; Bauer, H.J.; Whitney, C.F. Detached eddy simulation of film cooling performance on the trailing edge cutback of gas turbine airfoils. In Proceedings of the ASME Turbo Expo 2005: Power for Land, Sea, and Air, Reno, NV, USA, 6–9 June 2005.

26. Effendy, M.; Yao, Y.F.; Yao, J.; Marchant, D.R. DES study of blade trailing edge cutback cooling performance with various lip thicknesses. *Appl. Therm. Eng.* **2016**, *99*, 434–445. [CrossRef]

27. Mao, X.; Dal Monte, A.; Benini, E.; Zheng, Y. Numerical study on the internal flow field of a reversible turbine during continuous guide vane closing. *Energies* **2017**, *10*, 988. [CrossRef]

energies

MDPI

Article

Numerical Study on the Influence of Mass and Stiffness Ratios on the Vortex Induced Motion of an Elastically Mounted Cylinder for Harnessing Power

Vidya Chandran [1], Sekar M. [2], Sheeja Janardhanan [3],* and Varun Menon [4]

[1] Department of Mechanical Engineering, Karunya Institute of Technology and Sciences, Coimbatore, Tamil Nadu 600018, India; vidya.rudn@gmail.com
[2] Department of Mechanical Engineering, AAA College of Engineering and Technology, Sivakasi, Tamil Nadu 600018, India; mailtosekar@gmail.com
[3] Department of Mechanical Engineering, SCMS School of Engineering and Technology, Ernakulam, Kerala 673307, India
[4] Department of Computer Science and Engineering, SCMS School of Engineering and Technology, Ernakulam, Kerala 673307, India; varungmenon46@gmail.com
*   Correspondence: sheejajanardhanan@scmsgroup.org; Tel.: +91-828-194-3531

Received: 10 September 2018; Accepted: 25 September 2018; Published: 27 September 2018

Abstract: Harnessing the power of vortices shed in the wake of bluff bodies is indeed a boon to society in the face of fuel crisis. This fact serves as an impetus to develop a device called a hydro vortex power generator (HVPG), comprised of an elastically mounted cylinder that is free to oscillate in the cross-flow (CF) direction even in a low velocity flow field. The oscillatory motions in turn can be converted to useful power. This paper addresses the influence of system characteristics viz. stiffness ratio (k^*) and mass ratio (m^*) on the maximum response amplitude of the elastically mounted cylinder. Computational fluid dynamics (CFD) simulations have been used here to solve a two way fluid–structure interaction (FSI) problem for predicting the trend of variation of the non-dimensional amplitude Y/D with reduced velocity U_r through a series of simulations. Maximum amplitude motions have been attributed to the lowest value of m* with $U_r = 8$. However, the maximum lift forces correspond to $U_r = 4$, providing strong design inputs as well as indicating the best operating conditions. The numerical results have been compared with those of field tests in an irrigation canal and have shown reasonable agreement.

Keywords: computational fluid dynamics (CFD); flow around cylinder; fluid structure interaction (FSI); hydrodynamic response; numerical methods; simulation and modeling; vortex induced vibration (VIV) ratio

1. Introduction

As the sources of fossil fuels are depleting at a faster pace, energy scientists all over the world are keen on the search for new technologies that can provide renewable and clean energy. Hydroelectric power generation is of course a clean source of energy, but considering the capital investment and the effects of dams on natural ecosystems, the need for a much cleaner energy source becomes more important. This paper discusses the design and manufacture of the hydro vortex power generator (HVPG) model, which when scaled up can be viewed as one such cleaner source of electricity. Also, the paper discusses a numerical method to optimize the design parameters of HVPG model. The principle behind the working of HVPG is vortex shedding in the wake of bluff bodies in fluid flow. The phenomenon of vortex shedding behind bluff bodies has been an extensively researched topic [1,2]. The presence of such vortex shedding has been considered undesirable and researchers had been in search of methods to suppress vortex shedding [3]. Vortex power was proved useful to mankind by the researchers at Michigan University, who first converted vortex power into electricity [4,5]. HVPG works on the principle of vortex induced vibration of bluff bodies subjected to fluid current. The power of vortices shed in the wake of these bluff bodies are converted into vibration energy and then into electricity. HVPG can be made useful as a single standing power unit that can provide electricity to remote locations; and also as a multiunit module which can supply power to the grid [6].The paper discusses the design optimization of a single, power-generating, scaled-down module, harnessing power from vortices. In this paper, an attempt has been made to optimize the design based on the major influencing parameters, oscillating mass ratio (m^*) and stiffness ratio (k^*). Griffin consolidated experimental results and plotted them to show the dependence of maximum amplitude on system characteristics [7]. However, the drawback of Griffin's plot was its considerable scatter which could be attributed to the inclusion of mode shapes also as an influencing parameter. Later researchers could successfully reduce the scatter in Griffin's plot and establish a simple relationship between maximum amplitude and mass damping parameter by eliminating mode shapes from the list of variables [7,8]. The simplified mass damping parameter as in [8] was applicable for cylinders of high and low mass ratios equally. Many researchers have formulated empirical formulae to express the parametric relationship and experimentally verified the correlations [9,10]. Later in the experiments, however, it has been observed that with variation in the damping ratio, U_r and the Strouhal number (St) varies [11]. Also, these experiments revealed relatively larger response amplitude compared to other studies. Recently researchers have also succeeded in theoretically proving the effectiveness of harnessing VIV energy for powering underwater mooring platforms [12].

The present work studies the influence of k^* and m^* on maximum response amplitude of an elastically mounted cylinder with a single DOF. It also provides a detailed insight into the vortex shedding pattern at various U_r. The simulations are carried out at Re of the order 10^4 which corresponds to the realistic flow regime encountered by power generating vortices [13]. The numerical results have been verified using field tests conducted at Palissery irrigation canal (Palissery irrigational canal is one of the irrigation projects by Government of Kerala located in Thrissur district, Kerala, India).

Previous studies in the same domain had only considered the effect of complex and coupled parameters such as the Skop Griffin parameter [14]. The novelty in the present work is the effort made to represent the influence of tangible parameters viz the stiffness coefficient and mass of the cylinder. These parameters are easily controllable from a designer's point of view. A non-dimensional approach has been used here to generalize the results for the design of power harnessing devices of any scale.

2. The Concept of HVPG

HVPG works on the principle of vortex induced vibration (VIV). If a bluff body is not completely secured with at least one degree of freedom motion, and the frequency of vortex shedding matches the natural frequency of the structure, the structure begins to resonate, vibrating with harmonic oscillations of large amplitude. This phenomenon is known as 'lock-in'. During lock-in, vortex shedding frequency shifts to the natural frequency of the structure, leading to large amplitude vibrations. Vortex shedding in the wake of a cylinder is shown in Figure 1.

The vortex shedding occurs at a discrete frequency and is a function of the Reynolds number (*Re*), defined by Equation (1)

$$Re = \frac{\rho V D}{\mu}$$

(1)

The dimensionless frequency of the vortex shedding, $St = f_v D/V$, is approximately equal to 0.2 when the Reynolds number is greater than 1000 [15]. When vortices are shed from the cylinder, uneven pressure distribution develops around the upper and lower surfaces of the cylinder, generating an oscillating hydrodynamic lift force on the cylinder. This unsteady force given by Equation (2) can induce significant cross flow vibrations on a structure, especially if the resonance condition is met.

$$F_L = C_L \frac{1}{2} \rho A V^2.$$

(2)

where F_L is the lift force and C_L is the coefficient of lift. ρ is the density of water, A the projected area in the direction of flow, and V is the velocity of flowing water. The cylinder also experiences a net force along the flow direction and is called the drag force and is given by the Equation (3).

$$F_D = C_D \frac{1}{2} \rho A V^2 \tag{3}$$

where F_D is the drag force and C_D is the drag coefficient.

The oscillating lift force acting on the cylinder makes the cylinder oscillate in the cross flow (CF) direction at the frequency of vortex shedding. For the making of HVPG, the cylinder has been mounted elastically, which enables the entire module to be considered as a spring-mass system with the cylinder considered as the mass and the elastic supports as springs. When the natural frequency of spring-mass system matches the vortex shedding frequency, the cylinder oscillates with large amplitudes. The linear motion of the mass can then be converted to rotary motions through a slider-crank mechanism and the crank rotations can be used to drive a generator unit.

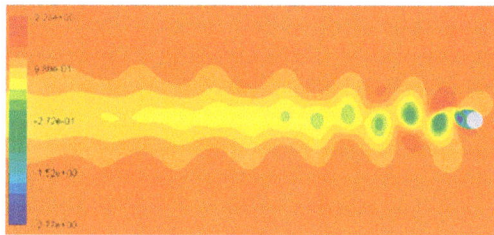

Figure 1. Shedding of alternate vortices behind a cylinder represented as pressure (N/m^2) contours. (Von Kármán Vortex Street).

3. Mathematical Model

A single power module of HVPG has been modeled as a spring mass system undergoing instability induced vibration. The instability is caused by the shedding of vortices in the wake of the cylinder when the flow encounters a bluff body. Alternate vortex shedding causes oscillatory forces that induce structural vibrations, where the rigid cylinder is now similar to a spring-mass system with a harmonic forcing term. This phenomenon is referred to as heave motion [14]. Equation of motion [16] for this system can be written as

$$m\ddot{Y} + c\dot{Y} + kY = F(t) \tag{4}$$

where Y is the displacement of the cylinder under VIV in the cross-flow direction; m is the sum of mass of the oscillating system or body mass, m_b and added mass of the system, m_a. m_a is defined as $m_a = C_A m_b$, where C_A is the added mass coefficient. c is the damping coefficient and k is the coefficient of stiffness of the spring mass system. $F(t)$ is the time varying force acting on the cylinder due the flow instability. Relatively small oscillation amplitudes are approximated by

$$F(t) = F_L sin(\omega_v\, t + \varphi) \tag{5}$$

where ω_v is the circular frequency of vortex shedding and φ the phase difference between the force and cylinder displacement. F_L is the maximum value of oscillating hydrodynamic lift force acting on the cylinder and is given by Equation (2).

The amplitude of oscillation of the system depends on the mass (m^*) ratio of the oscillating cylinder given by

$$m^* = \frac{m}{m_{fd}} \tag{6}$$

where m_{fd} is the mass of fluid displaced by the oscillating mass. The maximum possible response amplitude at any *Re*, Y_{max} can be calculated from the empirical relation between non-dimensional amplitude ($A_y = \frac{Y_{max}}{D}$) and Re [10] as given by Equation (7).

$$A_Y = -0.4435\left[log\frac{\alpha}{Re}\right] - 1.5 \tag{7}$$

where α is defined as

$$\alpha = (m^* + C_A)\zeta \tag{8}$$

where C_A is the added mass coefficient and ζ is the damping ratio.

Maximum amplitude of oscillation occurs when shedding frequency locks on to the natural frequency of the oscillating system (f_n). This condition is known as lock-in. Amplitude of oscillation of a spring mass system can also be obtained from Equation (9)

$$Y = \frac{F_L}{k}\left[\frac{1}{\sqrt{(1 - \eta^2)^2 + (2\zeta\eta)^2}}\right] \tag{9}$$

where η is the frequency ratio represented by Equation (10)

$$\eta = \frac{f_n}{f_v} \tag{10}$$

where f_v is the vortex shedding frequency. During lock-in $\eta = 1$ and Equation (9) simplifies as shown by Equation (11).

$$Y = \frac{F_L}{2k\zeta} \tag{11}$$

The paper discusses effect of mass and stiffness ratios of the oscillating cylinder on the maximum response amplitude. The structural damping variations are not considered as it is observed to be less significant compared to its inertia and elastic counterparts. Also, mass and stiffness of the system are more tangible parameters from design point of view compared to damping. Moreover, the study focuses on non-dimensionalizing the influencing parameters so that results hold applicable for prototypes and models equally. Many researchers have considered the combined effect of mass and damping through mass damping parameter $m^*\zeta$. In such analysis also ζ is kept constant and m^* is varied independently [17].

4. Numerical Determination of Hydrodynamic Lift Forces and Motions

Vortex induced vibration, a two-way fluid structure interaction phenomenon, is complex in nature due to the fact that the cylinder displacement is capable of changing the vortex shedding pattern behind it leading to a variation in the hydrodynamic load acting on the cylinder. Parameters k^*, m^*, and c have significant influence on the oscillation amplitude. A significant amount of research has been carried out to bring clarity on the influence of these parameterson the maximum cylinder response amplitude. Experiments conducted by [18] could capture lock-in phenomenon for a cylinder with a single degree of freedom. It was observed that cylinder oscillation frequency matches with vortex shedding frequency for cylinders having low mass ratios [19]. For mass ratios below critical mass ratio, m^*_{cr}, the range of U_r over which resonance occurs tend to extend towards infinity. The following section of the paper is an effort to understand the effect of mass and stiffness coefficient on the response of cylinder numerically in a simpler and economic way, and the results of the numerical study is verified using a field test.

4.1. Modeling the Flow

Numerically this problem has been treated as a case of two-way fluid structure interactions (two-way FSI). Modeling and meshing has been performed in ANSYS ICEM CFD (version 12) [20] and solving using ANSYS FLUENT (version 12) [21]. Flow around the cylinder is modeled using the transient, incompressible Navier–Stokes equation based RANS solver with k–ω SST as the turbulence model. The RANS solver does the virtual averaging of velocities over an interval of time and hence for a specific interval the velocity vector appears to be constant in a RANS solver. In the present work, an optimized fine grid is used to compensate for this drawback of the solver enabling it to capture the physics of von Kármán street eddies. RANS solver for transient two-dimensional analysis can be explained as follows [21]

$$\frac{\partial \rho}{\partial t} + \frac{\partial}{\partial x_i}(\rho u_i) = 0 \tag{12}$$

$$\frac{\partial}{\partial t}(\rho u_i) + \frac{\partial}{\partial x_i}(\rho u_i u_j) = -\frac{\partial p}{\partial x_i} + \frac{\partial}{\partial x_j}\left[\mu\left(\frac{\partial u_i}{\partial x_j} + \frac{\partial u_j}{\partial x_i} - \frac{2}{3}\delta_{ij}\frac{\partial u_l}{\partial x_l}\right)\right] + \frac{\partial}{\partial x_j}\left(-\overline{\rho u_i' u_j'}\right) \tag{13}$$

where u_i and u_i' are mean and fluctuating velocity components for i = 1, 2, and 3.

The velocities and other solution variables in the above equation represent the time averaged values. Equation (12) is solved by modeling Reynolds stresses $\overline{\rho u_i' u_j'}$, effectively using k–ω SST as turbulence model [22]. k–ω SST is capable of accurately predicting the commencement and the intensity of flow separation at fixed boundaries while the standard k–ε model has proved its efficacy in predicting the wake characteristics accurately. This fact has been established after extensive studies conducted by the authors. k–ωSST turbulence model demands a very high near wall grid resolution and hence the maximum element size is fixed to be less than 1×10^{-4}, which satisfies CFL criterion and near wall y+ values.

The above governing equations are discretized using finite difference method. Non iterative time advancement (NITA) scheme with fractional time stepping method (FSM) has been chosen for pressure-velocity coupling of the grid. A Least Squares Cell Based scheme is used for gradient in spatial discretization and a second order upwind scheme as convective scheme.

NITA

For capturing the physics of the flow with accuracy in the boundary as well as in the wake of the cylinder, the computational grid needs to be extremely fine. Solving a dynamic mesh case with such extreme grid fineness using iterative time advancement scheme demands a considerable number of iterations to be performed using a very small time step size to satisfy the dynamic mesh criteria. This in turn leads to huge computational cost and effort. As an alternative and computationally economic method, in the present work, NITA with FSM has been implemented. NITA scheme assures the same time accuracy by reducing the splitting error, which occurs while solving the discretized Navier–Stokes equation to the same order as the truncation error. Splitting error need not be reduced to zero in NITA scheme, saving a lot of computational effort.

4.2. Structural Modeling

An elastically mounted cylinder can be mathematically represented by Equation (4). This equation of motion is solved using six degrees of freedom solver (6DOF), an integral part of the main solver by defining the cylinder as an object with one degree of freedom (1DOF) in transverse direction. A user defined function (UDF) compiled in C programming language has been hooked to the cylinder dynamic boundary conditions. The governing equation for the motion of the center of gravity of the cylinder in the transverse direction is solved in the inertial coordinate system. Velocity in the transverse direction is obtained by performing integration on Equation (14).

$$\ddot{Y} = \frac{1}{m} \sum F \tag{14}$$

where $\ddot{Y}$, is the translational acceleration in the transverse direction, m is the mass of the cylinder and F, resultant fluid force acting on the cylinder. Position of the center of gravity of the cylinder (CG) is updated after solving the equation of motion of a spring mass system.

$$m\ddot{Y} + c\dot{Y} + kY = F(t) \tag{15}$$

The inertial force term on the left-hand side of Equation (15) is computed by the 6DOF solver for each time step from Equation (14) and the UDF hooked to the moving cylinder inputs the restoring force term as 6DOF load acting on the cylinder. Mass of the cylinder is given in the UDF as

$$m = m_b + m_a \tag{16}$$

$$m_a = (1 + C_A)m_b \tag{17}$$

Added mass coefficient C_A for the aspect ratio of the present model is found to be equal to 0.7 [13].

4.3. Mesh Deformation

Mesh motion to adapt to the movement of the cylinder is achieved by using displacement based smoothing algorithm. The governing equation for mesh motion is represented by Equation (18).

$$\nabla \cdot (\gamma \nabla \vec{u}) = 0 \tag{18}$$

where $\vec{u}$ is the velocity of mesh displacement. The boundary conditions for Equation (18) are computed by the 6DOF solver and the boundary mesh motion diffuses into the interior of the deforming mesh according to the Laplace equation, Equation (18). Diffusion coefficient, γ is calculated using boundary distance formulation given by Equation (19).

$$\gamma = \frac{1}{d^\tau} \tag{19}$$

where τ is the diffusion parameter and d is the normalized boundary distance. Diffusion parameter is set as unity to avoid excessive deformation of the near cylinder elements.

4.4. Fluid Structure Interaction

In this paper, a two-way implicit approach is used to study the effect of m^* and k^* on the response of cylinder under VIV. Flow equations and structural equations are solved simultaneously in iterations with a time step. A flow chart for the solution procedure is shown in Figure 2.

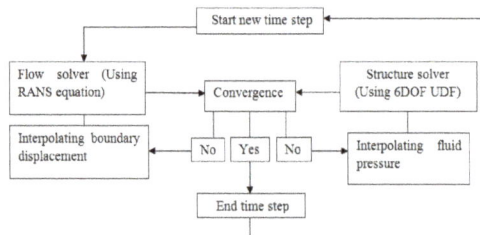

Figure 2. Flow chart of two-way implicit FSI solution procedure.

5. Problem Description

5.1. Simulation Parameters

In the present study, geometrically identical cylinders of different masses are considered for VIV of an elastically mounted horizontal cylinder. Based on the present study, a basic model of HVPG as in [5] was fabricated and a field study was carried out on it for verifying the numerical results. The main parameters of the model are summarized in Table 1. The model has an aspect ratio of 13.12 and an outer diameter of 0.0762 m. The structural damping is considered to be zero in the present study. The computational domain for the study is shown in Figure 3a. The representation of the present problem as a spring-mass system is depicted in Figure 3b. The mesh fineness in indicated in Figure 3c and the grid used for the present computation is represented by Figure 3d.

Table 1. Properties of the elastically mounted cylinder model.

Properties	Values	Units
Diameter of the cylinder (D)	0.0762	m
Aspect ratio of the cylinder (L/D)	13.12	-
Flow velocity (V)	0.5	m/s
Reynolds Number of flow (Re)	3.8×10^4	-
Mass ratio (m^*)	0.66	-

A simple representation of the two-dimensional computational domain for the cylinder is displayed in Figure 3a. The flow direction is parallel to the global *x*-axis and the flow velocity is set to be 0.5 m/s. Domain size is fixed based on previously published analysis and allowance is given to accommodate the vertical motion of cylinder boundary [23]. Simulations are performed for three different values of mass ratio, $m^* = 0.66, 1.32$, and 1.98 which correspond to moderate mass ratios above m^*_{cr}, for which maximum response in confined to the range of $U_r = 4$–12. For fixed mass, influence of stiffness ratio k^* is studied by varying the reduced velocity value which is defined as

$$U_r = \frac{V}{f_n D} \tag{20}$$

$$f_n = \frac{1}{2\pi}\sqrt{\frac{k}{m}} \tag{21}$$

Each case has been analyzed over a range of reduced velocity, $U_r = 4$–12, over which the cylinder is predicted to have maximum amplitude of oscillation [24]. $U_r = 5$ has also been analyzed since the case corresponds to $\eta = 1$ where one might expect resonance. The incoming flow velocity is fixed as 0.5 m/s to maintain the flow regime uniform at $Re = 3.8 \times 10^4$. The mass ratio, stiffness ratio, and other parameters for each case are summarized in Table 2. Stiffness coefficient of the cylinder has been non-dimensionalized to generalize the applicability of the analysis. Stiffness ratio is defined as

$$k^* = \frac{k}{mg/L} \tag{22}$$

Table 2. Reduced velocity, stiffness ratio, and frequency ratio at $Re = 3.8 \times 10^4$, for $m^* = 0.66, 1.32$, and 1.98.

U_r	k^*	η
4	11.17	1.3
5	6.9	1.0
6	4.81	0.84
8	2.7	0.63
10	1.73	0.51
12	1.21	0.42

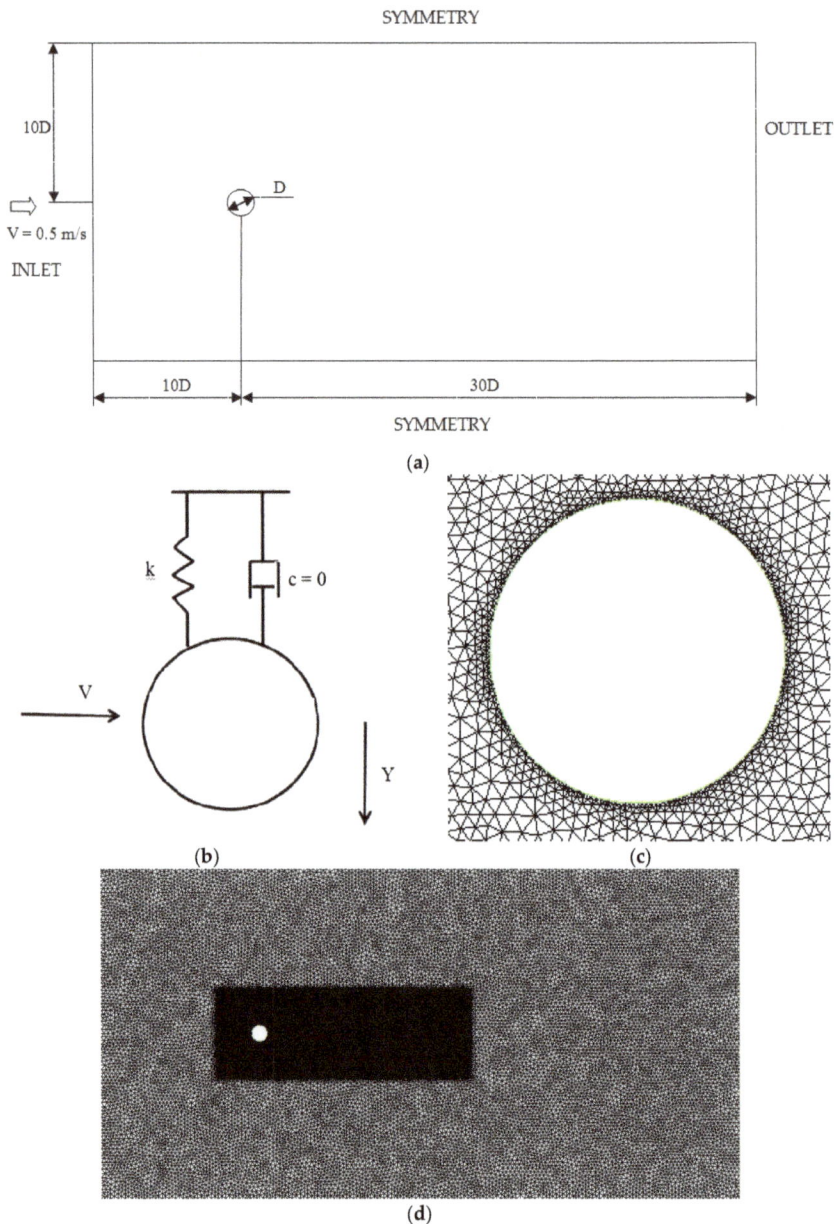

Figure 3. (**a**) Computational domain; (**b**) representation of elastically mounted cylinder model; (**c**) mesh around the cylinder; (**d**) computational mesh.

5.2. Fluid Domain and Boundary Conditions

Figure 3a shows the computational domain for the CFD simulation of VIV of an elastically mounted horizontal cylinder. The origin of the Cartesian coordinate system is located at the center of the cylinder. The length of the domain is 40D with the cylinder located at 10D away from the inlet boundary. The cross-flow width of the domain in 20D with the center of the cylinder at the middle. Detailed views of the mesh around the cylinder along with the computational domain after meshing have been shown in Figure 3c,d respectively. There are 307 nodes around the circumference of the cylinder and the minimum element size near the rigid wall boundary has been computed from boundary layer theory to be 0.0001*D*. The non-dimensional element size represented as *y*+, next to the cylinder surface is found to be less than unity. For the cylinder wall, a no slip boundary condition has been applied assuming the surface to be smooth. Inlet boundary has been treated has velocity-inlet with inflow velocity, *V* = 0.5 m/s. Outlet boundary has been treated as pressure outlet, the gradients of fluid velocity are set to zero and the pressure with zero reference pressure. On the two transverse boundaries a symmetry boundary condition has been applied.

5.3. Mesh Independence Study

An unstructured 2D mesh has been used in the present CFD simulation to facilitate computationally economic platform for dynamic mesh simulation. The three dimensionality of the wake reduces as a result of the motion of the cylinder [16]. Hence it is possible to get a reasonably accurate result from a 2D analysis saving much computational cost and effort. The meshing strategy is that finer mesh is used in the vicinity of the moving cylinder with extra fine meshing in the boundary layer. Boundary layer thickness and the near wall element size have been calculated from boundary layer theory. The thickness of laminar sub-layer is obtained from Equation (23) [25]

$$\delta' = \frac{11.6\vartheta}{V*} \tag{23}$$

where $V*$ is the frictional velocity given by

$$V* = \sqrt{\frac{\tau_0}{\rho}} \tag{24}$$

and τ_0, the wall shear stress is obtained as

$$\tau_0 = \frac{0.664}{\sqrt{Re_D}} \cdot \frac{\rho V^2}{2} \tag{25}$$

Increased mesh density has been adopted in the near cylinder and its wake in order to capture the physics of vortex shedding accurately. For ensuring that the results are independent of the grid size, a mesh independence study has been carried out. Three different grids have been used to simulate a specific case *m** = 2.45 and *U*ᵣ = 8 and has been verified using results of experiments conducted at a towing tank facility in the Department of Ocean Engineering, Indian Institute of Technology Madras, India [10]. The details of the mesh independency study are given in Table 3. The last three grids give almost similar results and experiments, and [10] shows a 4.9% deviation from the present results. This can be assumed to be due to not accounting for damping in the present study. It can be concluded that the variation in the numerical results given by Meshes II, III, and IV are in the acceptable range and considering the computational economy Grid II with 49,995 nodes has been chosen for further analysis.

Table 3. Mesh independency study results

	Re	$m*$	U_r	*Nodes*	Y_{max}/D
Grid I				35,487	1.241
Grid II	3.8×10^4	2.45	8	49,995	1.220
Grid III				70,857	1.219
Grid IV				98,475	1.219
Narendran et al. (2015)	$0.3–2.4 \times 10^5$	2.45	8	-	1.160

While due steps like mesh independence studies have been carried out in this investigation, we acknowledge that the use of eddy-viscosity based turbulence models introduces a small amount of discrepancy in the final results. The errors of the k–ωSST model in flow separation and wakes has been documented comprehensively in prior studies [26,27].

6. Results and Discussion

Numerical simulations have been carried out for three different mass ratios over reduced velocities ranging from 4 to 12. The cylinder is modeled to be having only single degree of freedom (SDOF) in the transverse or CF direction. Influence of m^* and k^* on both hydrodynamic force coefficient in the CF direction and the response of the cylinder have been studied in detail. The CF response results can be verified with experimental [17,28]. Time history of coefficient of lift, C_L and non-dimensional CF response Y/D over the range of U_r for $m^* = 0.66$, 1.32, and 1.98 are displayed in Figures 4–6 respectively.

6.1. Case I

Under Case I, the response of cylinder is studied for $m^* = 0.66$. The mass of the cylinder is taken as 3 kg and added mass coefficient $C_A = 0.7$. U_r is varied in the analysis by varying the coefficient of stiffness k and in turn the natural frequency f_n of the oscillating system. Details of the simulation parameters have been given in Table 2. For a stationary cylinder, it was observed from numerous experimental and numerical works [29–31] that C_L oscillates in a symmetrical fashion about zero due to vortex shedding [13]. In the present study for lower values of U_r, C_L is not oscillating about zero symmetrically. However, the response of the cylinder is observed to be symmetrical for all values of U_r at $m^* = 0.66$. At $U_r = 8$, C_L becomes almost symmetrical with an effective lift coefficient 0.18. Beat phenomenon is captured in the response of the cylinder at $U_r = 8$. Results for different cases are presented in Table 4. Maximum lift force is observed at $U_r = 4$ with $C_L = 1.12$ and maximum response at $U_r = 8$ with $Y/D = 1.26$. With increase in U_r response amplitude of the cylinder decreases beyond $U_r = 8$. Time histories of C_L and Y/D for $m^* = 0.66$ is presented in Figure 4.

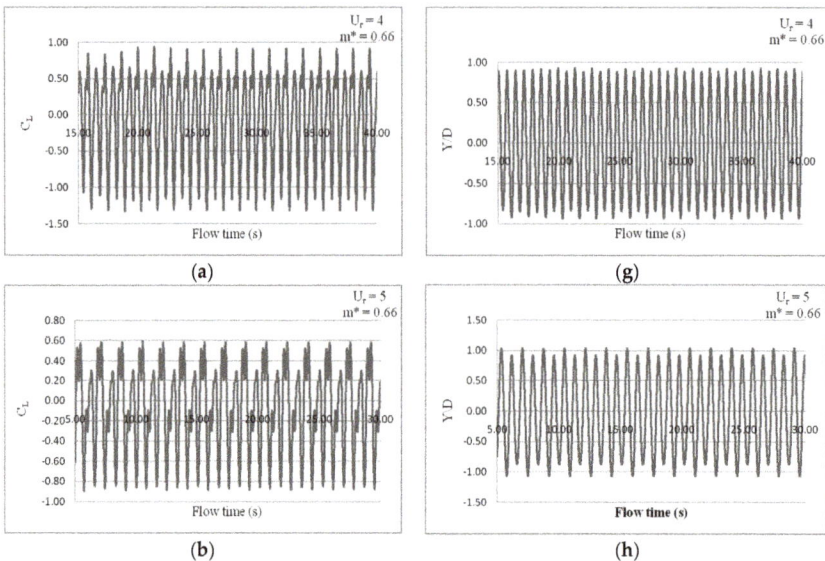

Figure 4. *Cont.*

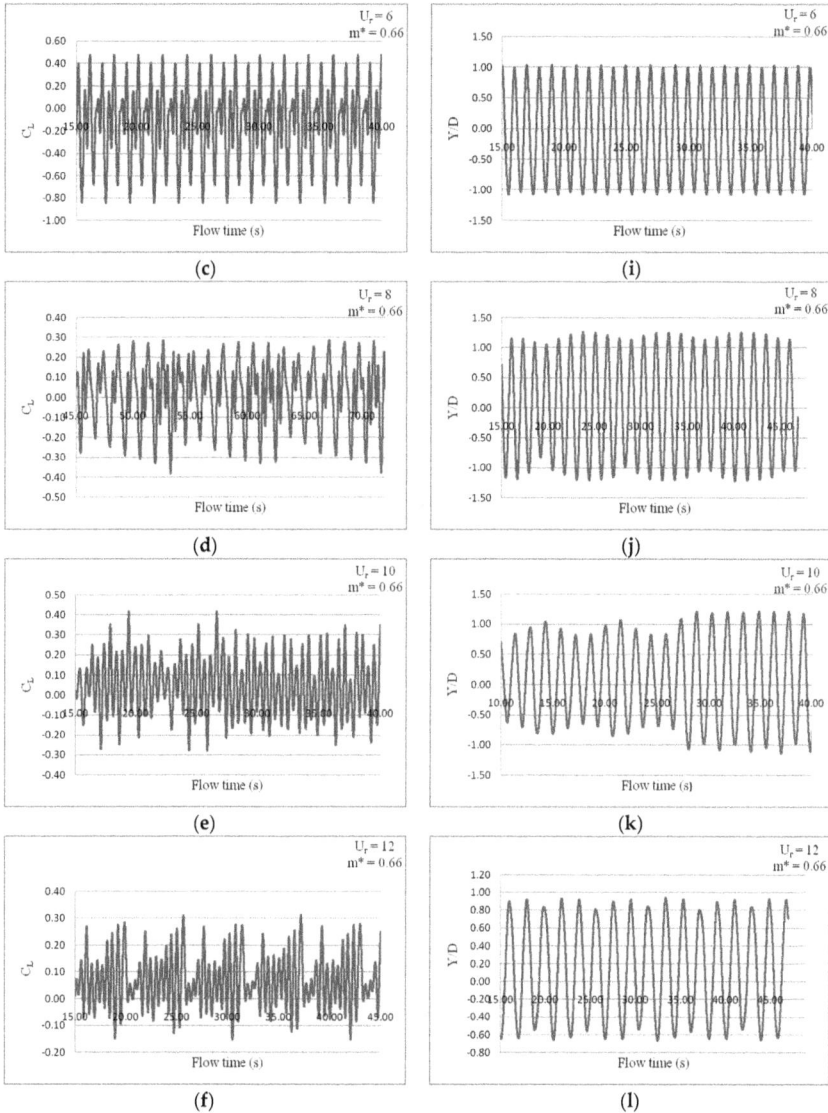

Figure 4. Time histories of hydrodynamic and structural response parameters at $m^* = 0.66$: (**a**) C_L for $U_r = 4$; (**b**) C_L for $U_r = 5$; (**c**) C_L for $U_r = 6$; (**d**) C_L for $U_r = 8$; (**e**) C_L for $U_r = 10$; (**f**) C_L for $U_r = 12$; (**g**) Y/D for $U_r = 4$; (**h**) Y/D for $U_r = 5$; (**i**) Y/D for $U_r = 6$; (**j**) Y/D for $U_r = 8$; (**k**) Y/D for $U_r = 10$; (**l**) Y/D for $U_r = 12$.

6.2. Case II

Under Case II, the response of cylinder is studied for $m^* = 1.32$. Mass of the cylinder is taken as 6 kg and added mass coefficient $C_A = 0.7$. Time histories of C_L and Y/D for $m^* = 1.32$ is presented in Figure 5. At $m^* = 1.32$, the cylinder exhibits similar response characteristics as in the previous case with maximum response at $U_r = 8$ with $Y/D = 1.17$. As the mass ratio increases, a slight decrease in the cross flow response amplitude is observed. Unlike the previous case, beat phenomenon is observed at $U_r = 10$.

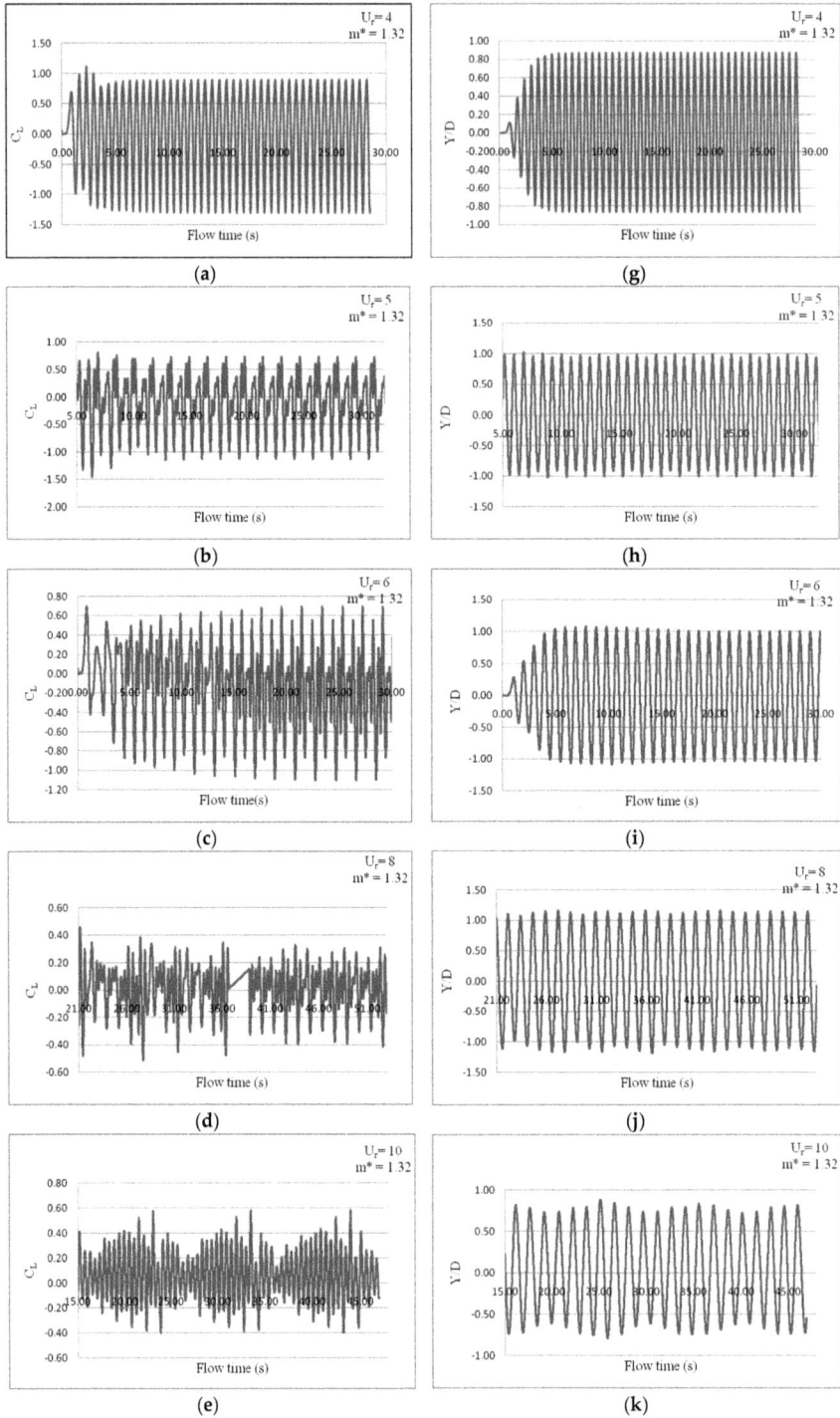

Figure 5. *Cont.*

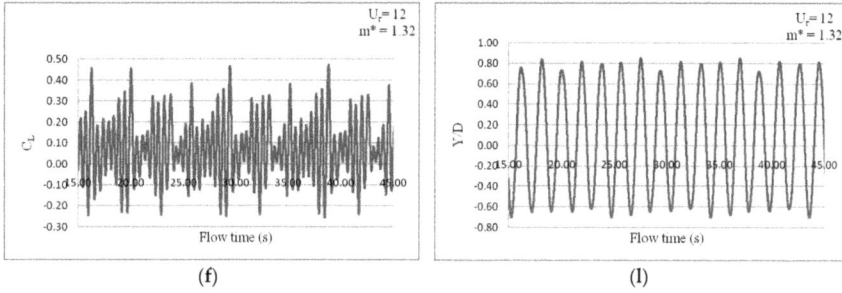

Figure 5. Time histories of hydrodynamic and structural response parameters at $m^* = 1.32$: (**a**) C_L for $U_r = 4$; (**b**) C_L for $U_r = 5$; (**c**) C_L for $U_r = 6$; (**d**) C_L for $U_r = 8$; (**e**) C_L for $U_r = 10$; (**f**) C_L for $U_r = 12$; (**g**) Y/D for $U_r = 4$; (**h**) Y/D for $U_r = 5$; (**i**) Y/D for $U_r = 6$; (**j**) Y/D for $U_r = 8$; (**k**) Y/D for $U_r = 10$; (**l**) Y/D for $U_r = 12$.

6.3. Case III

Under Case III, the response of cylinder is studied for $m^* = 1.98$. The mass of the cylinder is taken as 9 kg and added mass coefficient $C_A = 0.7$. Time histories of C_L and Y/D for $m^* = 1.98$ is presented in Figure 6. Maximum amplitude is observed at $U_r = 8$ with $Y/D = 1.13$ which is slightly less than the previous two cases. Here also prominent beat is observed at $U_r = 10$.

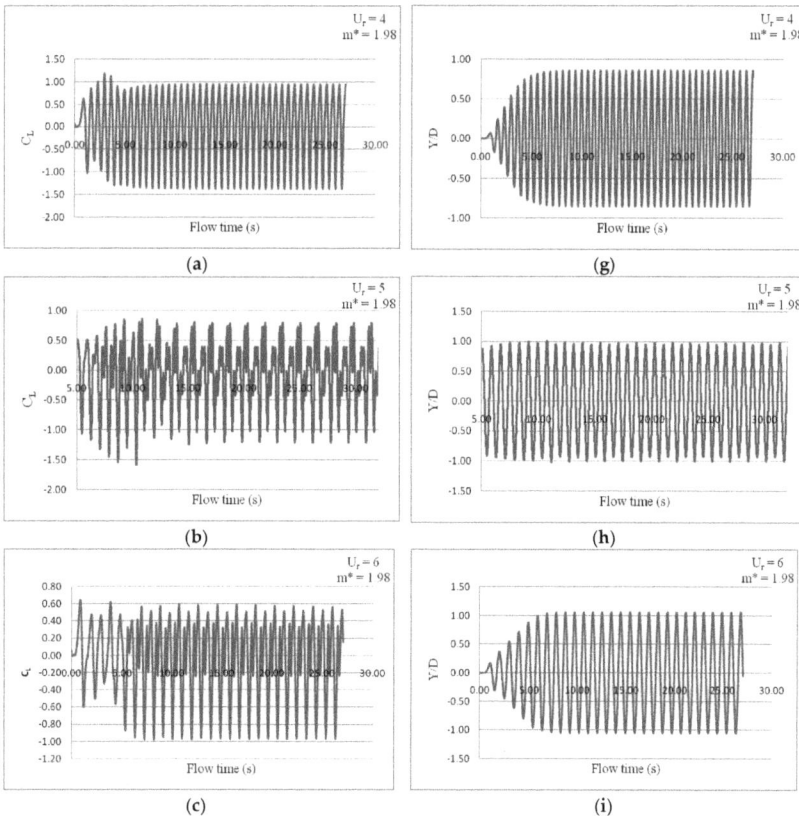

Figure 6. *Cont.*

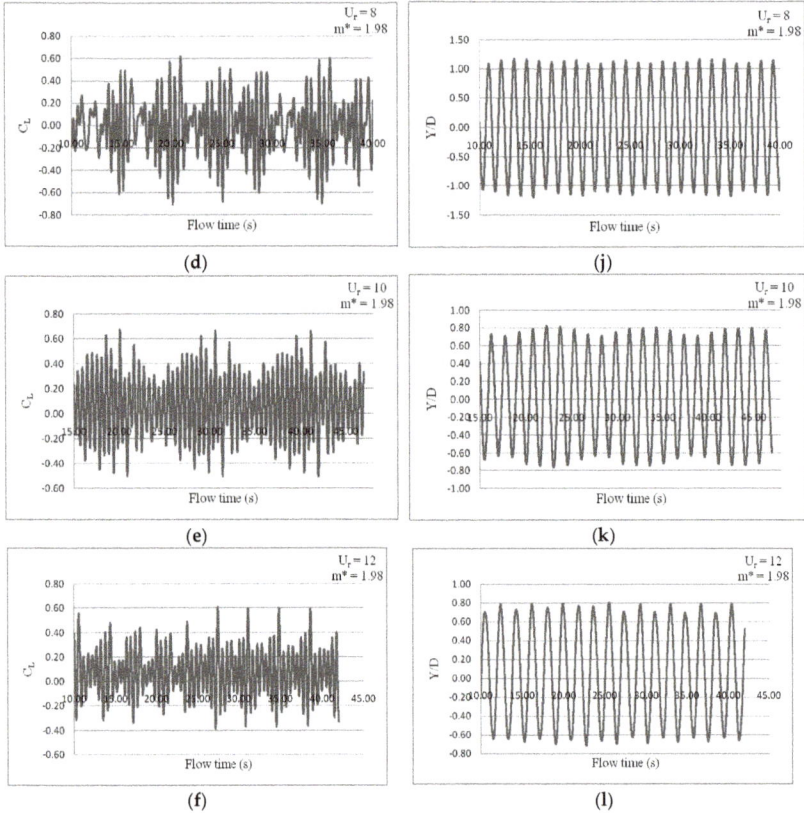

Figure 6. Time histories of hydrodynamic and structural response parameters at $m^* = 1.98$: (**a**) C_L for $U_r = 4$; (**b**) C_L for $U_r = 5$; (**c**) C_L for $U_r = 6$; (**d**) C_L for $U_r = 8$; (**e**) C_L for $U_r = 10$; (**f**) C_L for $U_r = 12$; (**g**) Y/D for $U_r = 4$; (**h**) Y/D for $U_r = 5$; (**i**) Y/D for $U_r = 6$; (**j**) Y/D for $U_r = 8$; (**k**) Y/D for $U_r = 10$; (**l**) Y/D for $U_r = 12$.

6.4. Shedding Characteristics

Numerical simulations for each case show that the shedding pattern and the characteristic variation of C_L strongly depend on the natural frequency of the oscillating mass. A more detailed history of C_L and Y/D for $m^* = 0.66$, 1.32, and 1.98 are presented in Figures 7–9 respectively.

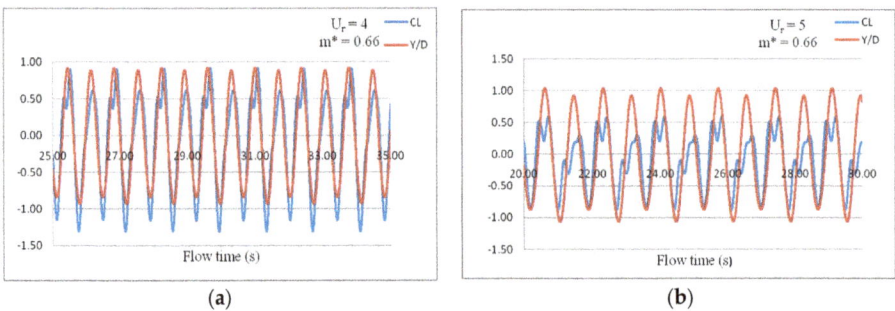

Figure 7. *Cont.*

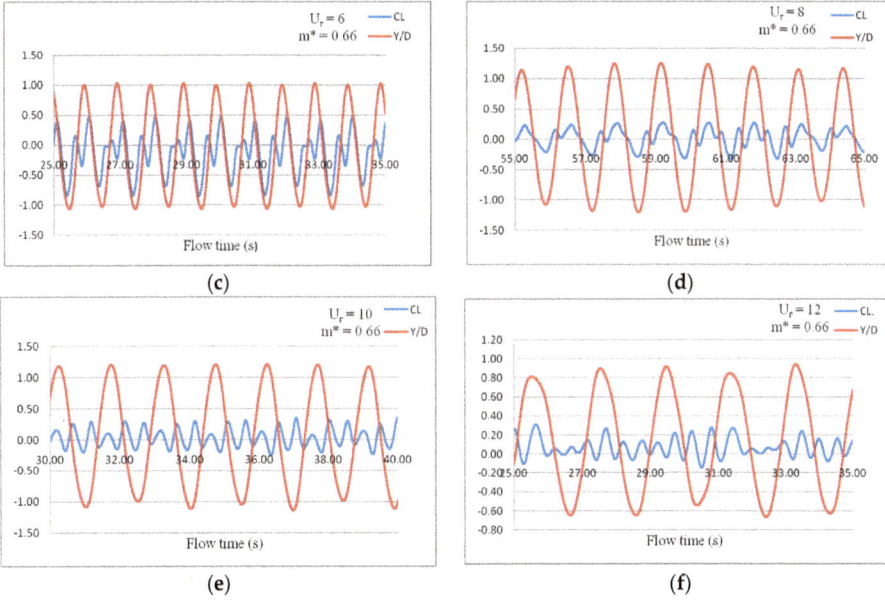

Figure 7. Variation of C_L with Y/D for $m^* = 0.66$ (**a**) $U_r = 4$; (**b**) $U_r = 5$; (**c**) $U_r = 6$; (**d**) $U_r = 8$; (**e**) $U_r = 10$; (**f**) $U_r = 12$.

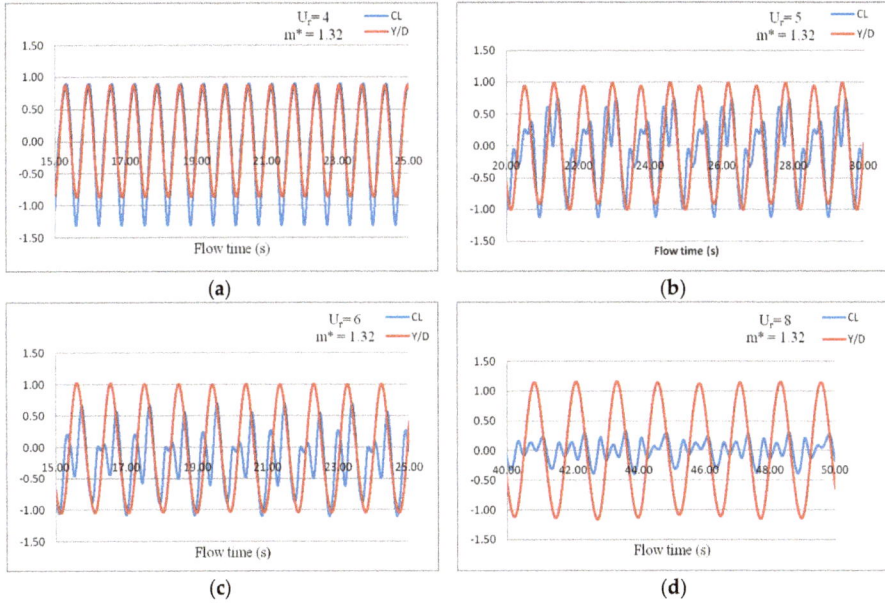

Figure 8. *Cont.*

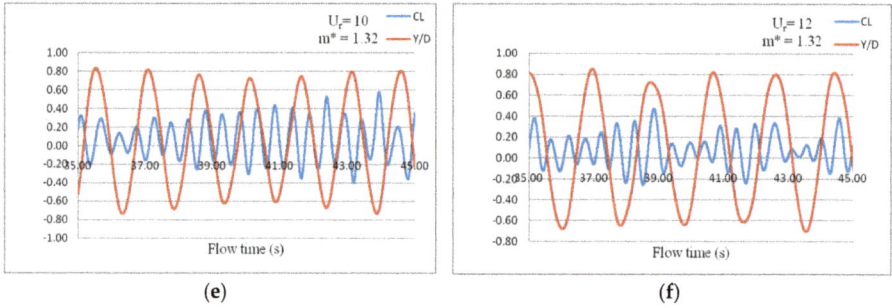

Figure 8. Variation of C_L with Y/D for $m^* = 1.32$: (**a**) $U_r = 4$; (**b**) $U_r = 5$; (**c**) $U_r = 6$; (**d**) $U_r = 8$; (**e**) $U_r = 10$; (**f**) $U_r = 12$.

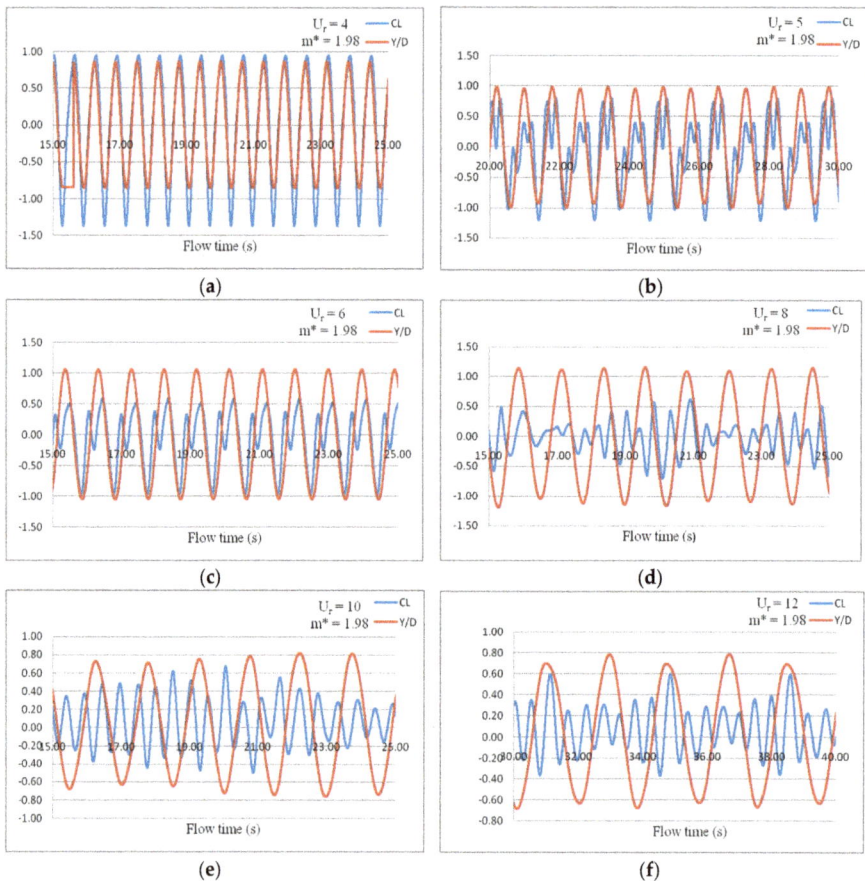

Figure 9. Variation of C_L with Y/D for $m^* = 1.98$: (**a**) $U_r = 4$; (**b**) $U_r = 5$; (**c**) $U_r = 6$; (**d**) $U_r = 8$; (**e**) $U_r = 10$; (**f**) $U_r = 12$.

From a detailed analysis of amplitude of variation of lift coefficient, it is observed that the characteristics of hydrodynamic load causing the cylinder to oscillate varies with U_r and in turn with frequency ratio, η. For all the three cases, a similarity can be observed in the shedding characteristics which are reflected in the pattern of lift coefficient variation. In each case, a single oscillation of the cylinder is actuated by a single oscillation of lift force when $U_r = 4$. For $U_r = 6$, double oscillation of lift coefficient is observed for each oscillation of Y/D. The first oscillation is of relatively low amplitude and the second one of larger amplitude. $U_r = 8$ and 10 which is more prone to beating phenomenon displays three separate C_L oscillations for each oscillation of the cylinder. $U_r = 12$ is observed to have an additional half oscillation. For all mass ratios, a 2P mode of vortex shedding is observed in the range of $U_r = 4$–12 which corresponds to synchronization.

A single beat sustains for a time period of 10.53 and 10.42 seconds for $m^* = 1.98$ and 1.32, respectively. Differently for $m^* = 0.66$, single beat sustains at $U_r = 8$ for 7.84 s which is significantly lesser than the other two cases, showing the highest value of cross flow response amplitude. Also, at the lowest mass ratio, $m^* = 0.66$, it is observed from the response amplitude history that the cylinder initially shows a tendency to beat at $U_r = 10$ with a significantly larger response amplitude with $Y/D = 1.08$ compared to 0.82 and 0.80 for $m^* = 1.32$ and 1.98 respectively at same U_r. Further study on the influencing parameters led to a trend showing a relationship between m^* and U_r, through which they have proved that the range of reduced velocity over which cylinder under VIV shows maximum cross flow response widens as m^* decreases [10]. The relationship put forward by [14] is shown in Figure 11. Results of the present study also show that Y/D increases with decreasing mass ratio and the range of U_r over which the maximum amplitude of response to be expected widens.

The effect of U_r on various simulations parameters is illustrated in Figure 10. A comparison of various hydrodynamic and structural parameters at different k* values for each case is given in Figure 11a,b. Present simulation results matches well with the values represented in Figure 11b. In all cases maximum response occurs at $U_r = 8$. Even though the response amplitude is high in the range $U_r = 4$–12, as it deviates away from $U_r = 8$, amplitude decreases. A similar result was reported by [19] is represented in the modified Griffin's plot.

The numerical simulation could successfully reproduce the response behavior depicted by the lower branch of response. Figure 11b shows the response of a cylinder of mass ratio $m^* = 2.4$ which is comparable with the present study.

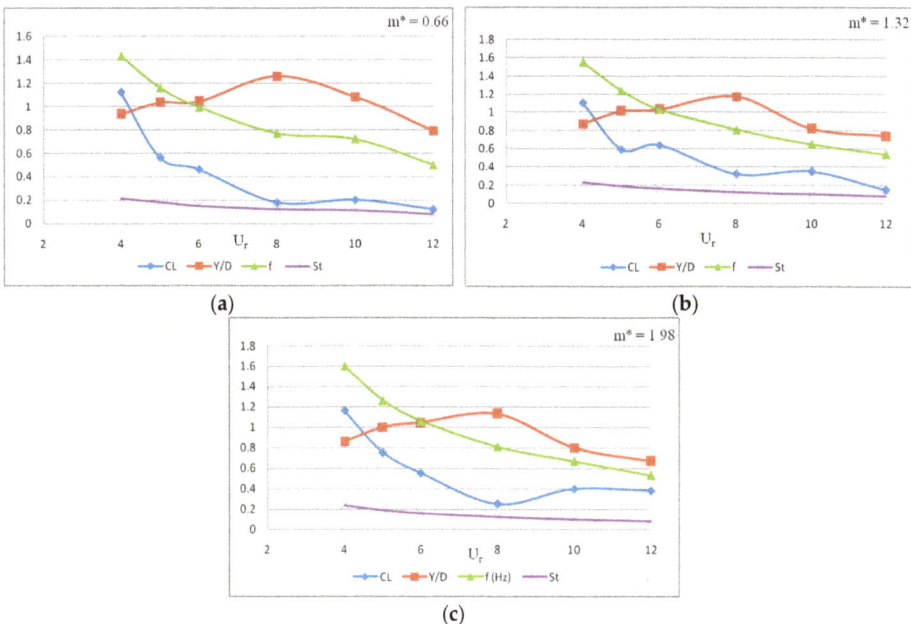

(a)

(b)

(c)

Figure 10. Effect of U_r on various hydrodynamic and structural parameters for VIV of a cylinder (a) $m^* = 0.66$; (b) $m^* = 1.32$; (c) $m^* = 1.98$.

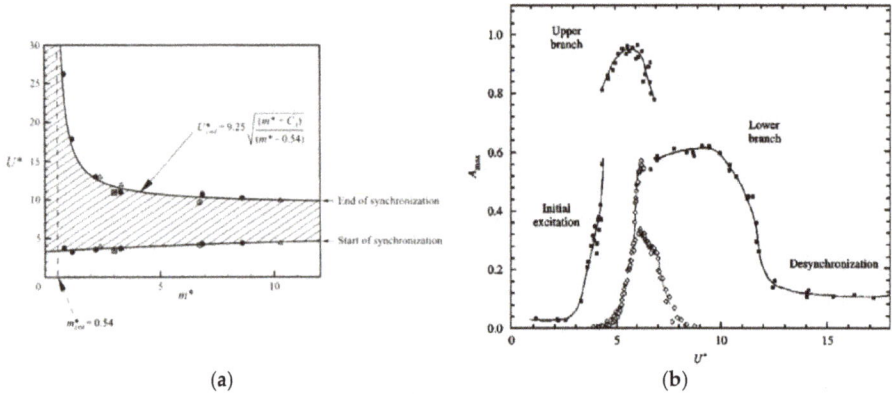

Figure 11. (a) Relationship between U_r and m^* presented by Williamson and Govardhan [23]; (b) maximum response amplitudes A_{max} as functions of the reduced velocity U_r for $m^* = 2.4$ and $m^* = 248$ [19].

It was previously observed that 2P mode shedding is the reason for synchronized response [20]. Since detailed analysis of physics of the flow is not in the scope of this paper, shedding pattern is not further discussed. A representation of 2P mode of vortex shedding in the synchronization range of U_r is shown in Figure 12. A detailed study to analyze this phenomenon is pending. Shedding characteristics at different m^* and k are listed in Table 4.

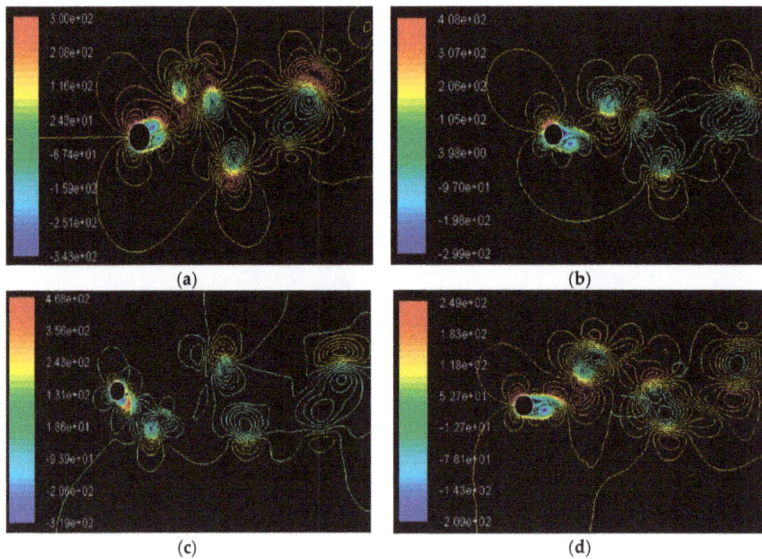

Figure 12. *Cont.*

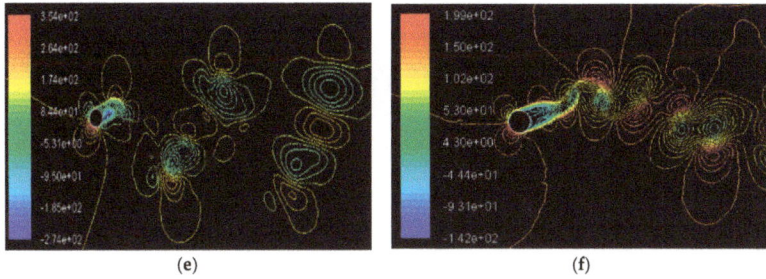

Figure 12. Representation of pressure (N/m^2) contours in 2P mode of vortex shedding for $m^* = 0.66$ (a) $U_r = 4$; (b) $U_r = 5$; (c) $U_r = 6$; (d) $U_r = 8$; (e) $U_r = 10$; (f) $U_r = 12$.

Table 4. Hydrodynamic and structural response characteristics for $m^* = 0.66$, 1.32, and 1.98.

Mass Ratio, m^*	U_r	C_L	Y/D	f_v (Hz)	η	St	Shedding Characteristics
	4	1.16	0.86	1.6	1.04	0.24	2P Lift force oscillates about zero value once during one time period of oscillation of the cylinder.
	5	0.75	1.0	1.26	1.04	0.19	2P Lift force oscillates twice during one time period of oscillation of the cylinder.
	6	0.55	1.05	1.06	1.03	0.16	2P Lift force oscillates twice during one time period of oscillation of the cylinder.
$m^* = 1.98$	8	0.25	1.13	0.81	1.01	0.12	2P Lift force oscillates thrice during one time period of oscillation of the cylinder.
	10	0.4	0.8	0.67	0.98	0.1	2P Lift force oscillates thrice during one time period of oscillation of the cylinder. Beat phenomenon is observed with time period 10.53 s.
	12	0.38	0.67	0.53	1.04	0.08	2P Lift force oscillates 3.5 times during one time period of oscillation of the cylinder.
	4	1.1	0.87	1.55	1.07	0.23	2P Lift force oscillates about zero value once during one time period of oscillation of the cylinder.
	5	0.59	1.01	1.24	1.06	0.19	2P Lift force oscillates twice during one time period of oscillation of the cylinder.
	6	0.64	1.03	1.03	1.06	0.16	2P Lift force oscillates twice during one time period of oscillation of the cylinder.
$m^* = 1.32$	8	0.32	1.17	0.81	1.01	0.12	2P Lift force oscillates thrice during one time period of oscillation of the cylinder.
	10	0.35	0.82	0.65	1.02	0.1	2P Lift force oscillates thrice during one time period of oscillation of the cylinder. Beat phenomenon is observed with time period 10.42 s.
	12	0.15	0.74	0.54	1.01	0.082	2P Lift force oscillates 3.5 times during one time period of oscillation of the cylinder.
	4	1.12	0.93	1.43	1.14	0.21	2P Lift force oscillates about zero value once during one time period of oscillation of the cylinder.
	5	0.56	1.03	1.16	1.13	0.18	2P Lift force oscillates twice during one time period of oscillation of the cylinder.
	6	0.46	1.04	0.99	1.10	0.15	2P Lift force oscillates twice during one time period of oscillation of the cylinder.
$m^* = 0.66$	8	0.18	1.26	0.77	1.06	0.12	2P Lift force oscillates thrice during one time period of oscillation of the cylinder. Beat phenomenon is observed with time period 7.84 s.
	10	0.2	1.08	0.72	1.1	0.11	2P Lift force oscillates thrice during one time period of oscillation of the cylinder. No beat is observed.
	12	0.12	0.79	0.5	1.1	0.08	2P Lift force oscillates 3.5 times during one time period of oscillation of the cylinder.

Energy possessed by a spring mass system undergoing oscillation can be represented as the sum of its kinetic and potential energies as given by Equation (26).

$$E = \frac{1}{2}m\dot{Y}^2 + \frac{1}{2}kY^2 \qquad (26)$$

When the position of the mass corresponds to maximum amplitude, the entire kinetic energy of the system will be converted into potential energy and Equation (26) reduces to

$$E = \frac{1}{2}kY^2 \qquad (27)$$

At zero amplitude position of the mass, the entire potential energy is converted into kinetic energy. Since the total energy of the system is conserved energy balance can be written as Equation (28).

$$E = \frac{1}{2}kY^2 = \frac{1}{2}m\dot{Y}^2 \qquad (28)$$

Hence the maximum possible velocity with which the system oscillates can be expressed as

$$\dot{Y} = \sqrt{\frac{k}{m}}Y \qquad (29)$$

Which may be also represented in non-dimensional form represented by Equation (30)

$$\dot{Y} = \sqrt{\frac{Dk^*g}{L/D}}\frac{Y}{D} \qquad (30)$$

For the optimum condition proven numerically, V_{max} is obtained as 0.5 m/s. Power associated with the oscillatory motion can be expressed by Equation (31)

Maximum velocity has been calculated using the above expression and a comparison of average power output estimated is represented in Table 5. Even though the maximum amplitude of oscillation is obtained for each m^* at $U_r = 8$, power output is maximum at $U_r = 4$, suggesting the best operating conditions.

$$P_{avg} = F_L\dot{Y} = C_L\frac{1}{2}\rho AV^2\dot{Y} \qquad (31)$$

Table 5. Cylinder velocity and calculated average power for different configurations.

Re	m^*	U_r	k^*	C_L	Y/D	$\dot{Y}$	P_{avg} (W)
3.8×10^4	0.66	4	11.17	1.12	0.93	0.74	7.90
		5	6.9	0.56	1.03	0.65	3.44
		6	4.81	0.46	1.04	0.54	2.38
		8	2.7	0.18	1.26	0.49	0.85
		10	1.73	0.2	1.08	0.34	0.64
		12	1.21	0.12	0.79	0.21	0.24
	1.32	4	11.17	1.1	0.87	0.69	7.26
		5	6.9	0.59	1.01	0.63	3.55
		6	4.81	0.64	1.03	0.54	3.28
		8	2.7	0.32	1.17	0.46	1.40
		10	1.73	0.35	0.82	0.26	0.86
		12	1.21	0.15	0.74	0.19	0.28
	1.98	4	11.17	1.16	0.86	0.69	7.57
		5	6.9	0.75	1	0.63	4.47
		6	4.81	0.55	1.05	0.55	2.87
		8	2.7	0.25	1.13	0.44	1.05
		10	1.73	0.4	0.8	0.25	0.95
		12	1.21	0.38	0.67	0.18	0.64

7. Field Test Validation of Numerical Results

The HVPG module with specifications as given in Table 1 has been used for a field test in the Palissery irrigation canal. The flow velocity in the canal was measured to be 0.5 m/s. The model has been tested for different k^* values. Cylinder displacement was measured by attaching a pantograph to the side vanes. As the cylinder oscillates, the marking pencil attached to the spring side of the module marks its impression on the paper attached on the vane side. The results are tabulated in Table 6. U_r = 4, 8, and 12, but the effect of k^* is observed to follow the same trend as predicted numerically. An average deviation of 30% is observed between numerical method and field tests. Over prediction of the response amplitude by the numerical method may be attributed to not accounting for structural damping. Also, friction in the sliding parts of the guide vane contributes to lowering the response. Components of the HVPG and field test set up are shown in Figure 13.

Table 6. Results of field test conducted at Palissery irrigation canal.

Re	m^*	U_r	k^*	Y_{field} (cm)	Y/D_{fileld}	$Y/D_{numerical}$
		4	11.17	5.5	0.72	0.93
3.8×10^4	0.66	8	2.7	7.5	0.98	1.26
		12	1.21	4.5	0.59	0.79

(a)

(b)

Figure 13. (a) Model of HVPG; (b) field test at Palissery irrigation canal for m^* = 0.66 and U_r = 4, 8, and 12.

8. Conclusions

Extensive research carried out to understand and interpret the intrinsic vortex shedding phenomenon has brought out several correlations to estimate response amplitude Y/D [32], but most of these expressions define the non-dimensional oscillation amplitude Y/D as a function of complex parameters like the Skop–Griffin parameter [14]. The present study is an effort to understand the hydrodynamic response of the cylinder from a designers' perspective by considering the effect of tangible system parameters only. Optimum response is obtained at m^* = 0.66 and U_r = 8, but optimum estimated power output for the same mass ratio is obtained at U_r = 4. It is observed that maximum power output can be derived from an HVPG operating at low mass ratio and in the lowest regime of reduced velocity in the synchronization range irrespective of amplitude of cylinder response. The reduced power output at U_r = 8 is due to the lower value of C_L which in turn is due to several oscillations of lift force within one time period of oscillation of the cylinder. The true values of frequency ratio, η obtained from the simulations are indicative of synchronized response for the range of U_r considered. Hence it is observed that the developed numerical method could successfully simulate the flow around an oscillating cylinder. The numerical method developed is capable of predicting the trend of variation of Y/D which is verified using the results of field test. At m^* = 0.66 where maximum response is observed, η significantly exceeds over unity compared to the higher mass ratio cases. It can be concluded that maximum amplitude of response is observed at mass ratios corresponding to η values greater than unity rather than at unity η values. Even though hydrodynamic lift force acting on the cylinder is proportional to the incoming flow velocity, C_L strongly depends on the natural frequency of the oscillating system and the shedding pattern. Response amplitude of the cylinder also depends on stiffness ratio k^* and in turn on natural frequency of the oscillating system f_n. Response amplitude increases with decreasing mass ratio and the range of U_r over which the response is high (resonance) widens. Occurrence of beat phenomenon also depends on m^* and k^* and the relationship is more pronounced at lower values of m^*. The mode of vortex shedding depends only on U_r at a constant flow velocity and is independent of m^*. On the whole, the work provides strong design inputs to the construction of the envisaged HVPG model.

9. Scope for Future Research

The discussions and conclusions from Sections 7 and 8 respectively indicate an immense scope for future research. Authors are in the process of improving the present design by incorporating structural damping in the studies. Inclusion of a greater number of rollers in the guide vanes is also being aimed at for eliminating the friction between sliding parts. The influence of these additional parameters in the equation of motion can reduce the errors in the subsequent numerical prediction, leading the design parameters closer to those of practical values. Improvised designs are also being devised for minimizing the transmission losses during power generation.

Author Contributions: Conceptualization, V.C., S.M., and S.J.; Data curation, V.C. and S.J.; Formal analysis, V.C. and S.J.; Funding acquisition, S.J. and V.M.; Investigation, V.C.; Methodology, S.M. and S.J.; Project administration, S.M. and S.J.; Resources, S.J. and V.M.; Software, V.M.; Supervision, S.M. and S.J.; Validation, V.C. and S.J.; Visualization, V.C. and S.J.; Writing—original draft, V.C. and S.J.; Writing—review & editing, V.C., S.J., and V.M.

Funding: This research was partially funded by Energy Management Center, Government of Kerala, India specifically for the development of Hydro Vortex Power Generator module with grant no. EMC/ET&R/18/R&D/SCMS/01.

Acknowledgments: The authors would like to thank the management and principal of SCMS School of Engineering and Technology for all the support offered for the research and development. The authors would also like to thank Energy Management Center, Government of Kerala for all the support.

Conflicts of Interest: The authors declare no conflict of interest.

References

1. Gerrard, J.H. The mechanics of the formation region of vortices behind bluff bodies. *J. Fluid Mech.* **1966**, *25*, 401–413. [CrossRef]
2. Gao, Y.; Fu, S.; Xiong, Y.; Zhao, Y.; Liu, L. Experimental study on response performance of vortex-induced vibration on a flexible cylinder. *Ships Offshore Struct.* **2016**, *12*, 116–134. [CrossRef]
3. Bimbato, A.M.; Pereira, L.A.; Hirata, M.H. Suppression of vortex shedding on a bluff body. *J. Wind Eng. Ind. Aerodyn.* **2013**, *122*, 16–18. [CrossRef]
4. Bernitsas, M.; Raghavan, K.; Ben-Simon, Y.; Garcia, E. VIVACE (Vortex Induced Vibration Aquatic Clean Energy): A New Concept in Generation of Clean and Renewable Energy from Fluid Flow. *J. Offshore Mech. Arct. Eng.* **2006**, *130*, 041101. [CrossRef]
5. An, X.; Song, B.; Tian, W.; Ma, C. Design and CFD Simulations of a Vortex-Induced Piezoelectric Energy Converter (VIPEC) for Underwater Environment. *Energies* **2018**, *11*, 330. [CrossRef]
6. Janardhanan, S.; Chandran, V.; Varghese, C.; Achuth, D.; Devassy, D.; Mathews, D.C. Hydro vortex power generator design and construction. In Proceedings of the Kerala Technological CONGRESS, KETCON 2018—Human Computer Interface, Thrissur, India, 24 February 2018.
7. Griffin, O.M. Vortex-Excited Cross-Flow Vibrations of a Single Cylindrical Tube. *ASME J. Press. Vessel Technol.* **1980**, *102*, 158–166. [CrossRef]
8. Khalak, A.; Williamson, C.H.K. Dynamics of a hydroelastic cylinder with very low mass and damping. *J. Fluids Struct.* **1996**, *10*, 455–472. [CrossRef]

9. Narendran, K.; Murali, K.; Sundar, V. Vortex-induced vibrations of elastically mounted circular cylinder at Re of the O(105). *J. Fluids Struct.* **2015**, *54*, 503–521. [CrossRef]
10. Govardhan, R.; Williamson, C.H.K. Defining the 'modified Griffin plot' in vortex-induced vibration: Revealing the effect of Reynolds number using controlled damping. *J. Fluid Mech.* **2006**, *561*, 147–180. [CrossRef]
11. Bernitsas, M. Out of the Vortex. *Mech. Eng.* **2010**, *132*, 22–27. [CrossRef]
12. Tian, W.; Mao, Z.; Zhao, F. Design and Numerical Simulations of a Flow Induced Vibration Energy Converter for Underwater Mooring Platforms. *Energies* **2017**, *10*, 1427. [CrossRef]
13. Khan, N.B.; Ibrahim, Z.; Tuan, L.; Javed, M.F.; Jameel, M. Numerical investigation of the vortex-induced vibration of an elastically mounted circular cylinder at high Reynolds number (Re = 104) and low mass ratio using the RANS code. *PLoS ONE* **2017**, *12*, e0185832. [CrossRef] [PubMed]
14. Williamson, C.H.K.; Govardhan, R. Vortex induced vibrations. *Annu. Rev. Fluid Mech.* **2004**, *36*, 413–455. [CrossRef]
15. Achenbach, E.; Heinecke, E. On Vortex Shedding from Smooth and Rough Cylinders in the Range of Reynolds Numbers 6 × 103 to 5 × 106. *J. Fluid Mech.* **1981**, *109*, 239–251. [CrossRef]
16. Blevins, R.D. *Flow-Induced Vibration*, 2nd ed.; Van Nostrand Reinhold: New York, NY, USA, 1990; pp. 163–164, ISBN 1-57524-183-8.
17. Gabbai, R.D.; Benaroya, H. An overview of modeling and experiments of vortex-induced vibration of circular cylinders. *J. Sound Vib.* **2005**, *282*, 575–616. [CrossRef]
18. Anagnostopoulos, P.W.; Bearman, P.W. Response characteristics of a vortex exited cylinder at low Reynolds number. *J. Fluids Struct.* **1992**, *6*, 39–50. [CrossRef]
19. Khalak, A.; Williamson, C.H.K. Investigation of the relative effects of mass and damping in vortex induced vibration of a circular cylinder. *J. Wind Eng. Ind. Aerodyn.* **1997**, *69–71*, 341–350. [CrossRef]
20. IcemCfd, A. *12.0 User's Ma*; Ansys Inc.: Canonsburg, PA, USA, 2009; Volume 5.
21. Fluent, A. *12.0 Theory Guide*; Ansys Inc.: Canonsburg, PA, USA, 2009; Volume 5.
22. Menter, F.R. Two-equation eddy-viscosity turbulence models for engineering applications. *AIAA J.* **1994**, *32*, 1598–1605. [CrossRef]
23. Anton, G. Analysis of Vortex-Induced Vibration of Risers. Master's Thesis, Applied Mechanics, Chalmers University of Technology, Gothenburg, Sweden, 2012.
24. Vandiver, J.K. Damping parameters for flow-induced vibration. *J. Fluids Struct.* **2012**, *35*, 105–119. [CrossRef]
25. Schlichting, H. *Boundary Layer Theory*, 8th ed.; McGraw-Hill Book Company: New York, NY, USA, 1979; ISBN 13 978-3540662709.
26. Iaccarino, G.; Mishra, A.A.; Ghili, S. Eigenspace perturbations for uncertainty estimation of single-point turbulence closures. *Phys. Rev. Fluids* **2017**, *2*, 024605. [CrossRef]
27. Mishra, A.A.; Gianluca, I. Uncertainty Estimation for Reynolds-Averaged Navier Stokes Predictions of High-Speed Aircraft Nozzle Jets. *AIAA J.* **2017**, *55*, 3999–4004. [CrossRef]
28. Govardhan, R.; Williamson, C.H.K. Modes of vortex formation and frequency response for a freely-vibrating cylinder. *J. Fluid Mech.* **2000**, *420*, 85–130. [CrossRef]
29. Chen, W.; Zhang, Q.; Li, H.; Hu, H. An experimental investigation on vortex induced vibration of a flexible inclined cable under a shear flow. *J. Fluids Struct.* **2015**, *54*, 297–311. [CrossRef]
30. Feng, C.C. The Measurements of Vortex Induced Effects in Flow Past a Stationary and Oscillating Circular and D-Section Cylinders. Master's Thesis, The University of British Columbia, Vancouver, BC, Canada, 1968.
31. Naudascher, E.; Rockwell, D. *Flow Induced Vibration—An Engineering Guide*; Dover Publications Inc.: Mineola, NY, USA, 2005; pp. 37–38, ISBN 13 978-0-486-44282-2.
32. Domal, V.; Sharma, R. An experimental study on vortex-induced vibration response of marine riser with and without semi-submersible. *J. Eng. Marit. Environ.* **2017**, *232*, 176–198. [CrossRef]

![energies logo] *energies*

MDPI

Article

Design and Optimization of Multiple Circumferential Casing Grooves Distribution Considering Sweep and Lean Variations on the Blade Tip

Weimin Song, Yufei Zhang and Haixin Chen *

School of Aerospace Engineering, Tsinghua University, Beijing 100084, China; songweimin1989@163.com (W.S.); zhangyufei@tsinghua.edu.cn (Y.Z.)
* Correspondence: chenhaixin@tsinghua.edu.cn; Tel.: +86-010-6278-9269

Received: 18 August 2018; Accepted: 9 September 2018; Published: 11 September 2018

Abstract: This paper focuses on the design and optimization of the axial distribution of the circumferential groove casing treatment (CGCT). Effects of the axial location of multiple casing grooves on the flow structures are numerically studied. Sweep and lean variations are then introduced to the blade tip, and their influences on the grooves are discussed. The results show that the ability of the CGCT to relieve the blockage varies with the distribution of grooves, and the three-dimensional blading affects the performance of both the blade and the CGCT. Accordingly, a multi-objective optimization combining the CGCT design with the sweep and lean design is conducted. Objectives, including the total pressure ratio and the adiabatic efficiency, are set at the design point; meanwhile, the choking mass flow and the near-stall performance are constrained. The coupling between the CGCT and the blade is improved, which contributes to an optimal design point performance and a sufficient stall margin. The sweep and lean in the tip redistribute the spanwise and chordwise loading, which enhances the ability of the CGCT to improve the blade's performance. This work shows that the present CGCT-blade integrated optimization is a practical engineering strategy to develop the working capacity and efficiency of a compressor blade while achieving the stall margin extension.

Keywords: circumferential groove casing treatment; sweep and lean; CGCT-blade integrated optimization

---

1. Introduction

The compressor of modern aero-engines has a high stage loading to fulfill the requirement of a high thrust–weight ratio. A high stage loading increases the potential risks of tip stall in some rotors [1]. A casing treatment is usually adopted to enhance the rotor's stability when the stall margin of the tip-critical rotor is insufficient. Two types of casing treatment, namely, axial slots and circumferential grooves (see Figure 1), are commonly used. Although the circumferential groove casing treatment (CGCT) typically generates less stall margin improvement, it has a smaller efficiency penalty and greater mechanical integrity than axial slots [2,3]. Consequently, a CGCT is more practical than axial slots if the efficiency cannot be sacrificed.

The mechanism of a CGCT for extending the stall margin is relevant to the alteration of the flow structures near the blade tip. Rabe and Hah [4] conducted experimental and numerical investigations on a rotor with three different CGCT configurations. They found that the CGCT could reduce the incidence angle and suppress the flow separation. Müller et al. [5] simulated four CGCT configurations under both design and off-design conditions, and they noticed that the grooves segmented the tip leakage vortex (TLV) and alleviated the blockage, which delayed the spillage of the low-energy fluid from the leading edge of the adjacent blade. Sakuma et al. [6] studied the effects of a single circumferential groove with different axial locations on NASA Rotor 37. They concluded that the CGCT reduced both the tip loading and the momentum of the tip leakage flow.

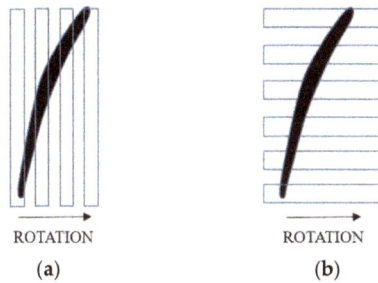

Figure 1. Illustration of axial slots and circumferential grooves in a top view: (**a**) axial slots; (**b**) circumferential grooves.

The axial location of the circumferential grooves significantly affects the rotor's performance. However, the design guidelines for the CGCT's axial distribution have not been unified. Houghton and Day [7] examined the effects of a single circumferential groove at different axial locations. They reported that whereas the groove located at the 8% and 50% points of the axial tip chord length ($C_{ax, tip}$) achieved the maximum stall margin improvement, the groove near the leading edge, near the trailing edge, and at the 18% point of C_{ax} were inefficient. The research carried out by Du et al. [8] showed that the groove located between the 50% and 60% points of C_{ax} was the most effective for the stall margin extension; however, the groove located at approximately 20% point of C_{ax} was actually harmful to the rotor's stability. Choi [9] tested a single casing groove at different positions. The groove installed near the leading edge was found to be the most effective for expanding the stable operating range with a small efficiency loss. The numerical study of Mao et al. [10] shows that a single groove located within the front 40% point of C_{ax} had obvious benefits to the efficiency and stall margin. Although the axial location of the CGCT plays a significant role in affecting the rotor's performance and stability, there is not an acknowledged conclusion on this issue. Nevertheless, most of the previous studies used a single circumferential groove to simplify the analysis of the effects of the CGCT's axial location on a stall. Those researchers [4,11,12] who used multiple grooves to seek a larger stall margin improvement and higher peak efficiency usually chose to adjust the depth and width of each groove while fixing their axial locations. Because the effects of multiple grooves are by no means a simple accumulation of the individual grooves' effects [13], it is necessary to fill the gap between the single groove study and the practical application of multiple grooves. In this paper, efforts are made to design and optimize the axial distribution of multiple grooves, which also contributes to the understanding of the CGCT's influences on the stall.

The benefits of a CGCT are obtained by altering the tip flow structures. The flow structures near the blade tip are also affected by the 3D shape of the blade. Researchers [14–18] have found that sweeps and leans can suppress the shockwave-boundary layer interaction, redistribute the spanwise loading, reduce the accumulation of the low energy fluids near the casing and impact the strength of the TLV and the secondary flow. Thus, the three-dimensional (3D) blade design is supposed to interact with the CGCT design. Houghton and Day [7] proposed the idea combining the casing treatment design with the blade design to maximize the effectiveness of the casing treatment. Recently, Hah [19] studied the inner workings of axial slots and suggested that casing treatments could be designed by optimizing the casing groove shape and the blade loading near the tip. However, the topic of the CGCT-blade integrated design or optimization has been rarely studied in the open literature. It is noteworthy that only Kim et al. [11] and Goinis and Nickle [20] have done some pioneer work in this field. Moreover, with the increasing use of 3D blading techniques in modern compressor rotors and fans, the CGCT design should consider the interactions between the grooves and a swept or leaned blade. With this motivation, effects of the sweep and lean on the CGCT design are studied in this paper. The CGCT-blade integrated optimization is further performed to achieve an ideal CGCT-blade combination.

This paper is organized as follows. First, the numerical methods are described and validated. Second, the flow fields of the baseline blade with different CGCTs are simulated and compared. Third, the sweep and lean are introduced to the blade tip. The performance of the modified blade with the CGCT is predicted, and the influences of the 3D blading on the CGCT design are discussed. Finally, the CGCT-blade integrated optimization is conducted. By this process, an optimal CGCT-blade combination is obtained. The mechanism by which the sweep and lean variations affect the effectiveness of CGCT is revealed.

2. Parameterization of the Blade and CGCT

The compressor studied in this paper is the Notre Dame Transonic Axial Compressor (ND-TAC) that has been experimentally tested at the University of Notre Dame. The details of the experimental setup can be found in the papers of Cameron et al. [21] and Kelly et al. [22]. The parameters of the original rotor blade are listed in Table 1.

Table 1. Design parameters of the Notre Dame Transonic Axial Compressor (ND-TAC) rotor.

Symbol	Value	Symbol	Value
N_c (rpm)	14,686	Axial Chord (mm)	35.56
U_{tip} (m/s)	352	R_{casing} (mm)	228.6
Solidity	1.21	R_{hub} (mm)	171.45
N_b	20	τ (mm)	0.762

At the full rotational speed, the corrected choking mass flow of the rotor is approximately 10.1 kg/s and the design mass flow rate is 9.67 kg/s. In addition, the actual mean tip clearance used in the computational fluid dynamics (CFD) simulation is 0.6858 mm. The ND-TAC rotor is tip-critical and it exhibits a spike-type stall behavior [23]. The application of a CGCT to enhance the stability of the ND-TAC [24,25] has demonstrated the effectiveness of CGCTs in this rotor. However, the variation of the design point performance resulting from CGCT was not discussed, or of concern. This paper focuses on the design and optimization of the axial distribution of a CGCT with consideration of the 3D blading design, which aims at developing a novel CGCT-blade integrated optimization strategy to improve the design point performance while extending the stall margin.

In this paper, the sweep and lean are introduced to the blade tip to investigate the effects of 3D blading on the CGCT design. The sweep is defined as the shift of the airfoil section along the local chord line. The upstream shift is called the forward sweep, and the downstream shift is the backward sweep. The movement of the section that is perpendicular to the local chord line is defined as the lean. If the direction of the lean is towards the pressure side, the lean is defined as a positive lean; otherwise, it is a negative lean.

The blade is generated by a third-order B-spline interpolation based on fourteen airfoils stacked from the hub to the casing. The sweep and lean are manipulated by two third-order Bezier curves. The values of the sweep and lean are defined by the percentage of the baseline-blade mean chord length L and the baseline-blade mean leading-trailing edge deviation $\Delta(r\cdot\theta)$, respectively. Initially, the leading edge (LE) and corresponding trailing edge (TE) for each baseline 2D airfoil are expressed in cylindrical coordinates (r, θ, z). Then, L and $\Delta(r\cdot\theta)$ are calculated by Equations (1) and (2), respectively:

$$L = \sqrt{(z_{TE} - z_{LE})^2 + [(r\cdot\theta)_{TE}^2 - (r\cdot\theta)_{LE}^2]}$$ (1)

$$\Delta(r\cdot\theta) = |(r\cdot\theta)_{TE} - (r\cdot\theta)_{LE}|$$ (2)

Next, the values of L and $\Delta(r \cdot \theta)$ of all of the 2D airfoils are averaged to obtain the mean value for the baseline blade. The above definitions of the sweep and lean adopted in this paper are essentially consistent with the conventional method that defines the sweep and lean by the absolute axial displacement and the circumferential angle. Still, the geometry changes defined by the relative variation can be more intuitive. Figure 2a,b illustrate the distributions of the sweep and lean along the blade span. The abscissa is the percentage of the mean chord length and the mean deviation. Both the sweep and the lean are regulated by five equally distributed control points from root to tip. In this study, only the control point at the blade tip is variable, which causes the 3D shape of the last 35% of the blade span to be affected by the sweep and lean variations.

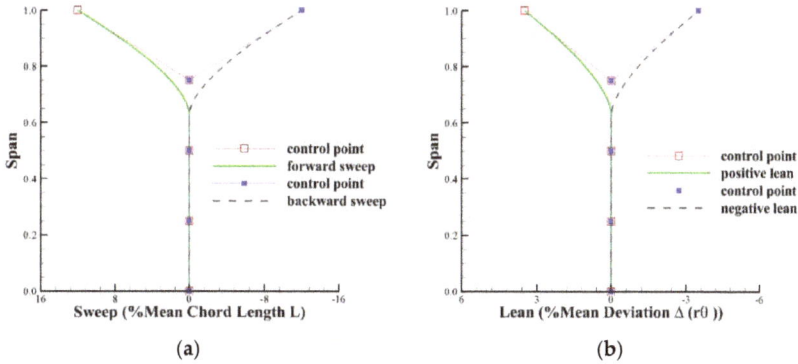

Figure 2. Illustration of the parameterization of the sweep and lean: (**a**) The sweep; (**b**) the lean.

With respect to the CGCT design, the groove number, width and depth of the grooves are fixed. Only the axial coverage of the CGCT is variable. The parameterization of the CGCT is shown in Figure 3. The groove number, along with the depth and width of each groove are given in the right side of the figure. The distance between any two adjacent grooves remains equal, to facilitate the manufacturing of the CGCT. In this figure, the abbreviation of "TG" means the tip gap. The depth of each groove is 4.04 times the tip gap size. The width of each groove is approximately equal to its depth. The dimension of the casing grooves refers to the previous research [24,25]. A few CFD tests, which are not provided in this paper, were also carried out to ensure the rationality of the parameter selection. Note that the width and depth design are not the interest of this paper; thus, only the axial coverage of the overall CGCT configuration installed on the casing can be controlled by two variables k and s, which correspond to the distance between the first groove and the blade's leading edge and the distance between two adjacent grooves, respectively. The variables k and s are non-dimensionalized by the axial chord length $C_{ax, \, tip}$.

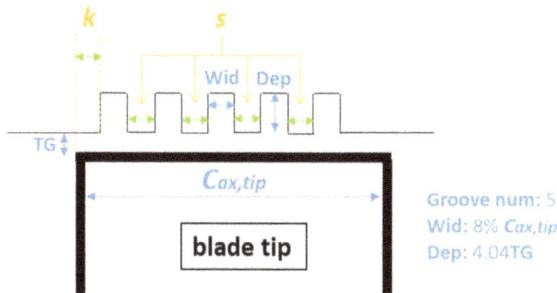

Figure 3. Illustration of the circumferential groove casing treatment (CGCT) design parameters.

3. Numerical Methods

An in-house CFD code called NSAWET (Navier-Stokes Analysis based on Window-Embedment Technique) is employed to solve the 3D compressible Reynold averaged Navier-Stokes (RANS) equation. An improved version of the HLL-Riemann solver by restoring the contact surface (HLLC) is selected to compute the convective flux [26]. The third-order monotone upstream-centred scheme for conservation laws (MUSCL) [27] with a Van Albada limiter [28] is adopted to ensure accuracy on non-uniform and skew cells. The k-ω shear stress transport (SST) turbulence model [29] is used to compute the turbulent viscosity. Moreover, the time marching is achieved by the Lower-Upper Symmetric-Gauss-Seidel (LU-SGS) implicit algorithm [30]. When the CGCT is adopted, the flux exchange between the groove-casing interfaces are dealt with using an overlap-area weighted reconstruction method [31]. The accuracy of NSAWET as applied to turbomachinery has been validated by previous studies [24,32,33].

The rotor is simulated in the single passage that is discretized by the multi-block structured mesh. The total pressure, total temperature, and axial flow angles are uniformly specified at the inlet boundary. With a given back pressure at the hub, a simplified radial equilibrium equation is employed to compute the radial distribution of the static pressure on the outlet plane. With a steady RANS, the stall point is approached by increasing the back pressure until a further 50 Pa increment will lead to an abrupt mass flow drop and the divergence of the computation. An automatic proportion-integration-differentiation (PID) conditioner [34] is embedded in the NSAWET code to adjust the back pressure until the mass flow converges to the target value at both the design point and the near-stall point.

The grid convergence study is conducted with three sets of meshes, corresponding to the coarse mesh, medium mesh, and fine mesh. The total grid numbers of the three sets of meshes are 73.6 million, 1.45 million, and 2.96 million, respectively. Figure 4 shows the details of the medium grid. The grid of circumferential grooves is shown in red, and the grid in the passage is shown in black. This set of mesh has 65 cells distributed along the whole blade span, and another 25 cells inside the tip gap in the radial direction. The wall unit y^+ of the first grid layer that is normal to the wall boundary, is strictly maintained below 1, in accordance with the requirements of the SST turbulence model. Each circumferential groove is discretized by a straight H block that has 40, 36, and 50 cells in the stream-, span- and pith-wise directions, respectively.

Figure 5 gives a comparison between the CFD results and the experimental data. The uncertainty in the measured total pressure ratio ($P_{t0,\ rat}$), the measured total temperature ratio ($T_{t0,\ rat}$), and the measured mass flow are 0.5%, 0.5%, and 1.0%, respectively. Figure 5a,b show the speed curve predicted by the three sets of meshes. All of the three meshes are capable of achieving a good agreement with the experimental data at the design point. The coarse mesh predicts a higher stall mass flow than the experimental data, due to the very sparse grid distribution in the passage. The speed curve obtained by the medium mesh is very close to the curve obtained by the fine mesh over the entire operating mass flow. The differences in the $P_{t0,\ rat}$ at the design point and the near-stall point between these two meshes are less than 0.15%, and this value reduces to 0.04% for the $T_{t0,\ rat}$ differences. The results of the medium and fine mesh show that the deviation of $P_{t0,\ rat}$ between CFD and experimental data occurs near the surge boundary. The present computation also slightly underestimates $T_{t0,\ rat}$. These numerical errors may be caused by the drawbacks of the turbulence model and the insufficient ability of the steady RANS for precisely capturing the large separation under near-stall conditions. Despite the numerical errors, the whole tendency of the speed curve predicted by the medium mesh and fine mesh coincides with the experimental data from choke to stall. In Figure 5c,d, the spanwise distributions of $P_{t0,\ rat}$ and $T_{t0,\ rat}$ predicted by the medium mesh at the downstream plane also show an acceptable agreement with the experimental data. To sum up, CFD results obtained by the medium mesh are close to the experimental data at the design point and the near-stall point. Because the optimization objectives are set at the design point, and the goal of the optimization is to gain a relative improvement based on the baseline blade, it is suitable to use the medium mesh to undertake CFD investigations to achieve a favorable balance between the accuracy and the computational cost.

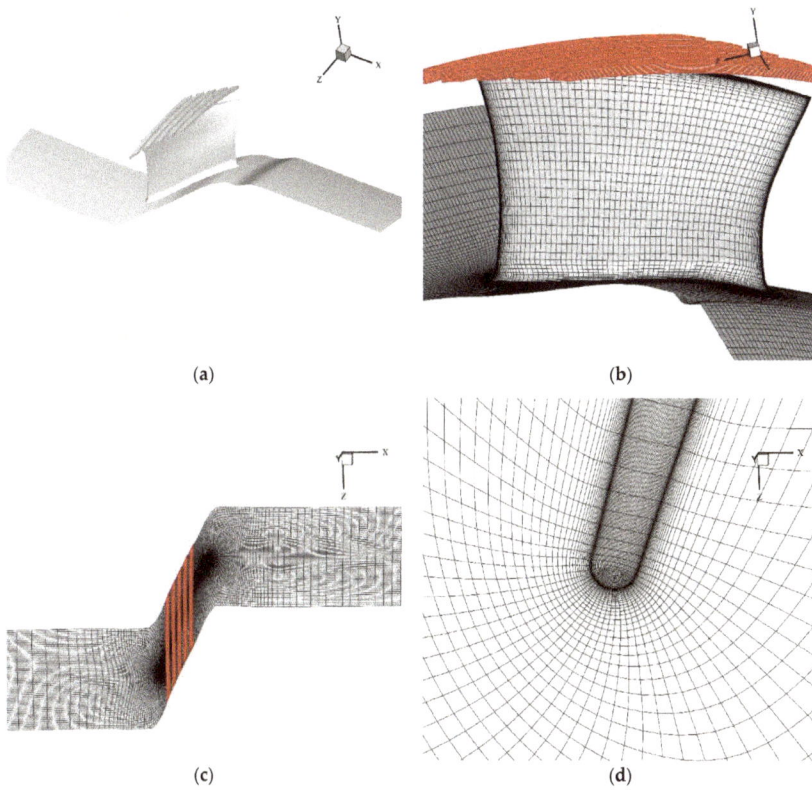

(a)

(b)

(c)

(d)

Figure 4. Structured mesh in the computational domain. (**a**) The geometry of the baseline rotor blade with five circumferential grooves installed on the casing; (**b**) the side view of the grid distribution on the blade surface; (**c**) the top view of the grid distribution on the casing; (**d**) the grid distribution near the leading edge of the blade tip.

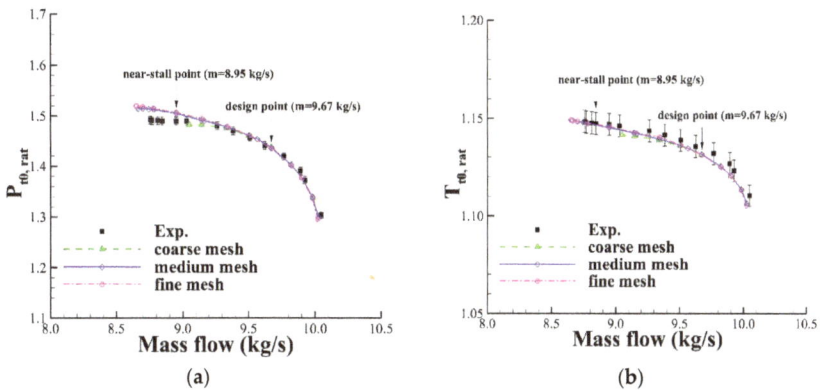

(a)

(b)

Figure 5. *Cont.*

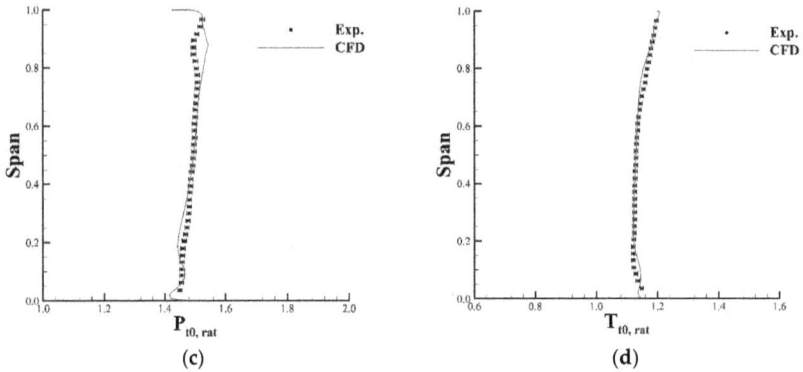

Figure 5. Validation of the computational results. (**a**) The total pressure ratio versus the mass flow rate; (**b**) the total temperature ratio versus the mass flow rate; (**c**) the exit total pressure ratio versus the spanwise location at the mass flow of 9.344 kg/s; (**d**) the exit total temperature ratio versus the spanwise location at the mass flow of 9.344 kg/s.

4. CGCT Application Tests

The effects of the CGCT and its axial coverage are first analyzed on the baseline blade that is installed with different CGCT configurations. Then, the sweep and lean are introduced to the blade tip. Influences of 3D blading on the performance of the CGCT are discussed. The mass flow rates of the design point and near-stall point are taken as 9.67 kg/s and 8.95 kg/s, respectively. The abbreviations adopted hereafter for each type of rotor are explained in Table 2.

Table 2. Definitions of the individual abbreviations.

Name	Definition
SW_*	solid casing rotor, including SW_0, SW_FS, SW_BS, SW_PL and SW_NL
CT_*	CGCT configuration, including CT_a, CT_b, CT_c, CT_d and CT_e
0	baseline ND-TAC rotor
FS	re-stacked rotor with forward swept tip
BS	re-stacked rotor with backward swept tip
PL	re-stacked rotor with positive leaned tip
NL	re-stacked rotor with negative leaned tip

4.1. Effects of the CGCT on the Performance of the Baseline Blade

By increasing the parameter k from 0 to 40% of $C_{ax,tip}$, five CGCT configurations with different axial coverage are generated. The parameter s is fixed as 5% of $C_{ax,tip}$ in all five of the CGCT configurations. Table 3 lists the axial coverages of these five CGCT configurations which are installed on the baseline blade. From CT_a to CT_e, the position of the CGCT is moved from the leading edge area to the trailing edge area.

Table 3. Five circumferential groove casing treatment configurations for testing.

Name	Axial Coverage (% $C_{ax,tip}$)
CT_a	0–60%
CT_b	10–70%
CT_c	20–80%
CT_d	30–90%
CT_e	40–100%

The speed curves of the ND-TAC rotor with and without installing the CGCT are shown in Figure 6. At both the design point and the near-stall point, all of the CGCT configurations improve both the total pressure ratio and the adiabatic efficiency (η). The stall mass flows of all of the grooved casing rotor are lower than that of the solid casing rotor. Specifically, the zoom-in of the stall point in Figure 6a shows that the stall mass flows of CT_a, CT_b and CT_c are very close. When the CGCT is moved to the trailing edge area, the ability of the CGCT to delay a stall dramatically decreases, which is reflected by the increased stall mass flows of CT_d and CT_e.

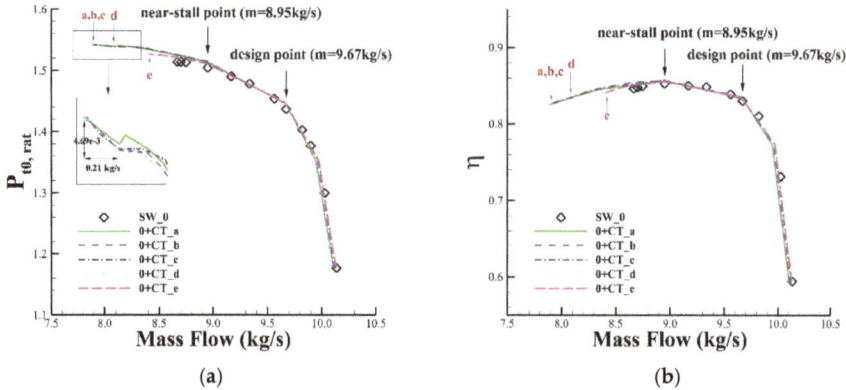

Figure 6. Speed curves of the baseline blade installed with five CGCT configurations. (**a**) The total pressure ratio versus the mass flow rate; (**b**) the adiabatic efficiency versus the mass flow rate.

At the design point, the spanwise distribution of the total pressure ratio, as well as the adiabatic efficiency at 25% of the tip chord length downstream of the trailing edge are shown in Figure 7. In Figure 7a, the CGCT decreases the total pressure ratio above 95% of the span but enhances the working capacity of the rest of the span, which demonstrates that the CGCT redistributes the spanwise loading. In Figure 7b, the increase of the efficiency concentrated in the blade tip indicates that the CGCT reduces the aerodynamic loss in the tip region. The baseline rotor installed with CT_e is found to have the highest total pressure ratio below 80% of the span and the highest efficiency above 90% of the span.

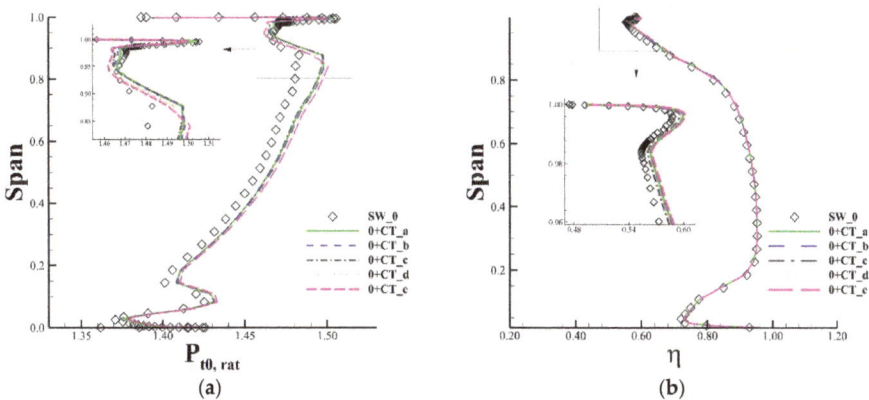

Figure 7. Spanwise distribution of (**a**) the total pressure ratio, and (**b**) the adiabatic efficiency at the design point.

The design point performance and the stall mass flow of the ND-TAC rotor with different CGCTs installed are summarized in Table 4. The performance improvement and the decrease of the stall mass flow due to the installation of different CGCTs are shown in Figure 8. Although the grooved casing rotor 0 + CT_e has the most significant improvement in the design performance, the stall mass flow of 0 + CT_e is obviously higher than the corresponding stall mass flows of the other CGCT cases. The greatest difference between the five grooved casing rotors is more than 0.1% for the design point performance, and is more than 6% for the stall mass flow. In the present CGCT tests, the optimal performance at the design point and near-stall point are not obtained by the same CGCT configuration. The effectiveness of the CGCT varies with its axial coverage.

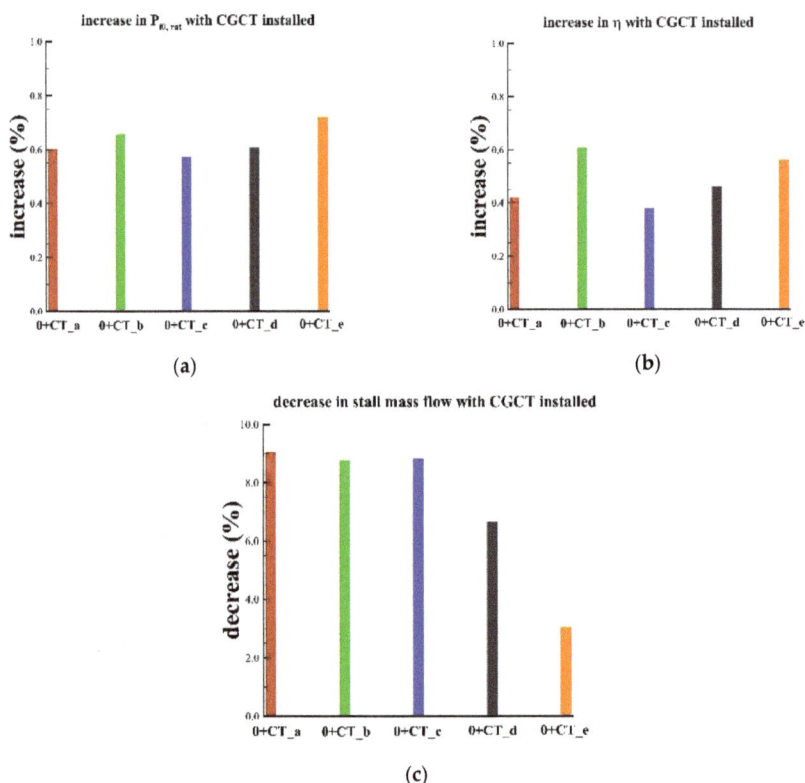

(a) (b)

(c)

Figure 8. Variation of the (a) total pressure ratio, (b) the adiabatic efficiency, and (c) the stall mass flow when different CGCTs are installed on the baseline blade at the design point.

Table 4. Effects of the CGCT on the rotor's design point performance ($P_{t0, rat}$, η) and the stall mass flow rate (M_s).

Name	$P_{t0, rat}$	Relative to SW_0	η	Relative to SW_0	M_s (kg/s)	Relative to SW_0
SW_0	1.43655	-	0.83008	-	8.666	-
0 + CT_a	1.44519	0.602%	0.83355	0.418%	7.882	−9.044%
0 + CT_b	1.44596	0.655%	0.83511	0.606%	7.907	−8.759%
0 + CT_c	1.44475	0.571%	0.83323	0.379%	7.901	−8.820%
0 + CT_d	1.44525	0.606%	0.83391	0.461%	8.089	−6.663%
0 + CT_e	1.44685	0.717%	0.83474	0.561%	8.403	−3.033%

As indicated by Figure 8, the differences between CT_c and CT_e in the design point performance and the stall mass flow are the most obvious. Thus, the flow fields of 0 + CT_c and 0 + CT_e are extracted for comparison and analysis.

In Figure 9, the blockage at 93% of the span formed by the TLV after passing the shockwave is characterized by the relative Mach number contours. Common effects of CT_c and CT_e on the flow fields in the tip region are alleviations of blockages at both the design point and the near-stall point. When the mass flow decreases with the variation of the operating point, the behaviors of the low Mach number region in 0 + CT_c are basically unchanged (see Figure 9c,d). By contrast, from the design point to the near-stall point, the low Mach number area in 0 + CT_e expands rapidly, and the average speed in this blockage region is also reduced (see Figure 9e,f). The rapid growth of the blockage in 0 + CT_e implies that CT_e that locates between 40% of $C_{ax,tip}$ and 100% of $C_{ax,tip}$ is less effective at delaying the stall than CT_c that locates between 20% of $C_{ax,tip}$ and 80% of $C_{ax,tip}$. However, at the design point, the blockage in 0 + CT_e is smaller than that in 0 + CT_c (see Figure 9c,e), which indicates that CT_e is more helpful at relieving the blockage under a high-mass-flow condition than CT_c is. Therefore, the Mach number contours plotted in Figure 9 demonstrates that 0 + CT_e has a stronger working ability at the design point, but a smaller stall margin than 0 + CT_c has. In addition, the blockage that originates downstream of the shockwave and then extends to the aft-part of the pressure side (PS) is supposed to affect the blade's aft-loading. If the blade's chordwise loading is redistributed due to the 3D blading, the effectiveness of CGCT can be changed.

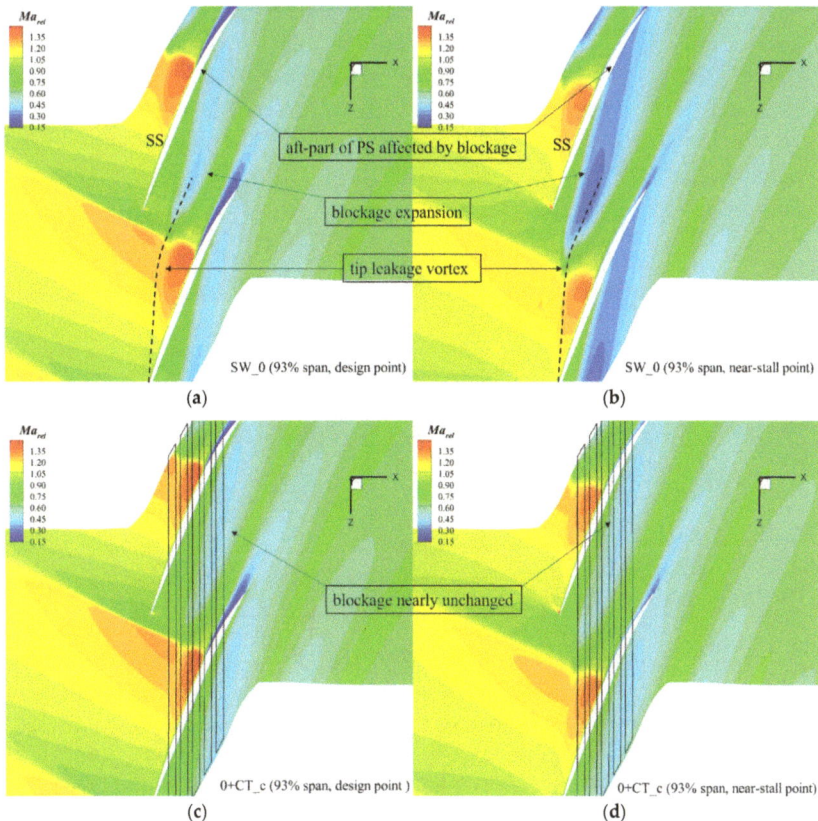

Figure 9. *Cont.*

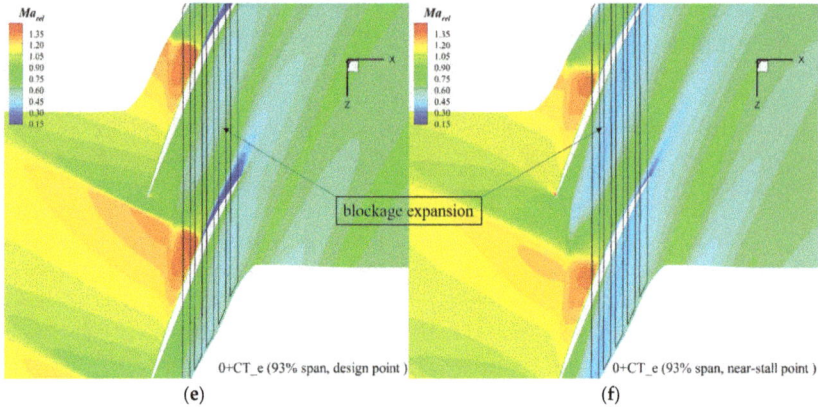

(e)   (f)

Figure 9. The relative Mach number contours on the S1 plane for (**a**) SW_0 at the design point; (**b**) SW_0 at the near-stall point; (**c**) 0 + CT_c at the design point; (**d**) 0 + CT_c at the near-stall point, (**e**) 0 + CT_e at the design point, and (**f**) 0 + CT_e at the near-stall point.

Figure 10 quantitatively compares the mass flow distribution along the span at the near-stall condition. It shows that both CT_c and CT_e increase the mass flow near the blade tip due to the reduction of the blockage. The mass flow of 0 + CT_c near the blade tip region is higher than that of 0 + CT_e, which demonstrates the conclusion drawn from Figure 9 that CT_c is more effective to alleviate the blockage at a low-mass-flow condition. It explains why 0 + CT_c presents a larger stall margin than 0 and 0 + CT_e. Moreover, Figure 10 also shows that the mass flow near the blade root is reduced by CGCT, which indicates that CGCT not only alters the tip flow field, but also changes the flow behaviors at the non-tip region.

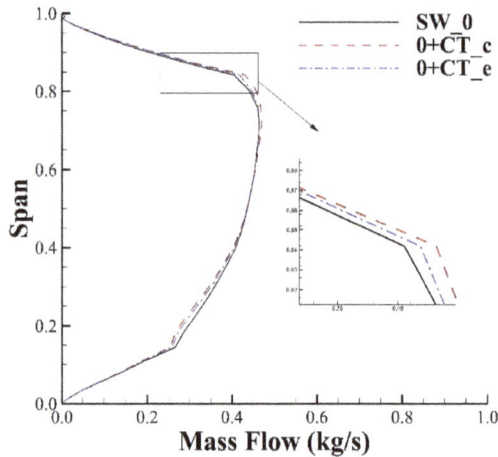

Figure 10. The spanwise mass flow distribution at the near-stall point.

Along the trajectory of the TLV, seven planes perpendicular to the Z-axis are extracted, and the entropy contours on them at the design point are shown in Figure 11. The yellow square in the figure encloses the cross-section where the core of the TLV interacts with the shockwave. Apparently, the entropy in this area of 0 + CT_e is less than the corresponding entropy of 0 + CT_c, which indicates that the strength of the shock-TLV interaction in 0 + CT_e is weaker than that in 0 + CT_c. Consequently, the high-entropy region marked by the dashed black square in the low-speed area of 0 + CT_e is smaller. The variations of the entropy distributions demonstrate that at the design point, CT_e is more effective at increasing the efficiency than CT_c is.

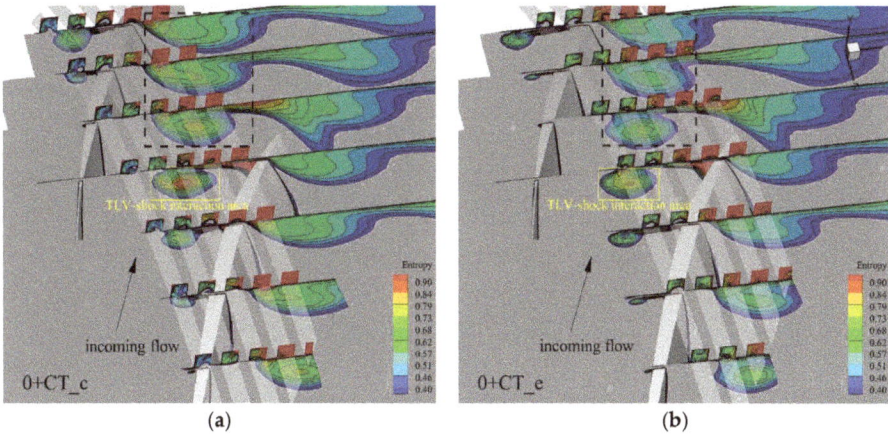

Figure 11. The entropy contours on the Z-planes at the design point in (**a**) 0 + CT_c and (**b**) 0 + CT_e.

The distributions of the relative total pressure ($P_{t0, rel}$), non-dimensionalized by the inlet total pressure in the tip clearance, are plotted in Figure 12. Figure 12a shows that the improvement of $P_{t0, rel}$ induced by CT_e after the 40% point of $C_{ax, tip}$ is higher than the corresponding improvement for CT_c at the design point; however, at the near-stall point, Figure 12b shows that CT_c is more effective at increasing $P_{t0, rel}$ after the 30% point of $C_{ax, tip}$ than CT_e is. The variation of $P_{t0, rel}$ in the tip clearance proves that the effectiveness of different CGCT configurations varies with the operating conditions.

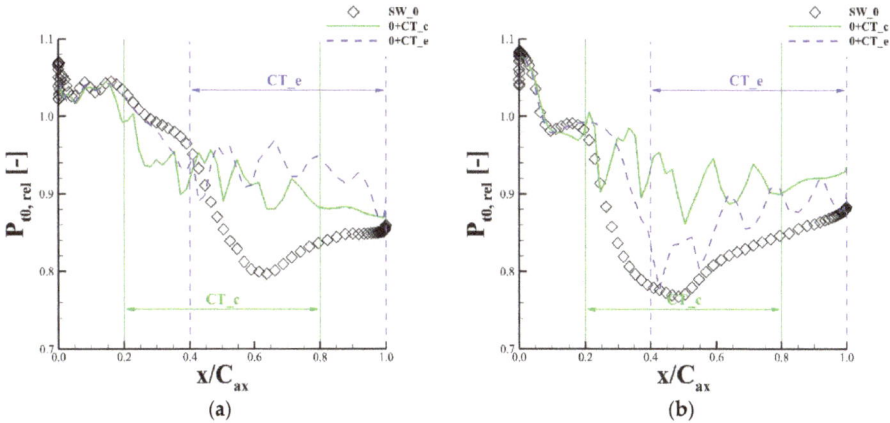

Figure 12. Distribution of the non-dimensional relative total pressure in the tip gap at (**a**) the design point and (**b**) the near-stall point.

4.2. Effects of the Sweep and Lean on the Effectiveness of the CGCT

As shown in Table 4, in terms of both the design point performance and the stall margin, the performance of CT_b is more satisfactory than the other CGCT configurations. In this section, CT_b is applied to four re-stacked rotor blades named FS, BS, PL, and NL to test the effects of the sweep and lean. The sweep and lean distributions adopted in this study are given in Figure 2. The 3D shapes of the four blades are plotted in Figure 13. Compared with the baseline ND-TAC blade, the control point of the sweep at 100% span is forward and backward shifted by 12% of the mean chord length to form the blades FS and BS, respectively. Moreover, the control point of the lean at 100% of the span is positively and negatively shifted by 3.5% of the mean leading-trailing edge deviation to produce the blades PL and NL, respectively. Although the position of the blade tip airfoil in the flow passage is changed in accordance with the sweep and lean, the geometry parameters of the 2D airfoil, including the thickness, camber and chord length, are unchanged. Additionally, the relative position of the circumferential grooves to the blade tip is fixed, regardless of how the blade is re-stacked.

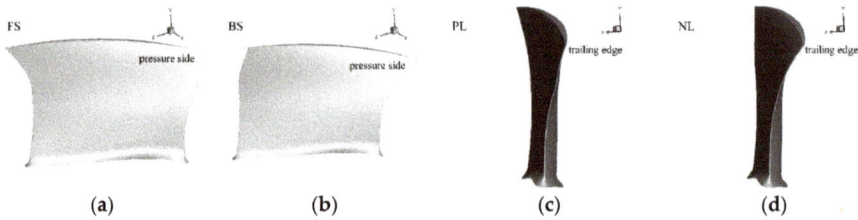

Figure 13. 3D shapes of the four re-stacked blades: (**a**) the tip of the forward swept blade FS; (**b**) the tip of the backward swept blade BS; (**c**) the tip of the positive leaned blade PL; (**d**) the tip of the negative leaned blade NL.

The CFD results show that when CT_b is installed, all four of the re-stacked blades are able to operate at the mass flow rate of 8.10 kg/s, which is approximately 0.60 kg/s lower than the stall mass flow rate of the baseline blade. Therefore, the stall margin of the re-stacked blades with CT_b installed is sufficient. Consequently, we focus on the performance of these blades instead of the stall margin.

Figure 14 shows the performance variations of re-stacked blades at both the design point and the near-stall point induced by CT_b. The forward and backward sweep result in a slight design performance deterioration, but the backward sweep leads to greater loss in $P_{t0, \, rat}$ and η than the forward sweep (see symbol Δ and ∇ in Figure 14a). After installing CT_b on the swept rotor's casing, the performance is improved significantly. The performance improvement induced by CT_b in blade FS is more than the corresponding improvement in the baseline blade 0 and blade BS (see Figure 14b). In addition, after introducing the sweep to the tip, the improvement of the near-stall performance induced by CT_b in FS and BS is more than twice the corresponding improvement in blade 0 (see Figure 14d). The positive-leaned blade PL has a slightly higher design performance than blade 0. By contrast, the negative lean causes a performance deterioration (see symbol $\diamond$ and $\bigcirc$ in Figure 14a). The performance improvement after installing CT_b in blade PL is less than the corresponding improvement in blade 0; however, the performance improvement of NL induced by CT_b is more than what we find in blade 0 (see Figure 14b,d).

From the above quantitative comparisons, it can be concluded that the sweep and lean introduced to the blade tip affect both the working capacity and the efficiency of the blade. The sweep and negative lean slightly decrease the solid casing rotor's performance. Moreover, the effectiveness of the CGCT is also altered by the sweep and lean. In these tests, the effectiveness of the CGCT at improving the solid casing rotor's performance is magnified by the forward sweep and negative lean in the tip.

Figure 15 shows the variations of the spanwise total pressure ratio distribution caused by the sweep and lean in the blade tip. In Figure 15a, whereas the forward sweep in the tip reduces the working ability of the blade tip, the backward sweep reduces the working ability of the rest of the blade span without changing the tip performance. When CT_b is installed on the casing, the total pressure ratio of the forward swept blade FS is greatly improved from the root to 90% of the span. The performance improvement is even larger than the corresponding improvement to the baseline blade with CT_b installed; however, for the backward swept blade BS, the performance improvement induced by CT_b is less than the corresponding improvement in FS at the same spanwise position.

In Figure 15b, the negative lean reduces the total pressure ratio along the entire span, whereas the positive lean slightly increases the total pressure ratio in the tip. The differences between the grooved leaned blades PL + CT_b and NL_CT_b are not as obvious as the differences between the grooved swept blades FS + CT_b and BS + CT_b.

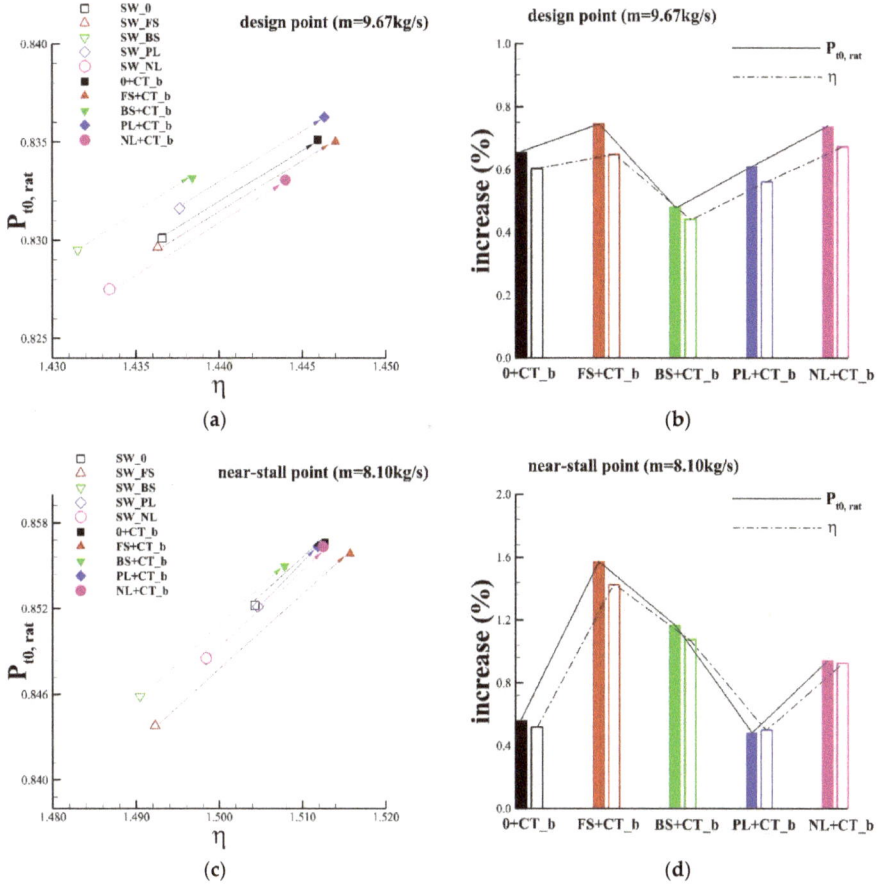

Figure 14. Performance variations of re-stacked blades induced by CT_b at the design point and the near-stall point: (**a**) $P_{t0,\,rat}$ and η at the design point; (**b**) the improvements in $P_{t0,\,rat}$ and η of the grooved casing rotors relative to the respective solid casing rotors at the design point; (**c**) $P_{t0,\,rat}$ and η at the near-stall point; (**d**) the improvements in $P_{t0,\,rat}$ and η of the grooved casing rotors relative to the respective solid casing rotors at the near-stall point.

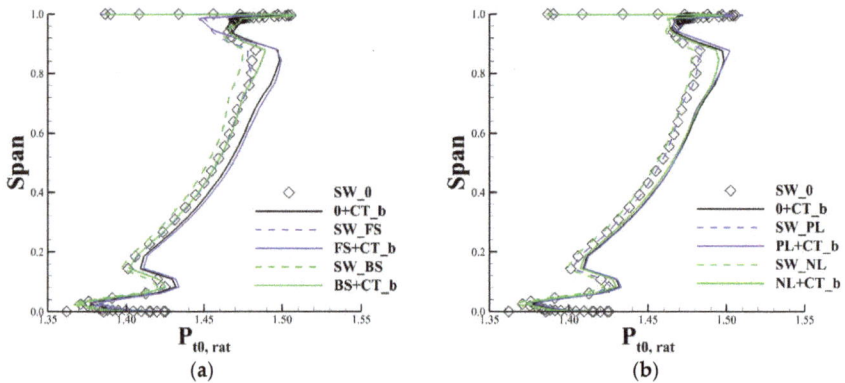

Figure 15. Spanwise distribution of the total pressure ratio at the design point for (**a**) swept blades and (**b**) leaned blades.

Figure 16 shows the radial relative velocities W_r at each groove-casing interface. W_r varies with the sweep in the tip (see Figure 16a–c). As shown in Figure 16d, compared with the baseline blade, the forward sweep reduces the aspiration of the front three grooves, and the backward sweep increases the aspiration of the front two grooves while obviously reducing the aspiration of the last one. Figure 16e) shows that the injection of fluids from the grooves to the passage is suppressed by the forward sweep in the front four grooves. By contrast, in the backward swept blade BS, the injection of the front four grooves is enhanced. To sum up, the forward sweep tends to decrease the injection and aspiration of grooves, while the backward sweep shows an opposite trend to the forward sweep in terms of affecting the fluid exchange at the groove–casing interfaces. The sweep is found to shift the spanwise loading as well as the chordwise loading [35]; therefore, the aspiration and injection of the grooves in the corresponding area vary with the changes of the blade loading.

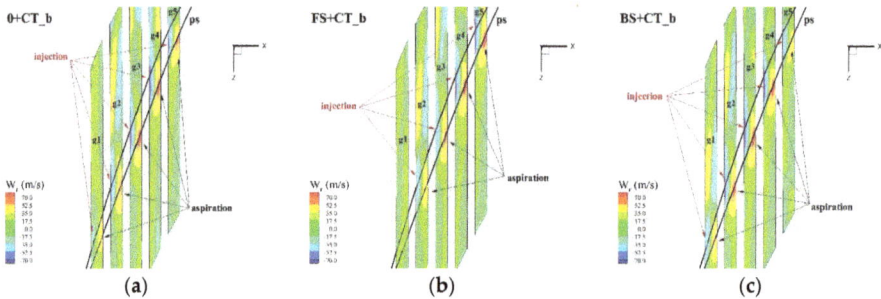

Figure 16. *Cont.*

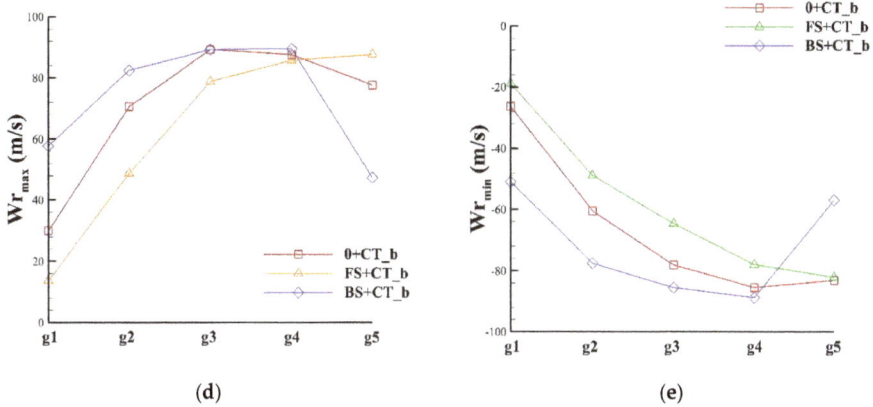

(d)　　　　　　　　　　　　　　**(e)**

Figure 16. Comparison of the radial relative velocity at the casing-groove interfaces (**a**) 0 + CT_b; (**b**) FS + CT_b; (**c**) BS + CT_b; (**d**) maximal radial velocity at each interface (aspiration); (**e**) minimal radial velocity at each interface (injection).

Although the relative position between the circumferential grooves and the tip airfoil is unchanged when the sweep and lean are introduced, the interactions between the blade and CGCT are altered, which impacts the performance of the blade and the effectiveness of the grooves. The effects of the 3D blading on the CGCT suggest that it is promising to achieve better design point performance while ensuring an adequate stable operating range by considering 3D blading in the process of the CGCT design.

5. CGCT-Blade Integrated Optimization

Because the CGCT-blade interactions are found to play a significant role in altering the rotor's performance and stability, we undertake an integrated optimization of the CGCT-blade combination. The goal is to verify the effectiveness of the CGCT design strategy that introduces 3D blading to the blade tip during the optimization of the grooves' axial distribution.

5.1. Optimization Setup

The sweep and lean of the blade tip and the axial distribution of the circumferential grooves are optimized together. The total number of design variables is four. The variation ranges of the sweep and lean at the blade tip are $[-12\%, +12\%]$ and $[-3.5\%, +3.5\%]$, respectively. The ranges of k and s controlling the grooves' location are $[-0.8\%, +20\%]$ and $[1.6\%, 12\%]$, respectively. A negative value of k means that the first groove is upstream of the leading edge. The variation range of s equals 0.2–1.5 times the width of the groove. Since the width of each groove is 8% of $C_{ax, tip}$, the start position of groove N can be expressed by Equation (3):

$$L_{start} = k + (8\% + s) \cdot (N - 1) \qquad (3)$$

where $N \in \{1, 2, 3, 4, 5\}$.

A hybrid aerodynamic optimization algorithm HSADE [36] is used in this paper to search a Pareto Front. HSADE combines the differential evolution (DE) algorithm with the radial basis function (RBF) response surface. Specifically, DE is a stochastic evolutionary optimization algorithm, well-known for its robustness, its strong global searching ability, and its suitability for high-dimensional problems [36]. RBF evaluates the similarity between the candidate design variables and known samples in the design space to predict their corresponding objective functions by interpolation. The results are used to construct the response surface for searching for potential optimal individuals. The Pareto Front is a group of optimal individuals that are not dominated by any other individuals obtained by the multi-objective optimization. To sum up, embedding RBF response surfaces into the basic DE improves the local searching ability of the basic DE, which helps to boost the convergence while obtaining an ideal Pareto Front.

The population size of each generation is nine. The total pressure ratio and the adiabatic efficiency at the design point are set as the two objectives for the optimization. Three operating conditions, namely, the choking point, design point, and near-stall point, are computed for each design in the optimization process. The operating conditions, design constraints, and objectives of the CGCT-blade integrated optimization problem are listed in Table 5. Note that the stall mass flow of the baseline ND-TAC rotor is 8.67 kg/s, and a baseline CGCT can decrease the stall mass flow by more than 0.7 kg/s; therefore, the near-stall condition of the rotor with the CGCT is set at 8.10 kg/s. Once the optimized blade with the CGCT is able to operate at 8.10 kg/s with a total pressure ratio of at least 1.511, the newly obtained CGCT-blade combination is considered to have a sufficient stall margin. The purpose of the CGCT-blade optimization is to obtain improvements to $P_{t0, rat}$ and η at the design point, as well as a favorable stall margin.

Table 5. Definition of the CGCT-blade integrated optimization problem.

Working Conditions	Constraints	Objectives
choking point	10.08 kg/s $\leq m \leq$ 10.34 kg/s	-
design point (m = 9.67 kg/s)	-	max $P_{t0, rat}$ and max η
near-stall point (m = 8.10 kg/s)	$P_{t0, rat} \geq$ 1.511	-

5.2. Results and Analysis

The optimization ran for 3400 CPU hours on a cluster with 300 cores in total, and it converged after 32 generations. The optimization histories of the two objectives are plotted in Figure 17. The performance differences between the individuals in the last few generations are small, and the objective values are concentrated on a narrow strip, which indicates that the optimization has converged.

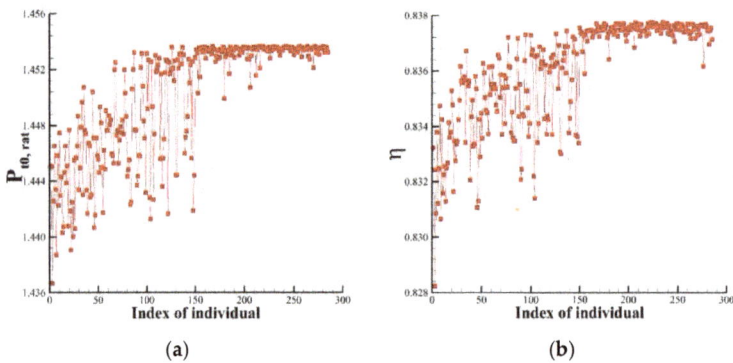

(a)   (b)

Figure 17. Convergence histories of the two design objectives: (a) the total pressure ratio, and (b) the adiabatic efficiency.

The values of the two design objectives of the individuals produced in the CGCT-blade optimization are plotted in Figure 18a. Performances of baseline blade 0, swept blade FS, BS, and leaned blade PL and NL, which have been discussed in Section 4.2, are also indicated in the figure with blue deltas. Compared with the tested blades, the CGCT-blade integrated optimization further improves the total pressure ratio $P_{t0, rat}$ and the adiabatic efficiency η. The result verifies the effectiveness of the CGCT-blade integrated optimization strategy. The Pareto Front of the optimization is shown in Figure 18b. The difference between these optimal individuals is less than 0.02% for $P_{t0,\,rat}$ and no greater than 0.05% for η.

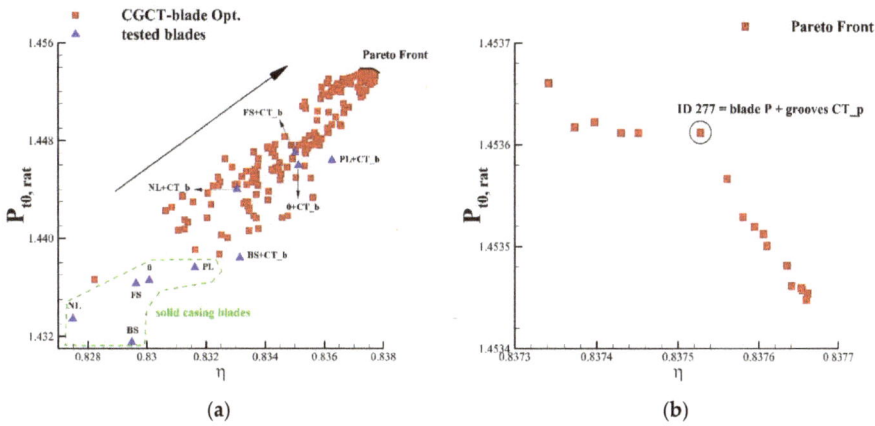

Figure 18. Design objectives of the CGCT-blade integrated optimization results for (**a**) all qualifying individuals, and (**b**) 18 optimal individuals on the Pareto Front.

Table 6 shows the average design performance of the grooved tested blades and the grooved optimized blades on the Pareto Front. The average performance of all of the grooved swept and leaned blades manually generated for the test is slightly lower than the performance of grooved baseline blade 0 + CT_b. Compared with 0 + CT_b, the CGCT-blade integrated optimization improves $P_{t0,\,rat}$ and η by 0.53% and 0.29%, respectively. When the optimization results are compared with the solid-casing baseline blade, the above two values increase to 1.19% and 0.90%, respectively.

Table 6. The average performance at the design point.

Type	$P_{t0,\,rat}$	η
SW_0	1.43655	0.83008
0 + CT_b	1.44596	0.83511
tested blades with CT_b	1.44393	0.83437
CGCT-blade integrated opt.	1.45354	0.83755

For the optimal individuals on the Pareto Front, the average values and variances of the four design variables are shown in Table 7. The differences between these optimal individuals are very small. The optimal blades are characterized by a forward swept and positive leaned tip. The distance between two adjacent grooves is approximately 7.7% of $C_{ax,tip}$, which is approximately 96% of the grooves' width. By substituting the average values of k and s into Equation (3), we find that the axial coverage of the CGCT obtained in the optimization is located between 11.7% and 82.5% of $C_{ax,tip}$.

Table 7. Averages and variances of the values of design variables of the individuals on the Pareto Front.

Statistics	3D Blading		CGCT Distribution	
	Sweep (%L)	Lean (%(r·˚))	k (%$C_{ax,tip}$)	s (%$C_{ax,tip}$)
average	11.9669	2.9675	11.6883	7.7067
variance	0.0419	0.3449	0.0028	0.0673

The CGCT-blade combination ID 277 marked by the black circle in Figure 18b is taken as the representative of the optimization for further analysis. The values of the sweep, lean, k, and s of ID 277 are 12.0000%, 2.8466%, 11.7790%, and 7.6247%, respectively. For the sake of simplicity, the blade of ID 277 is named "P", and the CGCT of ID 277 is named "CT_p" in the following discussion.

To eliminate the influence of the grid density and numerical errors on the prediction of the performance gains, the fine mesh used to conduct the grid convergence study is generated in the passage of blade P for validation. The design point performance predicted by the medium and fine mesh are compared in Table 8. Although the medium mesh slightly overestimates $P_{t0, rat}$ and η when compared with the fine mesh, the improvements in $P_{t0, rat}$ and η predicted by the medium mesh are close to the improvements given by the fine mesh. Therefore, it is proved that the refinement of the grid does not change the trend of the optimization, and the performance improvement predicted by the medium mesh in the optimizer is reliable. The analysis based on results of the medium mesh is performed to further investigate the inner mechanisms.

Table 8. Effects of the grid density on the predicted design point performance.

Performance and Variations	Medium Mesh (Used in the Optimization)	Fine Mesh (Used for Validation)
$P_{t0, rat}$ of baseline blade 0	1.43655	1.43449
$P_{t0, rat}$ of grooved blade 0 + CT_b	1.44596	1.44101
$P_{t0, rat}$ of optimized grooved blade P + CT_p	1.45361	1.45009
Improvement in $P_{t0, rat}$ (P+CT_p versus 0)	1.1876%	1.0875%
Improvement in $P_{t0, rat}$ obtained by optimization (P+CT_p versus 0+CT_b)	0.5291%	0.6301%
η of baseline blade 0	0.83008	0.82853
η of grooved blade 0 + CT_b	0.83511	0.83178
η of optimized grooved blade P +CT_p	0.83753	0.83626
Improvement in η (P+CT_p versus 0)	0.8975%	0.9330%
Improvement in η (P+CT_p versus 0+CT_b)	0.2898%	0.5386%

The speed curves of ID 277 with and without the CGCT are shown in Figure 19. The baseline blade installed with CT_p is also simulated for comparison. As shown by the blue solid squares and the black solid diamonds, the design point performance of the solid casing blade P is almost identical with the performance of the baseline blade, which indicates that the benefits of CGCT-blade integrated optimization are not obtained by improving the solid casing blade's performance. After installing CGCT, the stall mass flow is reduced below 8 kg/s in both rotors, and the near-stall performance of both are very similar, which shows that the ability of optimized grooves CT_p to extend the stall margin does not vary with the re-shape of the baseline blade. The zoom-in views at the design point show that the performance improvement induced by optimized grooves CT_p in the optimized blade P is larger than the corresponding improvement in the baseline blade. This result demonstrates that the CGCT-blade integrated optimization which alters the 3D shape of the blade and the axial distribution of the CGCT simultaneously, improves the CGCT-blade interactions to pursue an optimal CGCT-blade combination.

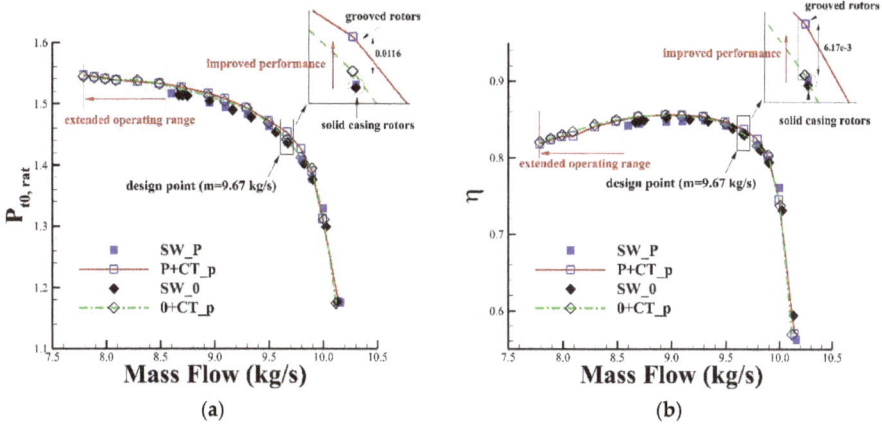

Figure 19. Speed curves of optimal blade ID 277 for (**a**) the total pressure ratio versus the mass flow rate; (**b**) the adiabatic efficiency versus the mass flow rate.

The working capacities and efficiencies of the baseline blade and optimized blade with and without installing CT_p are compared along the spanwise direction in Figure 20. In Figure 20a, the total pressure ratio of optimized blade P is reduced only in the tip region by the forward sweep and positive lean. When CT_p is installed, the total pressure ratio continues to decrease above 90% of the span, however, it increases below 90% of the span. Moreover, the total pressure ratio improvement induced by CT_p below 90% of the span in optimized blade P is obviously larger than the improvement in the baseline blade. Figure 20b shows that the efficiency improvement induced by CT_p is concentrated in the region above 85% of the span. The improvement of the efficiency in optimized blade P is also larger than the corresponding improvement in the baseline blade. The comparison of the diffusion factors in the spanwise direction is shown in Figure 20c. In the solid casing rotor, the forward sweep and positive lean in optimized blade P slightly reduce the tip loading; however, when CT_p is installed, optimized blade P has a more significant tip loading reduction than what the baseline blade has. Meanwhile, the enhancement of the loading in the non-tip region induced by CT_p is larger in optimized blade P than that in the baseline blade 0, which indicates that the forward swept and positive leaned blade tip improves the effectiveness of CT_p. These three plots demonstrate that the good performance of the optimal CGCT-blade combination is attributable to the redistributed spanwise loading caused by the CGCT-blade interactions in the tip region. Because the optimization does not change the design point performance of the solid casing rotor, the spanwise loading redistribution works to improve the CGCT-blade interactions, which is helpful to achieve a better overall performance of the CGCT-blade combination.

Figure 20a shows that the difference in the total pressure ratio between 0 + CT_p and P + CT_p is remarkable at 85% of the span; thus, the static pressure (P_s) distributions at the corresponding position are plotted in Figure 21 for comparison. The forward sweep and positive lean in the tip of blade P push the shockwave downstream, leading to a larger aft-loading in chordwise direction. After installing CT_p, P_s on the pressure side is increased. Compared with CT_p in the baseline blade, the zoom-in view in this figure shows that CT_p in the optimized blade is more effective at increasing P_s on the pressure side between 50% and 95% of the chord length. Note that P_s in this section of the optimized blade with a solid casing equals P_s of the baseline blade with a solid casing, which means the increase of P_s induced by CGCT is further improved after the CGCT-blade optimization.

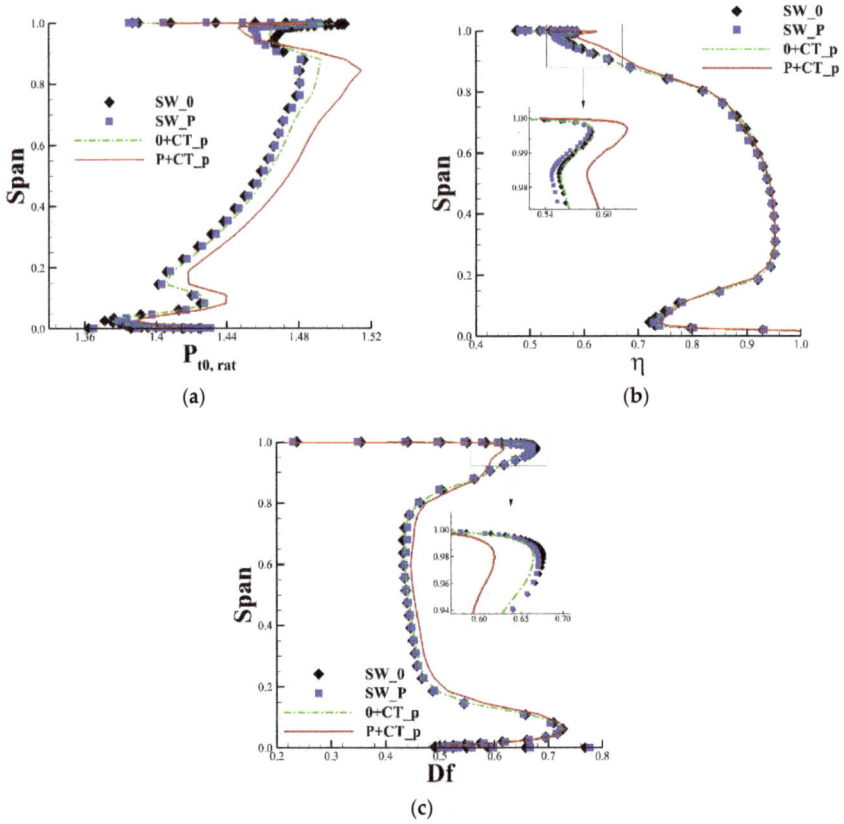

Figure 20. Spanwise distribution of (**a**) the total pressure ratio, (**b**) the adiabatic efficiency, and (**c**) the diffusion factor at the design point for optimal blade P and baseline blade 0 with and without CT_p.

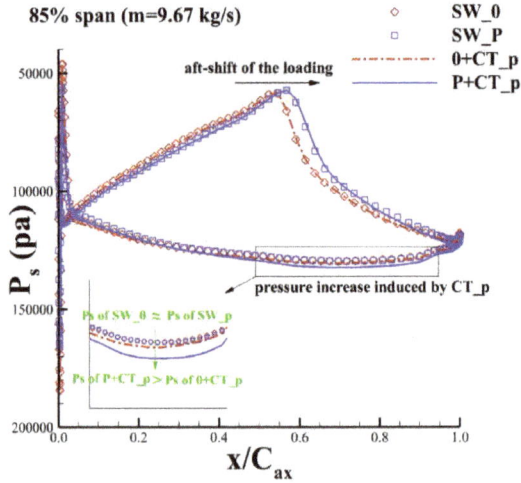

Figure 21. Static pressure distribution at 85% of the span in baseline blade 0 and optimized blade "P" with and without CT_p.

At 93% of the span, flow structures of 0 + CT_p and P + CT_p are shown in Figure 22. The main differences between the two flow fields are illustrated in this figure. The detached bow shock in the flow passage of blade P that has a forward swept and positive leaned tip is closer to the downstream. The area of the shock-induced separation on the suction side (SS) of blade P is also smaller. Moreover, because CGCT in blade P is more effective at relieving the blockage than CGCT in the baseline blade is, the flow tube sandwiched between PS and the blockage expands, which decelerates the near-wall flow inside the tube and increases the static pressure in the aft-part of the blade tip (see Figure 21). To sum up, when the airfoil is more aft-loaded in chordwise direction due to the forward sweep and positive lean, the effectiveness of CGCT to enhance the working capacity of the aft-part of the blade is improved.

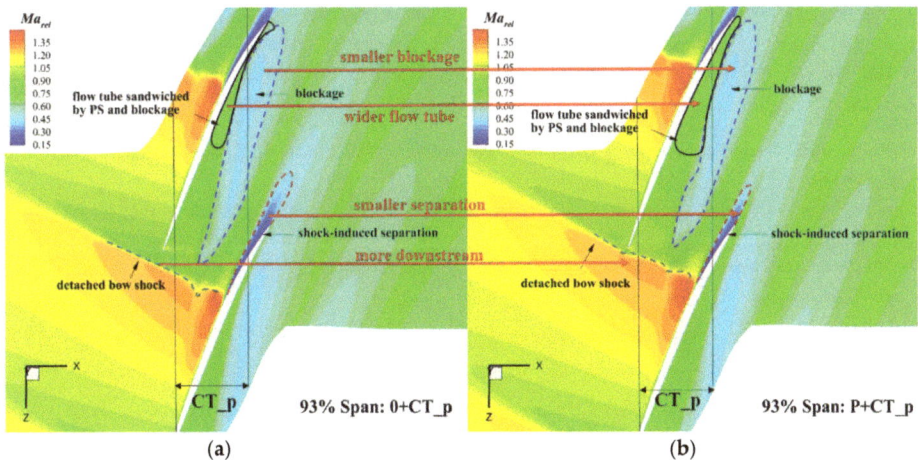

Figure 22. Comparison of flow structures between the baseline blade with CT_p and the optimized blade with CT_p at the design point: (**a**) baseline blade with optimized CGCT: 0 + CT_p; (**b**) optimized CGCT-blade combination: P + CT_p.

6. Conclusions

This paper studies multiple circumferential casing grooves' axial distribution design and optimization. The conventional CGCT design with a given blade is developed into a CGCT-blade integrated optimization. In the process of the optimization, the performance of CGCT-blade combinations is evaluated at the choking point, design point and near-stall point, which fulfills the constraints while improving the objectives. The work can be summarized as follows:

(1) A CGCT-blade combination with good design point performance does not necessarily results in good near-stall performance. For the ND-TAC rotor, the most effective CGCT that leads to the greatest working capacity improvement in the blade at the design point is located between 40% and 100% of the tip chord length; however, this CGCT configuration results in the smallest stall margin enhancement, because the grooves near the trailing edge are ineffective at suppressing the growth of the blockage in the tip region. Therefore, it is suggested that the CGCT design should coordinate the effects of CGCT under both the design and near-stall conditions.

(2) The sweep and lean introduced to the tip alter the CGCT-blade interactions, leading to the performance variations. The working ability of the blade at different span, as well as the fluid exchange between the passage and the grooves, are affected by the 3D blading, thus, it is promising to obtain a better CGCT-blade combination if the sweep and lean are introduced in the process of the CGCT design.

(3) The CGCT-blade integrated optimization improves the coupling between the CGCT and the blade tip, which strengthens the ability of the CGCT in improving the rotor's design point

performance. Compared with the baseline blade and all of the tested re-stacked blades with CGCTs, the CGCT-blade integrated optimization achieves the highest total pressure ratio and adiabatic efficiency while maintaining a sufficient stall margin. In the optimal CGCT-blade combination, a forward swept and positive leaned tip are generated, together with a CGCT located between 11.7% and 82.5% of $C_{ax,tip}$. In the spanwise direction, the blade loading is reduced in the tip region. In the chordwise direction, the detached bow shock is pushed downstream and the blade is more aft-loaded. These variations in loading are good for the CGCT to reduce the blockage and enhance the working capacity of the non-tip span and aft-part of the blade.

The newly developed CGCT-blade integration optimization method is helpful for advancing the understanding of CGCT-blade interactions. The combination of the CGCT and the 3D blading technique breaks the boundary between the CGCT design and the blade design. However, this design strategy is only verified in the ND-TAC rotor with the assistance of CFD. Efforts are needed to validate the effectiveness of this method in other types of rotors by both CFD calculations and experiments.

In future work, the CGCT-blade integrated design and optimization method will be refined by introducing more design variables to increase the design freedom of both the blade and CGCT.

Author Contributions: Formal Analysis, Software, Writing–Review & Editing, W.S.; Methodology, Data Curation, Project Administration, Y.Z.; Conceptualization, Resources, Supervision, H.C.

Funding: This work was supported by the National Natural Science Foundation of China (Grant No. 11872230).

Acknowledgments: The authors would like to thank the Notre Dame Turbomachinery Laboratory for providing the geometry and the experimental data of the ND-TAC.

Conflicts of Interest: The authors declare no conflict of interest.

Nomenclature

N_c	Rotational speed (rpm)
U_{tip}	Blade tip velocity (m/s)
N_b	Blade number
R_{casing}	Radius of casing (mm)
R_{hub}	Radius of hub (mm)
τ	Tip gap size (mm)
m	Mass flow (kg/s)
M_s	Stall mass flow (kg/s)
$P_{t0,\ rat}$	Total pressure ratio
$T_{t0,\ rat}$	Total temperature ratio
η	Adiabatic efficiency
Ma_{rel}	Relative Mach number
$P_{t0,\ rel}$	Non-dimensional relative total pressure
Df	Diffusion factor
W_r	Relative radial velocity (m/s)
C_{ax}	Axial chord length (mm)
P_s	Static pressure (Pa)
PS	Pressure side
SS	Suction side

References

1. Broichhausen, K.D.; Ziegler, K.U. Supersonic and Transonic Compressors: Past, Status and Technology Trends. In Proceedings of the ASME Turbo Expo 2005: Power for Land, Sea, and Air, Reno, NV, USA, 6–9 June 2005; pp. 63–74. [CrossRef]
2. Fujita, H.; Takata, H. A Study on Configurations of Casing Treatment for Axial Flow Compressors. *Bull. JSME* **1984**, *27*, 1675–1681. [CrossRef]

3. Hathaway, M.D. *Passive Endwall Treatments for Enhancing Stability*; NASA Rep. No. TM-2007-214409; NASA Glenn Research Center: Cleveland, OH, USA, 2007.

4. Rabe, D.C.; Hah, C. Application of Casing Circumferential Grooves for Improved Stall Margin in a Transonic Axial Compressor. In Proceedings of the ASME Turbo Expo 2002: Power for Land, Sea, and Air, Amsterdam, The Netherlands, 3–6 June 2002; pp. 1141–1153. [CrossRef]

5. Müller, M.W.; Schiffer, H.-P.; Hah, C. Effect of Circumferential Grooves on the Aerodynamic Performance of an Axial Single-Stage Transonic Compressor. In Proceedings of the ASME Turbo Expo 2007: Power for Land, Sea, and Air, Montreal, QC, Canada, 14–17 May 2007; pp. 115–124. [CrossRef]

6. Sakuma, Y.; Watanabe, T.; Himeno, T.; Kato, D.; Murooka, T.; Shuto, Y. Numerical Analysis of Flow in a Transonic Compressor with a Single Circumferential Casing Groove: Influence of Groove Location and Depth on Flow Instability. *J. Turbomach.* **2013**, *136*, 031017. [CrossRef]

7. Houghton, T.; Day, I. Stability Enhancement by Casing Grooves: The Importance of Stall Inception Mechanism and Solidity. *J. Turbomach.* **2012**, *134*, 021003. [CrossRef]

8. Du, J.; Li, J.; Gao, L.; Lin, F.; Chen, J. The Impact of Casing Groove Location on Stall Margin and Tip Clearance Flow in a Low-Speed Axial Compressor. *J. Turbomach.* **2016**, *138*, 121007. [CrossRef]

9. Minsuk Choi Effects of circumferential casing grooves on the performance of a transonic axial compressor. *Int. J. Turbo Jet-Engines* **2010**, *32*, 361–371. [CrossRef]

10. Mao, X.; Liu, B.; Tang, T.; Zhao, H. The impact of casing groove location on the flow instability in a counter-rotating axial flow compressor. *Aerosp. Sci. Technol.* **2018**, *76*, 250–259. [CrossRef]

11. Kim, J.H.; Choi, K.J.; Kim, K.Y. Aerodynamic analysis and optimization of a transonic axial compressor with casing grooves to improve operating stability. *Aerosp. Sci. Technol.* **2013**, *29*, 81–91. [CrossRef]

12. Zhao, Q.; Zhou, X.; Xiang, X. Multi-objective optimization of groove casing treatment in a transonic compressor. *Proc. Inst. Mech. Eng. Part A J. Power Energy* **2014**, *228*, 626–637. [CrossRef]

13. Houghton, T.O. *Axial Compressor Stability Enhancement*; University of Cambridge: Cambridge, UK, 2010.

14. Denton, J.D.; Xu, L. The Effects of Lean and Sweep on Transonic Fan Performance. In Proceedings of the ASME Turbo Expo 2002: Power for Land, Sea, and Air, Amsterdam, The Netherlands, 3–6 June 2002; Volume 1, pp. 23–32. [CrossRef]

15. Denton, J.D.; Xu, L. The exploitation of three-dimensional flow in turbomachinery design. *Proc. Inst. Mech. Eng. Part C J. Mech. Eng. Sci.* **1998**, *213*, 125–137. [CrossRef]

16. Wadia, A.R.; Szucs, P.N.; Crall, D.W. Inner Workings of Aerodynamic Sweep. *J. Turbomach.* **1998**, *120*, 671. [CrossRef]

17. Hah, C.; Puterbaugh, S.L.; Wadia, A. Control of Shock Structure and Secondary Flow Field inside Transonic Compressor Rotors through Aerodynamic Sweep. In Proceedings of the ASME 1998 International Gas Turbine and Aeroengine Congress and Exhibition, Stockholm, Sweden, 2–5 June 1998; pp. 1–15. [CrossRef]

18. Gallimore, S.J.; Bolger, J.J.; Cumpsty, N.A.; Taylor, M.J.; Wright, P.I.; Place, J.M.M. The Use of Sweep and Dihedral in Multistage Axial Flow Compressor Blading—Part I: University Research andMethods Development. *J. Turbomach.* **2002**, *124*, 521–532. [CrossRef]

19. Hah, C. The Inner Workings of Axial Casing Grooves in a One and a Half Stage Axial Compressor with a Large Rotor Tip Gap: Changes in Stall Margin and Efficiency. *J. Turbomach.* **2018**, accepted.

20. Goinis, G.; Nicke, E. Optimizing surge margin and efficiency of a transonic compressor. In Proceedings of the ASME Turbo Expo 2016: Turbomachinery Technical Conference and Exposition, Seoul, Korea, 13–17 June 2016; pp. 1–12. [CrossRef]

21. Cameron, J.D.; Cameron, J.D.; Morris, S.C.; Morris, S.C.; Corke, T.C.; Corke, T.C. A Transonic Axial Compressor Facility for Fundamental Research and Flow Control Development. *Mech. Eng.* **2006**. [CrossRef]

22. Kelly, R.; Hickman, A.R.; Shi, K.; Morris, S.C.; Jemcov, A. Very Large Eddy Simulation of a Transonic Axial Compressor Stage. In Proceedings of the 52nd AIAA/SAE/ASEE Joint Propulsion Conference, Salt Lake City, UT, USA, 25–27 July 2016; pp. 1–13. [CrossRef]

23. Cameron, J.D.; Morris, S.C. Spatial Correlation Based Stall Inception Analysis. In Proceedings of the ASME Turbo Expo 2007: Power for Land, Sea, and Air, Montreal, QC, Canada, 14–17 May 2007; pp. 433–444. [CrossRef]

24. Chen, H.; Huang, X.; Shi, K.; Fu, S.; Ross, M.; Bennington, M.A.; Cameron, J.D.; Morris, S.C.; McNulty, S.; Wadia, A. A Computational Fluid Dynamics Study of Circumferential Groove Casing Treatment in a Transonic Axial Compressor. *J. Turbomach.* **2013**, *136*, 031003. [CrossRef]

25. Shi, K.; Chen, H.X.; Fu, S. Numerical investigation of the casing treatment mechanism with a single circumferential groove. *Sci. China Phys. Mech. Astron.* **2013**, *56*, 353–365. [CrossRef]

26. Toro, E.F.; Spruce, M.; Speares, W. Restoration of the contact surface in the HLL-Riemann solver. *Shock Waves* **1994**, *4*, 25–34. [CrossRef]

27. Van Leer, B. Towards the ultimate conservative difference scheme III. Upstream-centered finite-difference schemes for ideal compressible flow. *J. Comput. Phys.* **1977**, *23*, 263–275. [CrossRef]

28. Van Albada, G.D.; Van Leer, B.; Roberts, W.W. A comparative study of computational methods in cosmic gas dynamics. *Astron. Astrophys.* **1982**, *108*, 76–84. [CrossRef]

29. Menter, F.R. Improved two-equation k-omega turbulence models for aerodynamic flows. *NASA Tech. Memo.* **1992**, 1–31. [CrossRef]

30. Yoon, S.; Jameson, A. Lower-upper Symmetric-Gauss-Seidel method for the Euler and Navier-Stokes equations. *AIAA J.* **1988**, *26*, 1025–1026. [CrossRef]

31. Chen, H.X.; Fu, S.; Li, F.W. Navier-Stokes simulations for transport aircraft wing-body combinations with deployed high-lift systems. *J. Aircr.* **2003**, *40*, 883–890. [CrossRef]

32. Huang, X.; Chen, H.; Fu, S. CFD Investigation on the Circumferential Grooves Casing Treatment of Transonic Compressor. In Proceedings of the ASME Turbo Expo 2008: Power for Land, Sea and Air, Berlin, Germany, 9–13 June 2008; pp. 1–9.

33. Chen, H.; Huang, X.; Fu, S. CFD Investigation on Stall Mechanisms and Casing Treatment of a Transonic Compressor. In Proceedings of the 42nd AIAA/ASME/SAE/ASEE Joint Propulsion Conference & Exhibit, Sacramento, CA, USA, 9–12 July 2006. [CrossRef]

34. Siller, U.; Voß, C.; Nicke, E. Automated Multidisciplinary Optimization of a Transonic Axial Compressor. In Proceedings of the 47th AIAA Aerospace Sciences Meeting Including The New Horizons Forum and Aerospace Exposition, Orlando, FL, USA, 5–8 January 2009; pp. 1–12. [CrossRef]

35. Ji, L.; Chen, J.; Lin, F. Review and Understanding on Sweep in Axial Compressor Design. In Proceedings of the ASME Turbo Expo 2005: Power for Land, Sea, and Air, Reno, NV, USA, 6–9 June 2005; pp. 1–9. [CrossRef]

36. Deng, K.; Chen, H. A Hybrid Aerodynamic Optimization Algorithm Based on Differential Evolution and RBF Response Surface. In Proceedings of the 17th AIAA/ISSMO Multidisciplinary Analysis and Optimization Conference, Washington, DC, USA, 13–17 June 2016. [CrossRef]

![energies logo]

MDPI

Article

Experimental Research on Hydraulic Collecting Spherical Particles in Deep Sea Mining

Guocheng Zhao [1,2], Longfei Xiao [1,2,*], Tao Peng [1,2] and Mingyuan Zhang [1]

[1] State Key Laboratory of Ocean Engineering, Shanghai Jiao Tong University, Shanghai 200240, China;
    guocheng.zhao@sjtu.edu.cn (G.Z.); Pengtao@sjtu.edu.cn (T.P.); mzbzmy@sjtu.edu.cn (M.Z.)
[2] Collaborative Innovation Center for Advanced Ship and Deep-Sea Exploration, Shanghai 200240, China
*   Correspondence: xiaolf@sjtu.edu.cn; Tel.: +86-21-34207051-2122

Received: 28 June 2018; Accepted: 16 July 2018; Published: 25 July 2018

Abstract: Hydraulic collecting is the key technology in deep sea mining and dredging engineering. It determines economic benefits of the project and environmental issues. However, mechanistic studies of hydraulic collecting are rarely described. In this study, the mechanism of collecting spherical particles is researched by dimensional analysis and experimental study. The experimental system is established to carry out three kinds of tests including 253 different test cases. The empirical model of collecting performance prediction is established by the tests of vertical force characteristics and vertical incipient motion characteristics of particles in suction flow field. The results show that the vertical suction force coefficient (C_{vs}) decreases exponentially with the ratio of bottom clearance to diameter of the particle (h/d), increases linearly with the ratio of diameter of the suction pipe to diameter of the particle (D/d), and is nearly independent of Reynolds number (Re). The empirical formula of vertical force and criterion-formula of vertical incipient motion of particles are obtained with the maximum tolerance less than 15%. The phenomenon that the vortex could help strengthen the suction force was observed in the tests. In addition, the characteristics of suction flow field were obtained by flow visualization tests, and applied to explain the force characteristics of particles in the suction flow field.

Keywords: deep sea mining; manganese nodules exploitation; hydraulic collecting; suction flow field; dimensional analysis

---

1. Introduction

Enormous mineral resources are deposited in the international seabed, of which polymetallic nodules have attracted more and more attention around the world for their high economic value and easy prospecting. With increasing demand for various mineral resources and gradual exhaustion of land resources in modern society and industry, research and development of seabed mineral resources is becoming increasingly urgent [1]. Therefore, deep sea mining is thought to be a key approach to the sustainable development of human beings and many countries around the world are stepping up exploration and exploitation of seabed mineral resources [2,3]. Most of the deep sea mining technologies and systems for exploration, recovery and processing were originally been done in the 1970s. Four subsystems are usually included in deep sea mining systems: surface systems, pipe systems, buffer systems and collector systems [4]. Chung [5,6] presented a brief summary of the developments in mining systems and technologies in the past years, and pointed out that the mobility, safety, collection efficiency and sweep efficiency of the collector system are the most important parameters in deep sea mining system design. The collecting part in the miner system is also considered as the key technology, because commercial production must achieve high sweep efficiency [7], and profitable deep sea mining exploitation is only feasible on the premise that there is a nodule collector with maximum collecting capacity of 140 kg of wet nodules per second [8]. Except for the high collecting efficiency, eco-friendly mining processes are also in demand because deep sea mining could be a new environmental challenge related to ocean biology [9,10]. To pick up nodules, a variety of collecting methods such as hydraulic methods, mechanical methods and hybrid collection methods have been developed. Ocean Management Incorporated (OMI's) sea trial results in 1978 showed that the hydraulic method has higher collection efficiency than the mechanical method [11]. Therefore, the problem of hydraulic collecting will be studied in this paper.

The mechanism of hydraulic collecting is fairly complicated because the collecting efficiency is influenced by a lot of factors such as particle size, bottom clearance, collector structure and flow rate. Several studies have been conducted to examine the process of hydraulic collecting. Hong et al. [12] dealt with experimental approaches for enhanced understanding of the hydraulic performance of a hybrid pick-up device. The experiments were conducted in a 2-D flume tank. By parametric experiments they found position and shape of baffle plates are significant factors for effective design of hydraulic nodule lifter. Yang et al. [13] discussed major parameters and their influences on the performance of the pick-up device by tests. The results showed that the hydraulic pick-up device with proper dimensions and parameters could get high pick-up rate and low content sediments. Tsutsui [14] had an experimental study of the flow field and the aerodynamic force on a sphere above a plane. The surface pressures on the sphere and the plane were measured by inclined multi-pipe manometers and the results were compared with photographs showing the flow visualization of the sphere. Lim et al. [15] analyzed flow field characteristics with outflow discharge from a collecting device in deep ocean by using software FLUENT. The study revealed seawater velocity and streamline distributions along with complicated flow characteristics downstream including nodule particles behavior. Zhao et al. [16] carried out studies of hydraulic collection of single spherical particle at various collecting bottom clearances and flow rates by both experiments and numerical simulations. It was verified that the vertical force prediction is feasible for single ore particle in collecting condition based on numerical simulation. The study revealed that the variation of the wake vortex is the dominant factor of force vibration.

The problems of hydraulic lifting have been studied, which could be helpful to understand the process of hydraulic collecting. Chung et al. [17] conducted experimental investigations to study the shape effect of solids on pressure drop in a 2-phase vertically upward transport. Sobota et al. [18] had an experimental study on nodule and water velocities at upward flow in a pipeline of $D = 150$ mm. The slip velocity values of 10-, 30- and 50-mm-dimeter nodules at the volume concentration 10% were measured. The results illustrated that the slip velocity values are of the same order as the fall velocity of these solids. Jiang et al. [19] had an investigation into the concentration characteristics of large size particles in vertical pipes for hydraulic lifting. Particles of different diameters (5–30 mm) were used for experimental studies. A formula for the radial distribution of particle concentration was developed and validated. Jiang et al. [20] did the experiments on single particle settling velocity in the static fluid, floating velocity for a single particle and critical velocity in a vertical hydraulic lifting system. They had found that the larger the size and concentration of particles were, the more possibly the particles moved to the pipe wall. Yang et al. [21] analyzed the influence of factors such as particle concentration, diameter, gradation and boundary conditions on single and group suspension velocity by doing numerous experiments. A relation between particle group and single particle suspension velocity ratio and volumetric concentration was given. Behavior of single particle in vertical lifting pipe is the foundation of the group particles in vertical lifting pipe. In addition, the studies of deep sea air-lift have been researched. Fan et al. [22] investigated the performance of an air-lift pump for artificial upwelling of deep water theoretically and experimentally with a vertical pipe of 0.4 m diameter and 28.3 m height. An empirical model of the air-lift artificial upwelling is presented. The performance of the model has been confirmed by the experimental findings. Pougatch et al. [23] presented a numerical model of the three-phase flow in the upward air-lift pipe. And the influence of the pipe diameter on the lifting efficiency has been investigated numerically. Ma et al. [24] assessed the technological feasibility and profitability analyses in terms of solid production rate, energy consumption per tonnage of mineral, and profitability per tonnage of mineral. The effects of submergence ratio, pipe diameter, particle diameter, mining depth, and gas flux rate are investigated.

The research of flow around a sphere is the foundation of study on hydraulic collecting. Achenbach [25] had experiments on the flow past spheres in the Reynolds number range $5 \times 10^4 \leq Re \leq 6 \times 10^4$. He compared his results with other available data and pointed out the dependence of friction forces on Reynolds number. Johnson and Patel [26] investigated the flow of an incompressible viscous fluid past a sphere at different Reynolds numbers by Detached Eddy Simulation (DES). Constantinescu and Squires [27] applied Large Eddy Simulation (LES) and DES to investigate the flow around a sphere at a Reynolds number of 10,000 in the subcritical regime.

Collecting efficiency and environmental issues from the collecting process are the biggest challenges for deep sea mining. Therefore, providing sufficient suction force and avoiding unnecessary disturbance by predicting the performance of collecting accurately is very important. In view of the previous studies including experiments and numerical simulations, however, there are few studies treated the mechanism study of hydraulic collecting and the research of collecting performance prediction.

In this study, mechanism of hydraulic collecting is researched by dimensional analysis and experiments. The characteristics of vertical force, vertical incipient motion and wake flow of spherical particles in suction flow field have been investigated. The empirical formula of vertical force of particles is obtained from measurements and further verified by the vertical incipient motion tests. Furthermore, the criterion-formula of vertical incipient motion of particles is derived to predict the collecting performance. The characteristics of flow field were observed by flow visualization tests. The results of this study will serve to further understand the hydrodynamic characteristics of the hydraulic collecting in deep sea mining and provide reference for the design of the deep sea miner collector.

2. Mechanism of Hydraulic Collecting

Firstly, the simplified models of manganese nodules and hydraulic collecting are founded. Then the method of dimensional analysis is used to study the mechanism of hydraulic collecting and instruct the design of the test program. At last, the empirical formula of vertical force and criterion-formula of vertical incipient motion of particles are derived to predict the hydraulic collecting performance.

2.1. Model of Manganese Nodules

Manganese nodules occur as potato-shaped concretions on the seafloor of abyssal plains in about 4000–6000 m water depth in all major oceans. They form two-dimensional deposits on the surface or within the first 10 cm of the deep-sea sediments (Figure 1).

Figure 1. High abundance of large nodules on the seafloor of the Peru Basin (ca. 4000 m water depth) [28].

Nodules in the eastern tropical Pacific and the central Indian Ocean are of special economic interest due to their high enrichment of metals such as Ni, Cu, Co, Mo, Li, rare earth elements (REEs), and Ga [28]. The dry bulk density of nodules ranges between 1.00 and 2.40 g/cm^3 [29,30]. They can have different sizes, shapes, and surface textures and can reach a size of up to 15 cm in diameter (general range from 2 cm to 10 cm). Some extremely large specimens of 21 cm have been found in the Peru Basin [31]. Moritani [32] did a study about the description and classification of manganese nodules. It is found that nodule occurrence is generally confined to the sediment-water interface of the sea bottom, in which nodules are buried in the sediments in different degrees. The relative position of occurrence seems to result in difference in surface texture, namely rough texture on the buried parts and smooth ones on the exposed parts. And the most common shapes of nodules are either ideal individual forms of spheroidal, ellipsoidal, discoidal, etc. or polylobate or intergrown aggregate form.

In terms of the studies, the assumptions are made to simplify the physical problems in this study:

(1) The shape of particles is spheroidal as a common shape of nodules.
(2) The surface texture of particles is smooth because the roughness of the exposed parts of nodules is usually small.
(3) The particles are on a flat surface because some parts of the mining area are flat and they are usually where we start deep sea mining; in addition, the sediments near the particles are so soft that the particles are always rolling out from sediments before being lifted by the collecting pipe.
(4) No deposit sediments surround particles because sediments contents are very low in the process of hydraulic collecting in engineering test.

2.2. Model of a Kind of Hydraulic Collecting

An underwater mining system was developed for operations and the flexible riser concept was validated in the Indian seas at 410 m water depth in 2000 jointly by National Institute of Ocean Technology (NIOT) and Institut für Konstruktion (IKS) of the University of Siegen (Germany). The underwater mining systems are shown in Figure 2 [33]. The collecting principle of these underwater mining systems is to suck the particles through a suction pipe. The suction flow field was formed near the bottom of suction pipe when water pump was running. The particles are lifted and sucked into the pipe by vertical suction force induced by pressure difference near the particles.

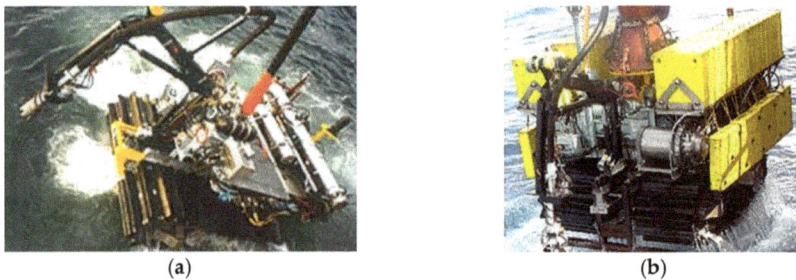

(a)   (b)

Figure 2. (a) Underwater mining system with manipulator and cutter for 500 m water depth; (b) Enhanced underwater mining system for 500 m water depth [33].

According to the collecting principle of these underwater mining systems, the physical model in this study is simplified as Figure 3.

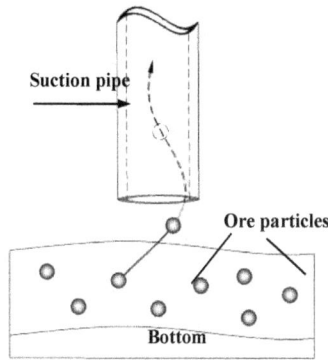

Figure 3. A Kind of hydraulic collecting model.

The main parameters are given in Figure 4, which represent various cases and determine the characteristics of flow field. The main parameters in relation with the vertical force of the particle in this study are listed in Table 1.

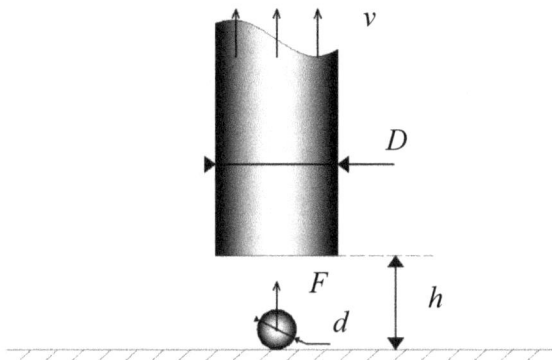

Figure 4. Main parameters of the model.

Table 1. Definition of main parameters.

Physical Parameter	Unit	Definition
d	m	Diameter of particle
h	m	Bottom clearance [1]
D	m	Diameter of suction pipe
F	N	Vertical suction force of particle
F_B	N	Buoyancy in water
G	N	Gravity of particle
ρ	kg/m³	Density of particle
ρ_w	kg/m³	Density of water
v	m/s	Flow velocity in pipe

[1] Bottom clearance h: Indicate the vertical height of the suction pipe; use the vertical distance from the pipe inlet to the bottom boundary.

The vertical suction force of the particle is similar to the drag force because both of them are induced by the pressure difference near the sphere or the velocity slippage between liquids and solids. The differences between them are: (1) different definitions, a fluid flowing past the particle exerts a force on it, the vertical force is the component of this force that is normal to ground (parallel to the gravity direction) while the drag force is the in-line component of the force parallel to the flow direction; (2) the drag force is usually defined in an oncoming flow with one direction while the vertical force is defined in the suction flow field which is clearly different with an oncoming flow. In this research, the flow velocity in pipe is used for 2 reasons: (1) the average velocity in pipe is stable along the whole pipe; (2) the flow velocity in pipe is one of the key parameters for engineering application because we could change the vertical force on particles by changing pump flow (the average velocity in pipe) to ensure collection efficiency in different cases.

2.3. Dimensional Analysis

2.3.1. Vertical Force F of Particles in Suction Flow Field

To indicate the vertical force of particles in suction flow field, vertical suction force coefficient C_{vs} is introduced. And the characteristics of flow field can be described by Reynolds number Re. Because of the complexity of the suction flow field, it is difficult to define Re and C_{vs} by using normal definition. In this study, the definition of Re and C_{vs} is proposed to be based on the flow velocity v in the collecting pipe as characteristic velocity.

The vertical force F of the particle fixed on the bottom boundary depends on the fluid density of water ρ_w, the viscosity coefficient of water μ, the velocity of the flow in pipe v, the diameter of the particle d, the diameter of the suction pipe D and bottom clearance h. As a result, the function of F is defined as:

$$F = f(\rho_w, \mu, v, D, d, h) \tag{1}$$

The dimensions of these parameters are as follows:

F	ρ_w	μ	v	d	D	h
MLT^{-2}	ML^{-3}	$ML^{-1}T^{-1}$	LT^{-1}	L	L	L

Therefore, we can get three basic dimensions of this problem with L, M and T, and then the theorem of dimensional analysis is used to obtain:

$$\Pi_1 = \frac{F}{\rho_w v^2 d^2} \tag{2}$$

Because d^2 of the spherical particle section has area dimension, the cross-section area A of the sphere is used instead:

$$\Pi_{1,\text{modified}} = \frac{F}{\frac{1}{2}\rho_w v^2 A} = C_{vs} \tag{3}$$

If the viscosity coefficient of fluid μ is used as dependent variable, then it can be achieved:

$$\Pi_2 = \frac{\mu}{\rho_w v d} \tag{4}$$

It can be derived by formula transformation:

$$\Pi_{2,\text{modified}} = \frac{\rho_w v d}{\mu} = Re \tag{5}$$

It is important to emphasize that the velocity v used in the Re definition is the flow velocity in the suction pipe, rather than the velocity around the particle.

In the same way, we can get the dimensionless variables:

$$\Pi_3 = \frac{D}{d} \tag{6}$$

$$\Pi_4 = \frac{h}{d} \tag{7}$$

The final dimensional relationships can be obtained:

$$\Pi_{1,\text{modified}} = f(\Pi_{2,\text{modified}}, \Pi_3, \Pi_4) \tag{8}$$

The vertical suction force coefficient is following:

$$C_{vs} = \frac{F}{\frac{1}{2}\rho_w v^2 A} = f\left(Re, \frac{D}{d}, \frac{h}{d}\right) \tag{9}$$

It is found that the vertical suction force coefficient C_{vs} could be related to the Reynolds number Re, D/d, and h/d.

2.3.2. Critical Bottom Clearance h for the Vertical Incipient Motion of Particles

According to the force balance analysis of the particle, we compare the theoretical wet weight mg-F_{buoy} of spherical particle to the vertical force calculated by C_{vs}. Therefore, the relationship between C_{vs} and Froude number Fr should be studied.

Based on the force analysis of the particles, the vertical incipient motion condition is as follows:

$$\vec{F} + \vec{F}_{buoy} = m\vec{g} \tag{10}$$

Using the functions of vertical force and volume of spherical particle:

$$F = \frac{1}{2}\rho_w A v^2 \times C_{vs} = \frac{1}{8}\pi\rho_w C_{vs} v^2 d^2 \tag{11}$$

$$mg - F_{buoy} = \frac{4}{3}g(\rho - \rho_w)\pi r^3 = \frac{1}{6}g(\rho - \rho_w)\pi d^3 \tag{12}$$

The following function can be obtained from Equation (10):

$$C_{vs} = \frac{4}{3}\frac{\rho - \rho_w}{\rho_w}\frac{gd}{v^2} \tag{13}$$

In terms of the definition of the Froude number Fr:

$$Fr = \frac{v}{\sqrt{gd}}\sqrt{\frac{v^2}{gd}} \tag{14}$$

The relationship between C_{vs} and Fr could be obtained:

$$C_{vs} = \frac{4}{3}\frac{\rho - \rho_w}{\rho_w}\frac{1}{Fr^2} \tag{15}$$

In consequence, if the vertical suction force coefficient determined by D/d and h/d is greater than the vertical suction force coefficient calculated by Fr, the particles will be lifted. Therefore, the criterion-formula is established to predict the vertical incipient motion of particles.

3. Experimental Description

In order to investigate the characteristics of ore particles in the suction flow field, the test system has been designed and constructed in the laboratory and three kinds of experiments are carried out: (1) vertical force tests: to measure the vertical force of particles in various cases; (2) vertical incipient motion tests: to measure the critical bottom clearances of particles and further validate the accuracy of the vertical force tests; (3) flow visualization tests: to observe the characteristics of flow field around the particle.

3.1. Test Set Up

Figure 5 is the sketch of vertical force tests. In Figure 5a, the spherical particle is fixed on the three-component force sensor, separated by an intermediate acrylic board. Tests aimed to measure the vertical force of the particle without relative velocity with the pipe. In Figure 5b, the spherical particle is free on the bottom to measure the vertical incipient motion clearance in suction flow field.

The particle is also still relative with the pipe before starting to lift in most tests. But in some exception, the particle will rotate around a vertical axis on the bottom and the phenomenon will be discussed later.

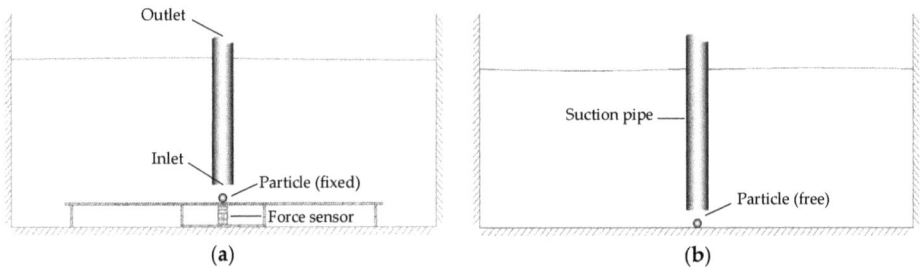

Figure 5. (a) Sketch of vertical force tests and (b) vertical incipient motion tests.

The whole test system consists of three parts: suction, movement and measurement as shown in Figure 6. The suction part is mainly composed of glass flume and pump circulation device. The size of glass flume is 2.5 m × 1.5 m × 1.0 m. The water in the flume is connected to the pump through a pipe, in which the rated power is 22 kW and the maximum flow rate is 100 t/h. The LDG-SIN-DN100 electromagnetic flow meter (made by MEACON, Hangzhou, Zhejiang, China) is installed at the outlet of the pump to measure the flow rate through the pump. Considering the fact that the flow rate at the inlet of the pipe is almost the same as the measured one, the average flow velocity is calculated and called flow velocity in pipe. In addition, a 0.7 m baffle is installed and two parallel honeycomb plates are placed to weaken the impact of water wave on force measurement. The movement part is composed of an ER50-C10 six-degree-of-freedom robot (made by Effort, Wuhu, Anhui, China), whose lower arm is fixed to the inlet of the pipe. By adjusting the motion of the mechanical arm to change the position of the pipe entrance, the precision can reach 0.01 mm and 0.01°. The measurement part adopts three-component force sensor. The top of the sensor is rigidly connected with the measured particle, and its bottom is rigidly connected to the stainless-steel plate placed on the bottom of the flume. In addition, an acrylic plate is installed under the particle, so that the acrylic plate can be considered as a boundary, and the force sensor has no effect on the flow field above the plate. In the experiment, the vertical distance between the inlet and the particle is measured as the vertical distance between the bottom of pipe and the acrylic plate, and it is defined as the bottom clearance h.

The force sensor is connected with the data acquisition system, and it can measure the data of the vertical suction force of the particles within a period of time. To ensure the accuracy of the measurement results, the layout structure of force sensor in the experiment has been improved many times, and the system tolerance is successfully maintained within 1%. In addition, in previous tests, it was found that if the measured particle is too heavy or too light, it will cause an oscillation of the measured value. Therefore, the particles used in the vertical force tests are designed to balance their buoyancy with the proper gravity, so that the measurement results can be more accurate.

In each of the test cases, the critical bottom clearance h was determined for the incipient motion of particles by accurately controlling the position of the ER50-C10 robot. The flow velocity is measured by the electromagnetic flow meter and shows the value of quantity of flow. Figure 7 shows the experimental value in the recorder of the flow in the pipe for a period of 200 s. The red line in the figure means that the target value of the flow quantity is 56.55 t/h in terms of the specified flow velocity. In comparison, the tolerance between the average of the experimental value and the target value is within 0.3%. The standard deviation is 0.1323 t/h, which is 0.2% of the target value of flow quantity. Thus the tolerance of flow velocity is small enough and can be acceptable in this study.

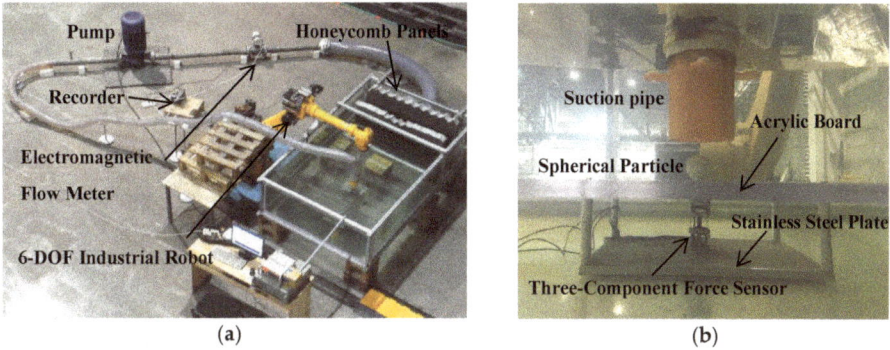

Figure 6. (a) Overlook of the test system; (b) Front view of the suction part.

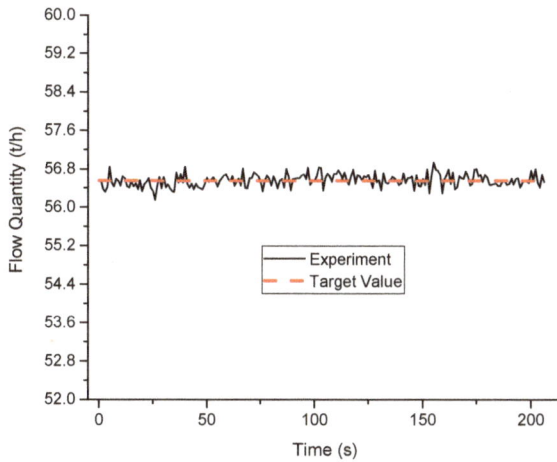

Figure 7. Target value (56.55 t/h) and experimental value of the flow quantity.

3.2. Test Program

To examine the effects on C_{vs} of Re, h/d, D/d and other parameters, a series of values of each factor has been chosen. The diameters of suction pipe are 0.075 m, 0.1 m and 0.125 m, and the diameter of spherical particles are 0.032 m, 0.036 m and 0.040 m. In total, 234 cases were carried out with 3~5 flow velocities in pipe ranging from 1 m/s to 2 m/s. In this range, the vertical forces of particles can be bigger and smaller than its weights in water. Moreover, each case was repeated three times with 3 min each time. The water temperature in this study is 10 °C and the dynamic viscosity coefficient μ equals 1.308×10^{-3} Pa·s. Therefore the range of dimensionless parameters can be determined. The range of Re is from 30,581 to 61,162, the range of D/d is from 1.875 to 3.906, and the range of h/d is from 1.389 to 2.375.

To obtain the incipient motion characteristics of spherical particles in suction flow field and further validate the accuracy of the vertical force tests, a total of 18 cases of vertical incipient motion tests were carried out with different parameters, as shown in Table 2. Each test was repeated ten times to measure the bottom clearance of vertical incipient motion of the particles when they were lifted into the suction pipe. The single particle is free on the bottom and right below the suction pipe.

Table 2. Parameters for vertical incipient motion tests of single particle.

Physical Parameter	Unit	Value
Particle density ρ	kg/m^3	2164.5, 2495.7
Particle diameter d	m	0.040, 0.036, 0.035
Pipe diameter D	m	0.100, 0.075
Ratio of diameter D/d		1.87, 2.50, 2.78, 2.89
Flow velocity in pipe v	m/s	2, 1.8, 1.6
Dynamic viscosity coefficient μ	Pa·s	1.308×10^{-3}

In addition, more vertical incipient motion tests of single and multiple particles were carried out to study the relations between the vertical incipient motion characteristics of single and multiple particles, such as the critical bottom clearance of each particle and the sequence of each particle to be lifted. Each test was also repeated ten times to measure the bottom clearance and the sequence of lifted particles. The parameters are listed in Table 3.

Table 3. Parameters for vertical incipient motion tests of multiple particles.

Physical Parameter	Unit	Value
Particle diameter d	m	0.019, 0.025, 0.030, 0.039
Particle weight G	N	0.092, 0.194, 0.352, 0.792
Pipe diameter D	m	0.1
Ratio of diameter D/d		5.21, 4.06, 3.33, 2.54
Flow velocity in Pipe v	m/s	2

Flow visualization tests were carried out to observe the characteristics of suction flow field. To compare with the results in references [16,34], the test cases are set as Table 4.

Table 4. Experimental parameters for flow visualization tests.

Physical Parameter	Unit	Case1	Case2
Bottom clearance h	m	0.07	0.07
Flow velocity in Pipe v	m/s	2.000	1.635
Reynolds number Re		61,162	50,000

4. Results and Discussion

Based on the measurements, the empirical formula of vertical force and criterion-formula of vertical incipient motion of particles are established separately. By comparing experiments of single and multiple particles, the correspondence of both incipient motion characteristics is obtained. The phenomenon that the vortex could help strengthen the suction force was found in both tests. In addition, the characteristics of flow field were observed by flow visualization tests, and applied to explain the force characteristics of particles in suction flow field.

4.1. Vertical Force Characteristics of Particles in Suction Flow Field

4.1.1. Influence of Re on C_{vs}

Figure 8 shows the test results of C_{vs} for different Re and h/d. It can be seen that the vertical suction force coefficient is nearly independent of Reynolds number when Re is in the range from 30,581 to 61,162. This range is subjected to the subcritical flow regime for flow around a sphere in a uniform flow, which also has the feature that the vertical suction force coefficient is nearly independent of Reynolds number [22]. Therefore, Equation (9) for C_{vs} can be simplified to:

$$C_{vs} = f(\frac{D}{d}, \frac{h}{d}) \qquad (16)$$

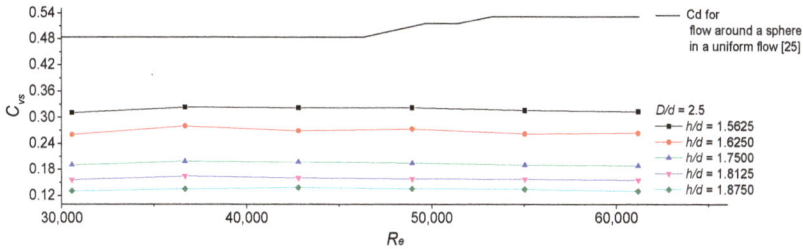

Figure 8. Relations between vertical suction force coefficients and Reynolds numbers.

In addition, the vertical suction force coefficients of spheres in suction flow are smaller than those in a uniform flow. One reason for this phenomenon is that the velocity close to the solid surface of particle is much smaller than that in the pipe. The velocity used to define C_{vs} is the velocity in pipe, however, the velocity used to define C_d is the oncoming flow velocity which is similar to that close to the solid surface of particles.

4.1.2. Influences of D/d and h/d on C_{vs}

The test results of C_{vs} for different D/d and h/d are shown in Figures 9 and 10. As shown in Figure 9, with the increase of D/d, the C_{vs} almost increases linearly. The coefficients of determination (R^2) of linear fitting functions for seven different fitting curves are over 0.9804.

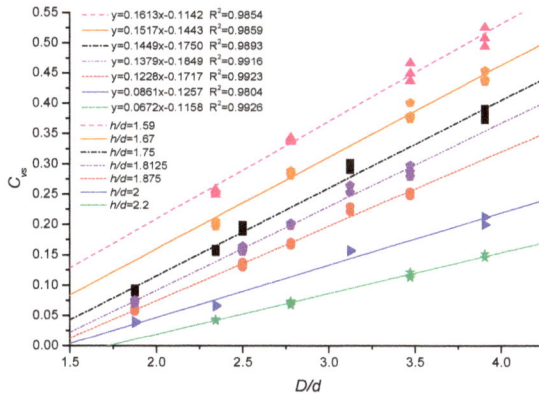

Figure 9. Relations between vertical suction force coefficients and ratios of D to d.

Unlike the influence of D/d, Figure 10 shows that the C_{vs} decreases exponentially with h/d in the hydraulic collecting process. The coefficients of determination (R^2) of exponential fitting functions are over 0.9996 and thus the fitting quality is extremely good. In consequence, h/d is a dominant factor to influence the C_{vs}. If the deep sea miner collector could work with a small h/d, large amount of collecting energy consumption will be saved and the disturbance to the sea floor will be minimized.

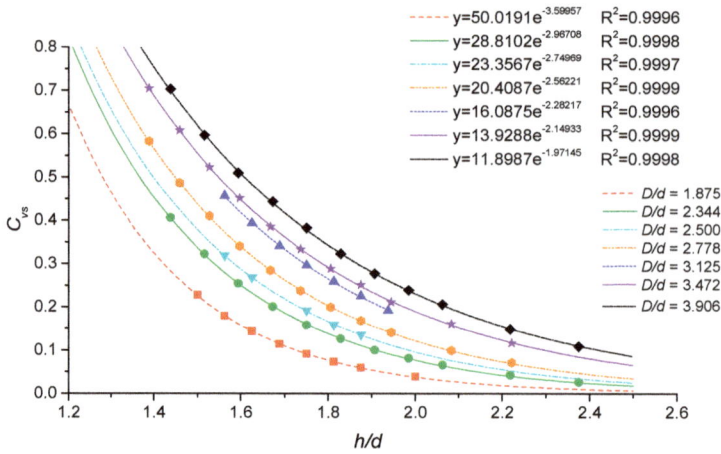

Figure 10. Relations between vertical suction force coefficients and ratios of h to d.

4.1.3. Empirical Formula among D/d, h/d and C_{vs}

The empirical formula among D/d, h/d and C_{vs} is required to predict the collecting performance under different cases. As a result, the functions between the natural logarithm of C_{vs} ($\ln C_{vs}$) and h/d for different D/d are drawn in Figure 11. An interesting phenomenon is observed that all the linear fitting lines tend to meet at one point.

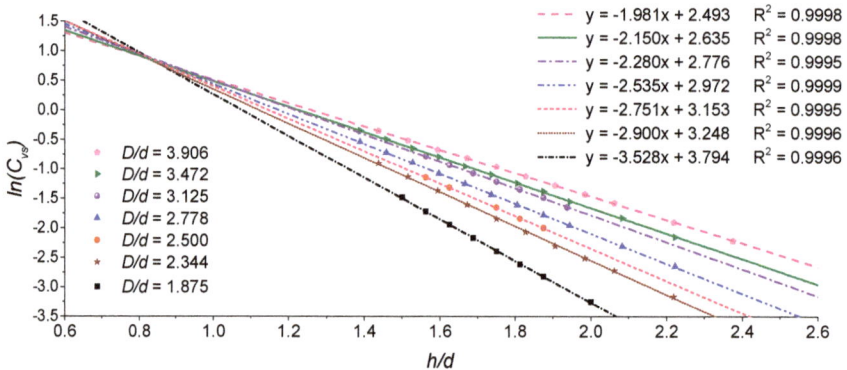

Figure 11. Relations between $\ln(C_{vs})$ and h/d.

By further analyzing the linear fitting parameters listed in Table 5 and making data fittings between the D/d and the two parameters, other two exponential fitting formulas for those parameters can be obtained respectively, as shown in Table 6.

Table 5. Linear fitting parameters of $\ln(C_{vs}) \sim h/d$.

D/d	Slope k	Intercept b	R^2
3.906	−1.981	2.493	0.9998
3.472	−2.15	2.635	0.9998
3.125	−2.28	2.776	0.9995
2.778	−2.535	2.972	0.9999
2.5	−2.751	3.153	0.9995
2.344	−2.9	3.248	0.9996
1.875	−3.528	3.794	0.9996

Table 6. Exponential fitting relations between parameters and D/d.

Fitting Formula	R^2
$k = -5.7005(D/d)^{-0.7876}$	0.9958
$b = 5.3305(D/d)^{-0.5669}$	0.9944

As a result, the empirical formula among D/d, h/d and $\ln C_{vs}$ is established:

$$\ln(C_{vs}) = -5.70\left(\frac{D}{d}\right)^{-0.788}\left(\frac{h}{d}\right) + 5.33\left(\frac{D}{d}\right)^{-0.567} \tag{17}$$

The graph of the empirical formula can be drawn as in Figure 12.

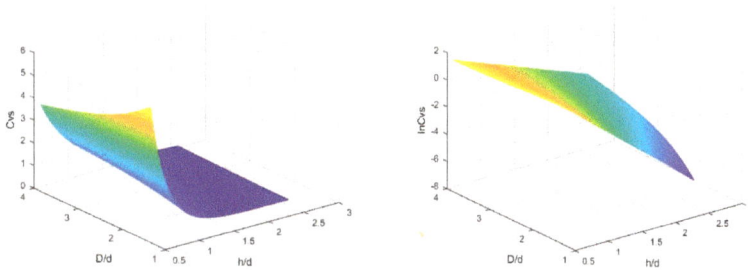

Figure 12. Relations among C_{vs}, h/d and D/d.

Figure 13 shows the tolerance between calculated value C_{vs}' by the empirical formula Equation (17) and measured value C_{vs} by experiments for 234 cases.

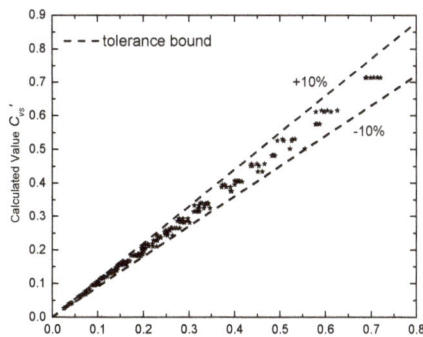

Figure 13. Tolerance between calculated C_{vs}' and measured C_{vs} for 234 cases.

The maximum tolerance is less than 10% and this indicates that the suction force prediction formula is sufficiently accurate to predict the vertical force of particles in different cases.

4.2. Vertical Incipient Motion Characteristics of Particles in Suction Flow Field

4.2.1. Criterion-Formula of Vertical Incipient Motion of Particles

In this study, the vertical suction force coefficient C_{vs} is used to represent the characteristics of vertical force of the particle in suction flow field. Because the equation $(F + F_{buoy} = mg)$ is the critical condition to trigger the vertical incipient motion of particles, therefore, if C_{vs} determined by D/d and h/d is bigger than that calculated by Equation (15), the particles will be lifted theoretically.

Figure 14 shows the tolerance of C_{vs} between the theoretical values by Equation (15) and the predicted values by Equation (17). The dash dotted line and solid line in the graph represent the theoretical values based on two kinds of densities: $\rho = 2164.5$ kg/m^3 and $\rho = 2495.7$ kg/m^3. Four dotted lines represent the upper and lower bounds with 15% tolerance of the two lines, respectively. The two kinds of data points are obtained based on the measurements of the cases. It can be seen that the measured data are distributed on both sides of the theoretical lines within the range of 15% tolerance. Therefore, the vertical force tests are further verified by the vertical incipient motion tests. Furthermore, the criterion-formula Equation (15) of vertical incipient motion of particles is obtained with the maximum tolerance less than 15% and can be used to predict the collecting performance in various cases. As a result, it is feasible to provide sufficient suction force and avoid unnecessary disturbance by predicting the performance of collecting accurately.

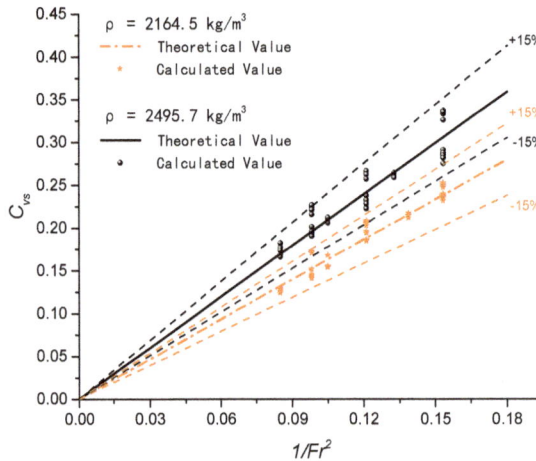

Figure 14. Relations between C_{vs} and $1/Fr^2$.

4.2.2. Vertical Incipient Motion Characteristics of Single Particles

Table 7 shows the test results of critical bottom clearance h in relation with D/d for the incipient motion of single particle. Each of the test cases for a single particle was repeated ten times and the recorded critical bottom clearance was represented by h_i ($i = 1, 2, \ldots , 10$). Then, the mean value $\bar{h}$ and standard deviation of the measurements were calculated. The test results show that the smaller particle has a higher standard deviation of h, which means the h of smaller particle exhibits more significant fluctuation. By observing the video taken by a high-speed camera, it is found that the particle d19.2 is always in circular motion with a large h. When particles remain almost still, however, the h is small. The reason for high standard deviation of particle d19.2 could be explained like this: the particle d19.2 tends to rotate around a vertical axis on the bottom before being sucked into the pipe because of the significant impact of vortex. The irregularity of vortex devotes to the high standard deviation of h and the vortex could help strengthen the suction force.

Table 7. Results of single particle tests.

D/d	5.21	4.05	3.32	2.54
h_i ($i = 1, 2, \ldots , 10$)	d19.2	d24.7	d30.1	d39.4
h_1 (mm)	59.81	60.76	65.00	69.32
h_2 (mm)	62.8	61.16	66.44	69.46
h_3 (mm)	69.18	63.00	67.45	70.45
h_4 (mm)	84.53	63.83	68.51	71.86
h_5 (mm)	61.18	60.94	66.23	69.39
h_6 (mm)	63.57	61.82	66.52	69.57
h_7 (mm)	65.28	61.86	66.86	69.71
h_8 (mm)	73.77	63.58	67.63	70.79
h_9 (mm)	65.77	62.23	66.92	69.81
h_{10} (mm)	67.81	62.6	66.98	69.89
$\bar{h}$ (mm)	67.37	62.18	66.85	70.03
Standard deviation (mm)	6.90	1.02	0.93	0.80

4.2.3. Vertical Incipient Motion Characteristics of Multiple Particles

Table 8 lists the results of the sequence of lifting particles and the critical bottom clearances of test cases for four particles, which are right below the fixed pipe. Similarly, each of the test cases for multiple particles was repeated 10 times. It is clear that for most cases the particles d39.39 and d19.2 would be lifted firstly, and this is consistent with the results in Table 7, where the $\bar{h}$ of particles d39.39 and d19.2 are larger than the other particles.

Table 8. Results of multiple particle tests.

D/d	5.21	4.05	3.32	2.54
No.i, h_i ($i = 1, 2 \ldots ,10$)	d19.2	d24.7	d30.1	d39.4
No.1	2nd	3rd	4th	1st
h_1 (mm)	70.32	63.90	60.33	70.32
No.2	1st	2nd	4th	3rd
h_2 (mm)	88.16	88.16	61.54	68.23
No.3	2nd	4th	3rd	1st
h_3 (mm)	65.79	62.12	62.12	67.84
No.4	2nd	3rd	3rd	1st
h_4 (mm)	64.24	64.24	64.24	68.17
No.5	4th	2nd	3rd	1st
h_5 (mm)	63.82	66.25	66.25	67.50
No.6	1st	3rd	4th	2nd
h_6 (mm)	83.38	64.29	61.06	68.47
No.7	1st	3rd	3rd	4th
h_7 (mm)	69.60	61.22	61.22	59.78
No.8	2nd	4th	3rd	1st
h_8 (mm)	68.92	63.82	63.82	68.92
No.9	2nd	4th	3rd	1st
h_9 (mm)	67.59	59.63	61.03	67.59
No.10	1st	3rd	4th	2nd
h_{10} (mm)	91.85	60.72	60.72	70.65

Especially, in Nos.1, 3, 4, 8 and 9, the particle d39.39 is lifted firstly and the particle d19.2 is lifted secondly. Moreover, the relevant experiment video in Figure 15 shows that particles remain almost still before lifted. In Nos.2, 6, 7 and 10, on the other hand, the particle d19.2 is lifted firstly and the video shows that particles are in circular motion before lifted, as shown in Figure 16. As a result, the smallest particle is influenced by the circular motion significantly. The instability of the fluid field generally causes the smallest particle to be sucked firstly, and this supports the supposition in Section 4.2.2 that the vortex could help strengthen the suction force. Therefore, the characteristics of collecting single particle would be meaningful for further research on collecting multiple particles.

Figure 15. A snapshot of multiple particles without circular motion.

Figure 16. A snapshot of multiple particles with circular motion.

4.3. Flow Visualization Results

Flow visualization experiments were carried out to observe the characteristics of suction flow field. The visualizations in Figure 17b are similar to the numerical results [16]. The streamlines gather in the middle and the flow field exhibit cohesive characteristics.

(a)

(b)

Figure 17. (a) Velocity vector distribution from numerical solution (h = 0.07 m, v = 2 m/s, Re = 61162) [16]; (b) Flow visualization results (h = 0.07 m, v = 2 m/s, Re = 61162).

Furthermore, by comparing the flow visualization test results in Figure 18, it is evident that the wake vortex separation point in suction flow is much more near the top of the particle than that in uniform flow. The closer the separation point of the wake vortex near the top of the particle, the smaller area on the particle is influenced by the wake vortex, leading to much smaller vertical suction force coefficient. Therefore, this phenomenon also contributes to the result that the vertical suction force coefficients of spheres in suction flow are smaller than those in a uniform flow in Figure 8. In addition, this physical phenomenon could explain why tests could have good repeatability and the prediction for both vertical force and incipient motion of particles in suction flow field could be accurate.

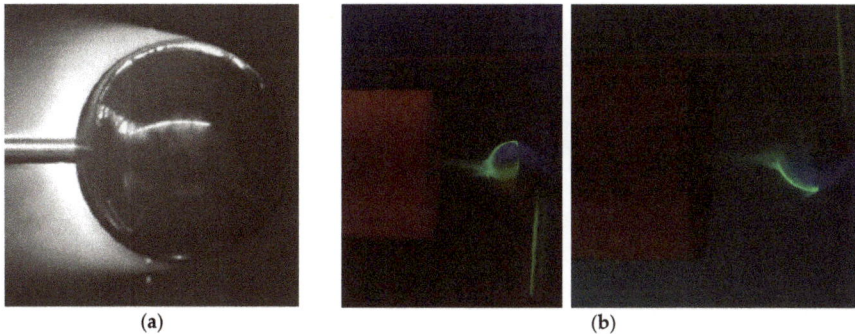

(a)   (b)

Figure 18. Flow visualization results for flow around a sphere in the uniform flow (Re = 50,000) [34] (**a**); and the suction flow (Re = 50,000) (**b**).

5. Conclusions

The mechanism of hydraulic collecting is researched by dimensional analysis and experimental studies. The experimental system is established in laboratory to carry out three kinds of tests including 253 different test cases. The results show that the vertical suction force coefficient (C_{vs}) decreases exponentially with the ratio of bottom clearance to diameter of the particle (h/d, range from 1.875 to 3.472), increases linearly with the ratio of diameter of the suction pipe to diameter of the particle (D/d, range from 1.875 to 3.472), and is nearly independent of Reynolds number (Re, range from 30,581 to 61,162). The empirical formula of vertical force of particles is obtained with respect to h/d and D/d, and the maximum tolerance is less than 10%.

The empirical formula of vertical force of particles in suction flow field is further verified by the vertical incipient motion tests. The criterion-formula of vertical incipient motion of particles is obtained to predict the collecting performance with the maximum tolerance less than 15%. As a result, it is feasible to provide sufficient suction force and avoid unnecessary disturbance by predicting the performance of collecting accurately. By comparing experiments on single and multiple particles, the correspondence of both incipient motion characteristics is obtained, and thus the research on a single particle could reveal some principles of multiple particles. In addition, the phenomenon that the vortex could help strengthen the suction force is observed in the tests.

The characteristics of suction flow field were observed by flow visualization tests. It is found that the wake vortex separation point is near the top of the particle and very small area on the particle is influenced by the unstable wake vortex. This physical phenomenon could contribute to the result that the C_{vs} in a suction flow is much smaller than the C_d in a uniform flow and explain why the flow field around the particle is stable enough to guarantee the good repeatability in tests.

Author Contributions: Conceptualization, G.Z. and L.X.; Data curation, G.Z. and M.Z.; Formal analysis, G.Z. and M.Z.; Investigation, G.Z., L.X., T.P. and M.Z.; Methodology, G.Z., L.X., T.P. and M.Z.; Resources, L.X. and T.P.; Supervision, L.X. and T.P.; Validation, G.Z., L.X., T.P. and M.Z.; Visualization, G.Z.; Writing—original draft, G.Z., L.X. and M.Z.

Funding: This research was funded by the National Key Research and Development Program of China, grant number 2016YFC0304103 and the APC was funded by the National Key Research and Development Program of China.

Acknowledgments: The present work is supported by the National Key Research and Development Program of China (2016YFC0304103). The support is gratefully acknowledged.

Conflicts of Interest: The authors declare no conflict of interest.

Abbreviations

The following abbreviations are used in this article:

C_{vs} vertical suction force coefficient
D diameter of suction pipe
DES   Detached Eddy Simulation
d diameter of particle
F vertical suction force of particle
F_B buoyancy in water
Fr Froude number
G gravity of particle
h bottom clearance
IKS   Institut für Konstruktion
LES   Large Eddy Simulation
NIOT   National Institute of Ocean Technology
OMI   Ocean Management Incorporated
Re Reynolds number
REE   rare earth element
v flow velocity in pipe
ρ density of particle
ρ_w density of water
μ dynamic viscosity coefficient

References

1. Bury, J. Mining mountains: neoliberalism, land tenure, livelihoods, and the new Peruvian mining industry in Cajamarca. *Environ. Plan. A* **2005**, *37*, 221–239. [CrossRef]
2. Mero, J.L. The mineral resources of the sea. *Elsevier Oceanogr. Ser.* **1965**, *1*, 1–5.
3. Willums, J.O.; Bradley, A. MIT's deep sea mining project. In Proceedings of the 6th Offshore Technology Conference, Houston, TX, USA, 6–8 May 1974; pp. 1072–1076.
4. Chung, J.S. Deep-Ocean Mining Issues and Ocean Mining Working Group (OMWG). In Proceedings of the 3rd ISOPE Ocean Mining Symposium, Goa, India, 8–10 November 1999.
5. Chung, J.S. Deep-ocean Mining Technology: Development II. In Proceedings of the 6th ISOPE Ocean Mining Symposium, Changsha, China, 9–13 October 2005; pp. 1–6.
6. Chung, J.S. Deep-ocean mining technology III: Developments. In Proceedings of the 8th ISOPE Ocean Mining Symposiumth, Osaka, Japan, 20–24 September 2009.
7. Chung, J.S. Advances in manganese nodule mining technology. *Mar. Technol. Soc. J.* **1985**, *19*, 39–44.
8. Herrouin, G.; Lenoble, J.P.; Charles, C.; Mauviel, F.; Bernard, J.; Taine, B. A Manganese Nodule Industrial Venture Would Be Profitable: Summary of a 4-Year Study in France. In Proceedings of the Offshore Technology Conference, Houston, TX, USA, 1–4 May 1989.
9. Boetius, A.; Haeckel, M. Mind the seafloor. *Science* **2018**, *359*, 34–36. [CrossRef] [PubMed]
10. Ma, W.; Schott, D.; Lodewijks, G. A new procedure for deep sea mining tailings disposal. *Minerals* **2017**, *7*, 47. [CrossRef]
11. McFarlane, J.; Brockett, T.; Huizingh, J.P. *Analysis of Mining Technologies Developed in the 1970s and 1980s*; International Seabed Authority: Kingston, Jamaica, 2008.
12. Hong, S.; Choi, J.S.; Kim, J.H.; Yang, C.K. Experimental study on hydraulic performance of hybrid pick-up device of manganese nodule collector. In Proceedings of the Third ISOPE Ocean Mining Symposium, Goa, India, 8–10 November 1999; pp. 69–77.
13. Yang, N.; Tang, H. Several considerations of the design of the hydraulic pick-up device. In Proceedings of the Fifth ISOPE Ocean Mining Symposium, Tsukuba, Japan, 15–19 September 2002; pp. 119–122.
14. Tsutsui, T. Flow around a sphere in a plane turbulent boundary layer. *J. Wind Eng. Ind. Aerodyn.* **2008**, *96*, 779–792. [CrossRef]
15. Lim, S.J.; Kim, J.W.; Jung, S.T.; Cho, H.Y.; Lee, S.H. Deep Seawater flow Characteristics around the Manganese Nodule Collecting Device. *Procedia Eng.* **2015**, *16*, 544–551. [CrossRef]

16. Zhao, G.; Xiao, L.; Lu, H.; Chen, Z. A Case Study of Hydraulic Collecting a Single Spherical Particle. In Proceedings of the 27th International Ocean and Polar Engineering Conference, San Francisco, CA, USA, 25–30 June 2017; pp. 30–38.

17. Chung, J.S.; Yarim, G.; Savasci, H. Shape effect of solids on pressure drop in a 2-phase vertically upward transport: silica sands and spherical beads. In Proceedings of the Eighth International Offshore and Polar Engineering Conference, Montreal, PQ, Canada, 24–29 May 1998; Volume 1, pp. 58–65.

18. Sobota, J.; Boczarski, S.; Petryka, L.; Kotlinski, R.; Stoyanova, V. Slip Velocity in Nodules Vertical Flow-Experimental Results. In Proceedings of the Fourth ISOPE Ocean Mining Symposium, Szczecin, Poland, 23–27 September 2001; pp. 127–131.

19. Jiang, L.; Li, P.; Tian, L.; Han, W. Concentration Characteristics of Large-Size Particles in Vertical Pipes for Hydraulic Lifting. In Proceedings of the Sixth ISOPE Ocean Mining Symposium, Changsha, China, 9–13 October 2005; pp. 106–109.

20. Jiang, L.; Li, P.; Tian, L.; Han, W. Experiment Study on Critical Velocity in Vertical Pipes for Hydraulic Lifting. In Proceedings of the Sixth ISOPE Ocean Mining Symposium, Changsha, Hunan, China, 9–13 October 2005; pp. 115–118.

21. Yang, N.; Chen, G.; Tang, D.; Jin, X.; Xiao, H. Behavior of single particle and group particles in vertical lifting pipe in China. In Proceedings of the Ninth ISOPE Ocean Mining Symposium, Maui, HI, USA, 19–24 June 2011; pp. 153–157.

22. Fan, W.; Chen, J.; Pan, Y.; Huang, H.; Chen, C.T.A.; Chen, Y. Experimental study on the performance of an air-lift pump for artificial upwelling. *Ocean Eng.* **2013**, *59*, 47–57. [CrossRef]

23. Pougatch, K.; Salcudean, M. Numerical modelling of deep sea air-lift. *Ocean Eng.* **2008**, *35*, 1173–1182. [CrossRef]

24. Ma, W.; van Rhee, C.; Schott, D. Technological and Profitable Analysis of Airlifting in Deep Sea Mining Systems. *Minerals* **2017**, *7*, 143.

25. Achenbach, E. Experiments on the flow past spheres at very high Reynolds numbers. *J. Fluid Mech.* **1972**, *54*, 565–575. [CrossRef]

26. Johnson, T.A.; Patel, V.C. Flow past a sphere up to a Reynolds number of 300. *J. Fluid Mech.* **1999**, *378*, 19–70. [CrossRef]

27. Constantinescu, G.S.; Squires, K.D. LES and DES investigations of turbulent flow over a sphere at Re = 10,000. *Flow Turbul. Combust.* **2003**, *70*, 267–298. [CrossRef]

28. Hein, J.R.; Koschinsky, A. Deep-ocean ferromanganese crust and nodules. In *Earth Systems and Environmental Sciences, Treatise on Geochemistry*, 2nd ed.; Holland, H., Turekian, K., Eds.; Elsevier: Amsterdam, The Netherlands, 2013; pp. 273–291.

29. Hein, J.R.; Mizell, K.; Koschinsky, A.; Conrad, T.A. Deep-ocean mineral deposits as a source of critical metals for high- and green-technology applications: comparison with land-based resources. *Ore Geol. Rev.* **2013**, *51*, 1–14. [CrossRef]

30. Blöthe, M.; Wegorzewski, A.V.; Müller, C.; Simon, F.; Kuhn, T.; Schippers, A. Manganese-cycling microbial communuties inside deep-sea manganese nodules. *Environ. Sci. Technol.* **2015**, *49*, 7692–7700. [CrossRef] [PubMed]

31. Von Stackelberg, U. Growth history of manganese nodules and crusts of the Peru Basin. In: Manganese mineralization: Geochemistry and mineralogy of terrestrial and marine deposits. *Geol. Soc. Spec. Publ.* **1997**, *119*, 153–176. [CrossRef]

32. Moritani, T. Description, classification, and distribution of manganese nodules. *Geol. Surv. Japan Cruise Rep.* **1977**, *8*, 136–158.

33. Atmanand, M.A.; Shajahan, M.A.; Deepak, C.R.; Jeyamani, R.; Ravindran, M.; Schulte, E.; Panthel, J.; Grebe, H.; Schwarz, W. Instrumentation for underwater crawler for mining in shallow waters. In Proceedings of the International Symposium of Autonomous Robots and Agents, Singapore City, Singapore, 26 May 2000.

34. Bakić, V.; Perić, M. Vizualization of Flow around a Sphere for Reynolds Numbers between 22,000 and 400,000. *Therm. AeroMechanism* **2005**, *12*, 307–315.

Article

Impeller Optimized Design of the Centrifugal Pump: A Numerical and Experimental Investigation

Xiangdong Han [1,2,3], **Yong Kang** [1,2,3,4,*], **Deng Li** [1,2,3] and **Weiguo Zhao** [5]

1   Key Laboratory of Hydraulic Machinery Transients, Ministry of Education, Wuhan University,
    Wuhan 430072, China; hanxiangdong@whu.edu.cn (X.H.); 2008lee@whu.edu.cn (D.L.)
2   Hubei Key Laboratory of Waterjet Theory and New Technology, Wuhan University, Wuhan 430072, China
3   School of Power and Mechanical Engineering, Wuhan University, Wuhan 430072, China
4   Collaborative Innovation Center of Geospatial Technology, Wuhan 430079, China
5   School of Energy and Power Engineering, Lanzhou University of Technology, Lanzhou 730050, China;
    hanromeolut@163.com
*   Correspondence: kangyong@whu.edu.cn; Tel.: +86-27-6877-4442

Received: 22 March 2018; Accepted: 17 May 2018; Published: 4 June 2018

Abstract: Combined numerical simulation with experiment, blade wrap angle, and blade exit angle are varied to investigate the optimized design of the impeller of centrifugal pump. Blade wrap angles are 122°, 126°, and 130°. Blade exit angles are 24°, 26°, and 28°. Based on numerical simulation, internal flow of the centrifugal pump with five different impellers under 0.6, 0.8, 1.0, 1.2, and 1.5 Q_d are simulated. Variations of static pressure, relative velocity, streamline, and turbulent kinetic energy are analyzed. The impeller with blade wrap angle 126° and blade exit angle 24° are optimal. Distribution of static pressure is the most uniform and relative velocity sudden changes do not exist. Streamlines are the smoothest. Distribution scope of turbulent kinetic energy is the smallest. Based on performance experiments, head and efficiency of the centrifugal pump with the best impeller are tested. The values of head and efficiency are higher than that of the original pump. Centrifugal pump with the best impeller has better hydraulic performance than the original centrifugal pump.

Keywords: centrifugal pump; impeller; blade wrap angle; blade exit angle; optimized design

1. Introduction

A centrifugal pump is one kind of general machinery [1] which is widely and fully utilized in the industrial and agricultural fields [2], such as irrigation and water supply. It normally includes four different parts: suction pipe, impeller, volute, and exit pipe. The impeller is the core part and it converts the mechanical energy into pressure energy [3], which directly determines the transport capacity and the hydraulic performances of centrifugal pump. So, optimized design of the impeller is essential and significant for the efficient operation of a centrifugal pump [4,5].

One-dimensional flow theory, two-dimensional flow theory, and three-dimensional flow theory are three basic theories to optimally design the centrifugal pump [6,7]. Two-dimensional flow theory regards the distribution of meridian velocity across the cross-section as symmetrical. The flow is one kind of potential flow. Optimized design of hydraulic machinery in engineering fields, which is based on two-dimensional flow theory, is rare. It is only employed to optimally design the higher specific speed impellers of mixed-flow pump and runners of mixed-flow turbine. Three-dimensional flow theory, proposed by Wu [8], is also called S_1 and S_2 relative flow surface theory. In this method, the three-dimensional flow is converted into two two-dimensional flows in the surfaces S_1 and S_2. Only in the ideal conditions of flow being inviscid, incompressible, and unsteady, can this method be applied to design the impeller successfully. This method is difficult and complex. Thus, it is not always employed to optimally design an impeller for a centrifugal pump in the engineering field. However, one-dimensional flow theory, which is based on Euler equations, is convenient and simple. Flow in the meridional surface is symmetrical. Engineers coming from home and abroad factories widely employ one-dimensional flow theory to optimally design an impeller for a centrifugal pump [9–11].

Blade exit angle (Φ) and blade wrap angle (β_2) are two significant parameters in the impeller optimized design process according to one-dimensional flow theory [12–15]. The hydraulic performance of a centrifugal pump is mainly determined by these parameters. In the actual optimized design process, blade exit angle often varies from 15° to 40° and blade wrap angle often varies from 90° to 130° [16]. Some scholars studied the effects of blade exit angle and blade wrap angle on the hydraulic performances of the centrifugal pump [17–19]. The selection of blade wrap angle and blade exit angle mainly depend on the specific speed of the centrifugal pump and the actual optimized design experience. If the blade wrap angle is too large, the impeller friction area becomes large, which is bad for the improvement of efficiency. If the blade wrap angle is too small, the impeller cannot control the flow of water effectively and the impeller cannot operate stably. For the blade exit angle, a moderate value could effectively improve the hydraulic performances of the centrifugal pump. Comprehensively considering the specific speed of the target centrifugal pump (n_s = 112) and the effects of blade wrap angle and blade exit angle on the hydraulic performances of the centrifugal pump, the variation scope of the blade wrap angle and blade exit angle was selected carefully. The blade wrap angle was tested in the range of 120° to 130° and the values were Φ = 122°, 126°, and 130°, respectively. The blade exit angle was tested in the range of 20° to 30° and the values were β_2 = 24°, 26°, and 28°, separately.

2. Basic Parameters of the Impeller

The inlet diameter of the impeller D_1 = 200 mm and the outlet diameter D_2 = 420 mm. The exit width b_2 = 34 mm. The original blade wrap angle Φ = 120°. The original blade exit angle β_2 = 26°. The basic parameters of the impeller are shown in Figure 1.

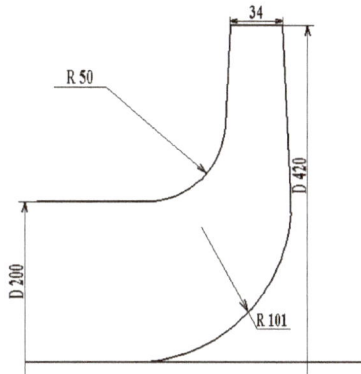

Figure 1. Two-dimensional model of the impeller.

3. Method to Modify the Impeller

The impeller shape is directly determined by the blade profiles. In the impeller optimization design process, all blade profiles must be smooth. The bend of the blade profiles should be in one single direction. An 'S' shape should not exist [16,20,21]. All blade profiles should be symmetrical. The blade profile differential equation is

$$d\theta = \frac{ds}{r \tan \beta} \qquad (1)$$

Here, s is the axial streamline. θ denotes the axial angle. r represents the radius of one arbitrary point in the blade profile. β is the blade angle.

The speed triangle is shown in Figure 2. Here, c represents absolute velocity, w is the relative velocity, and u is the following velocity. c_m and w_m are the component of absolute velocity and relative velocity in the meridional plane, respectively. c_u is the component in the circumferential plane.

Based on the speed triangle, $\tan\beta$ could be described as

$$\tan\beta = \frac{c_m}{u - c_u} \tag{2}$$

Thus,

$$\frac{ds}{rd\theta} = \frac{c_m}{u - c_u} \tag{3}$$

So,

$$ds = \frac{c_m}{u - c_u} rd\theta \tag{4}$$

According to the above equations, blade profile optimized design has a closed relation with blade inlet angle (β_1), blade exit angle (β_2), and blade wrap angle (Φ). Three different methods are usually employed to modify the blade profiles. They are that

(1) β_1 is the constant. β_2 and Φ are changed.
(2) β_1 and β_2 are fixed. Φ is varied.
(3) β_2 and Φ are invariant. β_1 is altered.

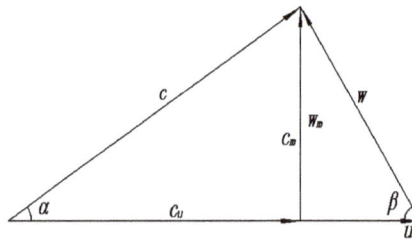

Figure 2. Speed triangle.

Method 1 (β_1 is the constant. β_2 and Φ are changed.) was always employed to optimally design the impeller. As shown in Table 1, the blade exit angle of Schemes 1–3 was identical and the blade wrap angles varied from 122° to 130°, which were employed to investigate the effects of different blade wrap angles on the optimized design of the centrifugal pump. For Schemes 2, 4, and 5, the blade wrap angle was the same and the blade exit angle was changed, which was used to discuss the effects of diverse blade exit angles on the optimized design of the centrifugal pump.

Table 1. Optimized schemes of the blade profile.

Scheme	1	2	3	4	5
Φ	122°	126°	130°	126°	126°
β_2	26°	26°	26°	28°	24°

The impeller optimized design procedure is that, via the analysis of effects of blade wrap angle (Schemes 1–3) on the hydraulic performances of the centrifugal pump, one optimal scheme could be determined via computational fluid dynamics (CFD). According to the discussion of effects of blade exit angle (Schemes 2, 4, and 5) on the variations of head and efficiency, one optimal scheme could be got by numerical simulation. Then, these two different schemes are compared and the best one could be obtained. The optimized procedure includes two main parts, as shown in Figure 3.

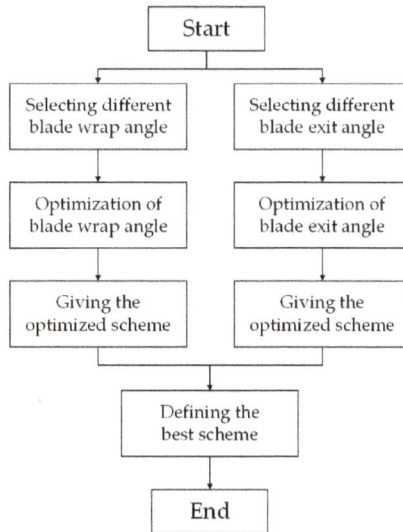

Figure 3. Flow chart of the optimized design of the impeller.

The physical models of the blades under different blade wrap angle and blade exit angle conditions are given in Figure 4.

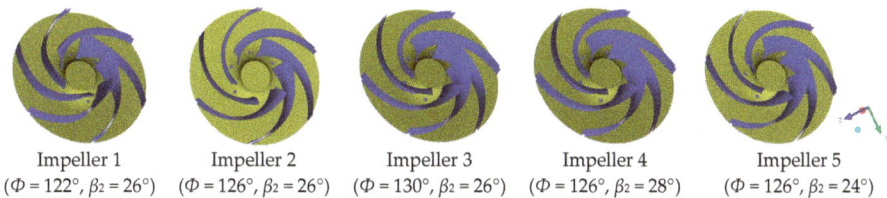

Impeller 1	Impeller 2	Impeller 3	Impeller 4	Impeller 5
($\Phi = 122°$, $\beta_2 = 26°$)	($\Phi = 126°$, $\beta_2 = 26°$)	($\Phi = 130°$, $\beta_2 = 26°$)	($\Phi = 126°$, $\beta_2 = 28°$)	($\Phi = 126°$, $\beta_2 = 24°$)

Figure 4. Physical model of the blades.

4. Numerical Method

4.1. Fundamental Equations

The fundamental equations [22] are employed to describe the flow characteristics in the centrifugal pump, which include two main parts: continuity equation and motion equation, corresponding with mass conservation law and momentum conservation law.

$$\frac{\partial \rho}{\partial t} + \frac{\partial (\rho u_i)}{\partial x_i} = 0 \tag{5}$$

$$\rho \frac{\partial u_i}{\partial t} + \rho u_j \frac{\partial u_i}{\partial x_j} = \rho f_i - \frac{\partial p}{\partial x_i} + \frac{\partial}{\partial x_j}\left[\mu\left(\frac{\partial u_i}{\partial x_j} + \frac{\partial u_j}{\partial x_i}\right)\right] - \frac{2}{3}\frac{\partial}{\partial x_i}\left(\mu \frac{\partial u_j}{\partial x_j}\right) \tag{6}$$

Here, ρ is the density of water. t represents the time. u denotes the velocity. x is the Cartesian coordinate. f_i is the body force vector. p is the pressure. μ is the dynamic viscosity.

4.2. Turbulent Model

RNG k-ε turbulent model proposed by Yakhot and Orzag [23] was employed to deal with turbulent flow. In the centrifugal pump, the impeller is the rotating part whose rotation effects could be fully dealt by the turbulent dissipation rate (ε) equation in this turbulent model. On the other hand, the RNG k-ε turbulent model has high-precision, which could guarantee the accuracy of the numerical results.

$$\frac{\partial(\rho k)}{\partial t} + \frac{\partial(\rho k u_i)}{\partial x_i} = \frac{\partial}{\partial x_j} \times \left[(\alpha_k(\mu + \mu_t)) \frac{\partial k}{\partial x_j} \right] + G_k + \rho\varepsilon \tag{7}$$

$$\frac{\partial(\rho\varepsilon)}{\partial t} + \frac{\partial(\rho\varepsilon u_i)}{\partial x_i} = C_{1\varepsilon}\frac{\varepsilon}{k}G_kC_{2\varepsilon}\rho\frac{\varepsilon^2}{k} + \frac{\partial}{\partial x_j}\left[\alpha_\varepsilon(\mu + \mu_t)\frac{\partial\varepsilon}{\partial x_j}\right] \tag{8}$$

$$\mu_t = \rho c_\mu \frac{k^2}{\varepsilon} \tag{9}$$

Here, k is the turbulent kinetic energy. ε is the turbulent dissipation rate. μ_t is the turbulent viscosity. The five terms, $C_{1\varepsilon}$, $C_{2\varepsilon}$, α_k, α_ε, and c_μ are empirical coefficients and the values are 1.42, 1.68, 1.39, 1.39, and 0.09, separately. G_k is one generation term of turbulent kinetic energy which is caused by the mean velocity gradient.

5. Numerical Simulation Setup

5.1. Physical Model

Three-dimensional (3D) single-stage and single-suction centrifugal pump was employed, as shown in Figure 5. The suction pipe was employed to keep the uniform of the flow. Exit pipe was utilized to avoid the backflow. The performance parameters were designed flow rate Q_d = 550 m^3/h; designed head H_d = 50 m; rated motor power P = 110 KW; and rated rotational speed n = 1480 rpm. The geometric parameters of the impeller and the volute were shown in Table 2.

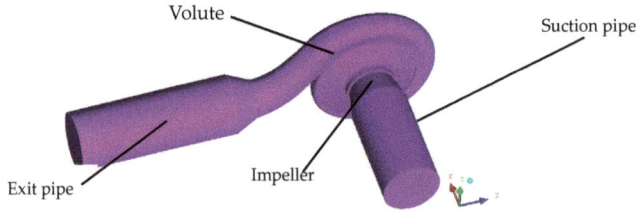

Figure 5. Physical model of the centrifugal pump.

Table 2. Geometric parameters of the impeller and volute

Parameter	Value
Impeller inlet diameter D_1, mm	200
Impeller outlet diameter D_2, mm	420
Impeller exit width b_2, mm	34
Number of blade Z	6
Base diameter of the volute D_3, mm	435
Volute inlet width b_3, mm	72
Volute outlet diameter D_4, mm	250

To get accurate numerical simulation results a balance hole, front chamber, and back chamber were added to the 3D physical model of the centrifugal pump, as displayed in Figures 6 and 7. Diameter of the balance hole was 10mm and it was mainly employed to balance the axial force [24].

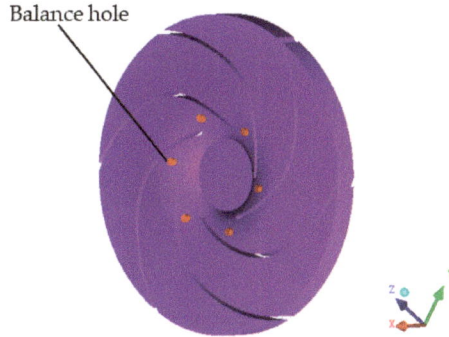

Figure 6. Physical model of the impeller.

Figure 7. Chamber of the centrifugal pump.

5.2. Mesh Generation

ANSYS-ICEM (15.0, ANSYS, Inc., Canonsburg, PA, USA) is employed to discrete the centrifugal pump computational domains. To guarantee the uniformity of the flow, structured meshes were employed to discretize the computational domains of the suction pipe and exit pipe, as displayed in Figure 8d–e. Fully considering the complex structure of the impeller and the volute, to get the better adaptability of the flow, unstructured meshes were employed to discrete the computational domains of impeller, front and back chamber, and volute, as shown in Figure 8a–c. Meshes in the leading edges of the impeller and balance hole were refined.

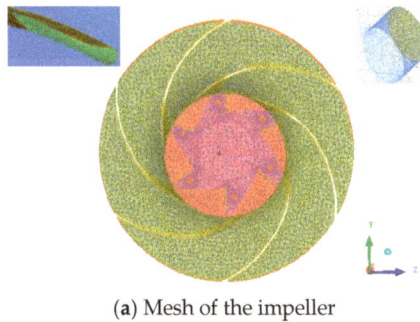

(**a**) Mesh of the impeller

Figure 8. *Cont.*

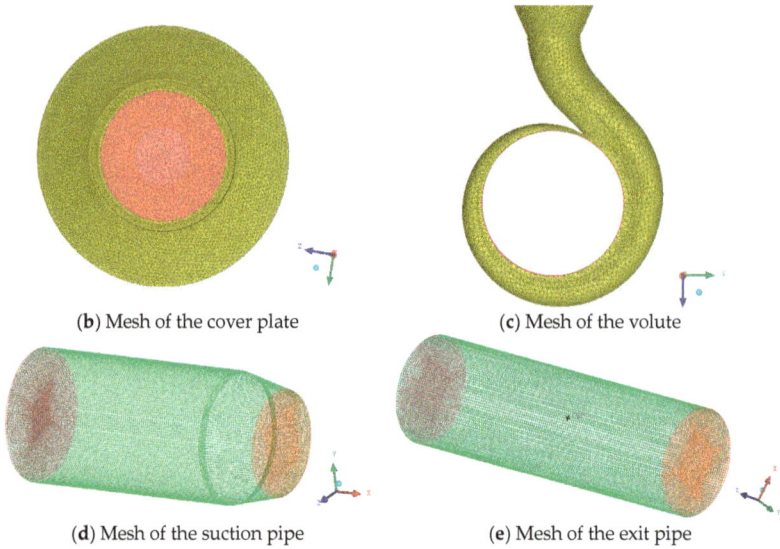

(**b**) Mesh of the cover plate (**c**) Mesh of the volute

(**d**) Mesh of the suction pipe (**e**) Mesh of the exit pipe

Figure 8. Meshes of the computational domain of the centrifugal pump.

Mesh numbers of the five different impellers were 3,472,923, 3,473,052, 3,471,896, 3,472,816, and 3,473,011, respectively. All mesh quality was higher than 0.3. Mesh number and quality of front and back chamber, volute, suction pipe, and exit pipe were displayed in Table 3.

Table 3. Mesh number and quality

	Front and Back Chamber	Volute	Suction Pipe	Exit Pipe
Mesh number	934,607	2,238,179	498,506	327,624
Mesh quality	>0.4	>0.4	>0.8	>0.8

Head under designed flow rate condition was calculated to verify mesh independence. Results indicated that with the increase of mesh number, head increased firstly. Then, the differences of head were slight, as shown in Figure 9. Total mesh numbers of the five different centrifugal pumps were 7,139,826, 7,146,497, 7,143,016, 7,145,682, and 7,147,024, separately.

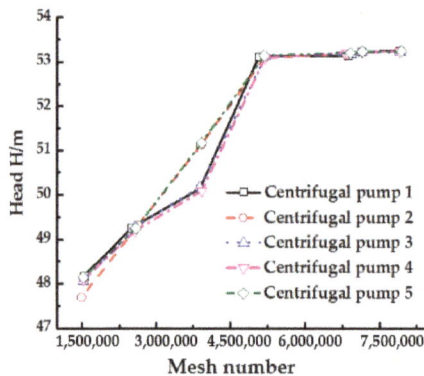

Figure 9. Meshes independence.

5.3. Boundary Conditions

Reynolds averaged Naiver–Stokes (RANS) method was employed to simulate the internal flow of the centrifugal pump. At inlet, the velocity was set. At outlet, the free outflow was set. Near wall flow was treated by standard wall function. The interface was set between suction pipe and impeller, impeller and volute, volute and exit pipe. The SIMPLE scheme was used to solve the coupled equations of velocity. The PRESTO! was employed to compute pressure. The second order upwind scheme was used for the solving of momentum, turbulent kinetic energy, and turbulent dissipation rate. All residuals were less than 1.0×10^{-5}.

5.4. Verification of the Algorithm

The above-mentioned algorithm is employed to numerically simulate the internal flow of the original centrifugal pump and optimized centrifugal pump. To verify the reasonableness of the algorithm used, flow in one single-stage and single suction centrifugal pump which is shown in Figure 10 was simulated under different flow rate conditions. The rated flow rate is $Q_d = 500 \text{ m}^3/\text{h}$ and the rated head is $H_d = 53$ m. Rated rotating speed is $n = 1480$ rpm.

Figure 10. Experimental centrifugal pump.

The numerical results of head and efficiency had a good agreement with the experimental results, as displayed in Figure 11. The maximum relative error of the head was less than 3.7%. The largest fractional error of the efficiency was less than 2.2%. The algorithm designed was reasonable.

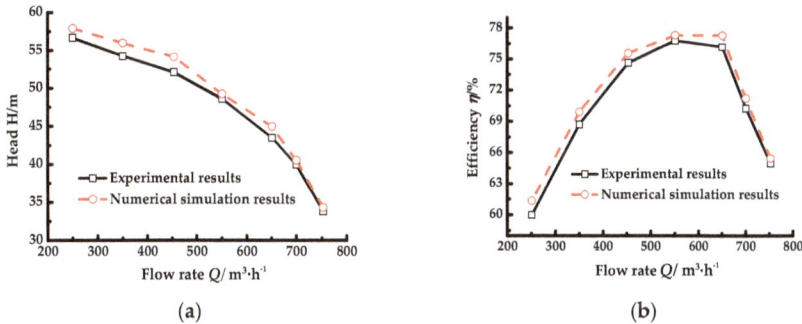

Figure 11. Comparison of the hydraulic performances of centrifugal pump. (**a**) Head (H)-flow rate (Q); (**b**) Efficiency (η)-flow rate (Q).

6. Numerical Simulation Results Analysis

To measure the effects of blade wrap angle and blade exit angle on the performances of centrifugal pump, variations of static pressure, relative velocity, streamlines, and turbulent kinetic energy in the middle span of the impeller were analyzed under the typical flow rate conditions (low flow rate 0.6 Q_d and 0.8 Q_d, rated flow rate 1.0 Q_d, and high flow rate 1.2 Q_d and 1.5 Q_d). Also, the variations of head and efficiency under the above flow rate conditions were compared.

6.1. Variation of Static Pressure

The overall static pressure variation law is that static pressure in different passages distributed uniformly. Differences of static pressure in all passages were relatively small, as shown in Figure 12. At the inlet of the impeller, static pressure was the lowest. Static pressure at the outlet of the impeller was the highest. For Impellers 1–5, static pressure increased overall with the growing of flow rate. Under 0.8 Q_d condition, the increase of static pressure was more manifest. At outlet, static pressure attained the maximum under 1.5 Q_d condition.

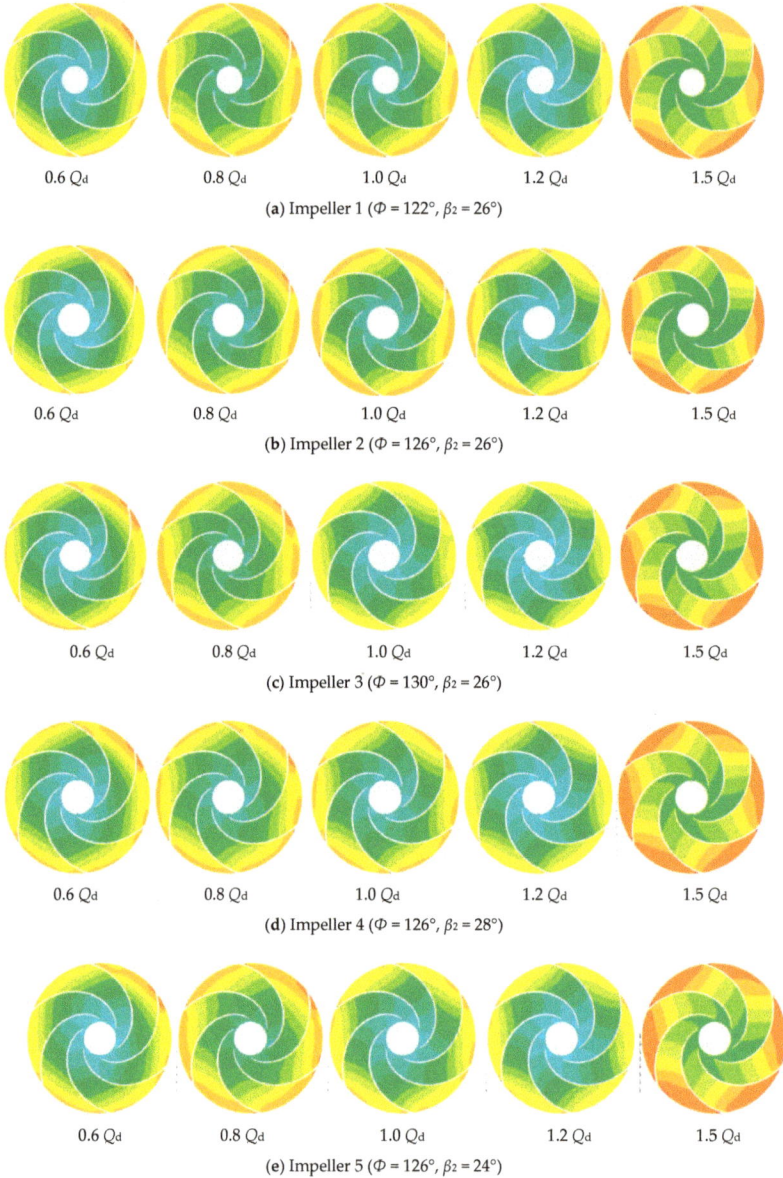

0.6 Q_d 0.8 Q_d 1.0 Q_d 1.2 Q_d 1.5 Q_d

(**a**) Impeller 1 ($\Phi = 122°$, $\beta_2 = 26°$)

0.6 Q_d 0.8 Q_d 1.0 Q_d 1.2 Q_d 1.5 Q_d

(**b**) Impeller 2 ($\Phi = 126°$, $\beta_2 = 26°$)

0.6 Q_d 0.8 Q_d 1.0 Q_d 1.2 Q_d 1.5 Q_d

(**c**) Impeller 3 ($\Phi = 130°$, $\beta_2 = 26°$)

0.6 Q_d 0.8 Q_d 1.0 Q_d 1.2 Q_d 1.5 Q_d

(**d**) Impeller 4 ($\Phi = 126°$, $\beta_2 = 28°$)

0.6 Q_d 0.8 Q_d 1.0 Q_d 1.2 Q_d 1.5 Q_d

(**e**) Impeller 5 ($\Phi = 126°$, $\beta_2 = 24°$)

Figure 12. *Cont.*

Static pressure /Pa

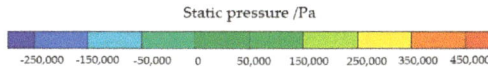

-250,000   -150,000   -50,000   0   50,000   150,000   250,000   350,000   450,000

Figure 12. Variation of static pressure in different impellers.

Static pressure in Impellers 1, 2, and 4 fluctuated more obviously and the pressure gradient was higher, which was caused by the intense rotor-stator interaction of the impeller and the volute [25]. Volumetric loss of the centrifugal pump is severe. Thus, the centrifugal pump could not operate efficiently. For Impeller 2, under low flow rate conditions, the distribution scope of lower pressure was much larger than that of other impellers. Centrifugal pumps operating under these conditions were unstable. Compared with Impellers 1, 2, 4, and 5, the lower pressure region in Impeller 3 was much larger under 0.8 Q_d and 1.2 Q_d conditions, mainly caused by secondary flow. Cavitation in Impeller 3 could occur easily, which could make the performances of the centrifugal pump decrease sharply [26]. In Impeller 5, the pressure gradient was the smallest and the lower pressure region was the smallest under all flow rate conditions, proving that the optimization of Impeller 5 is the best. Impeller with $\Phi = 126°$ and blade exit angles $\beta_2 = 24°$ could guarantee the safety operation of the centrifugal pump.

6.2. Variation of Relative Velocity

At the inlet of the impeller, the relative velocity was small. At the outlet of the impeller, the relative velocity attained the maximum, which was in agreement with the experimental and theoretical results [27]. Relative velocity in the pressure surface was much smaller than that of suction pressure. With the increase of flow rate, the low relative speed zone became smaller, as shown in Figure 13.

Under low flow rate conditions, low relative velocity zone in Impeller 5 was smaller than that of other impellers, which could guarantee the stable operation of the centrifugal pump. Velocity gradient in Impellers 1–4 was larger and the flow in these impellers was disorderly. Under rated flow rate conditions, the low speed velocity zone in Impeller 1 was the most obvious. Although differences of low speed zones of Impellers 2–5 were smaller, the zone of Impeller 5 was smaller than that of Impellers 2–4. Under high flow rate conditions, the low velocity speed zone of Impeller 5 was the smallest. Velocity sudden change did not appear. The distribution of relative velocity was uniform, which reflected that blade wrap angle $\Phi = 126°$ and blade exit angle $\beta_2 = 24°$ could let the centrifugal pump operate stably. The hydraulic performances of Impeller 5 were better than others.

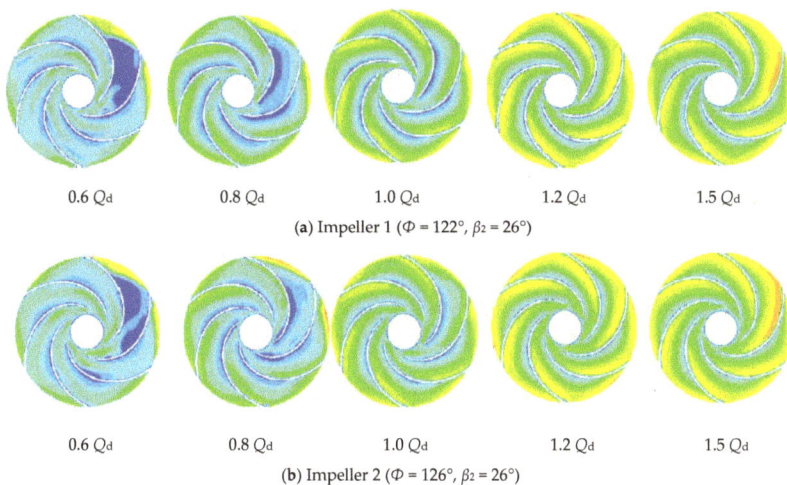

0.6 Q_d 0.8 Q_d 1.0 Q_d 1.2 Q_d 1.5 Q_d

(a) Impeller 1 ($\Phi = 122°$, $\beta_2 = 26°$)

0.6 Q_d 0.8 Q_d 1.0 Q_d 1.2 Q_d 1.5 Q_d

(b) Impeller 2 ($\Phi = 126°$, $\beta_2 = 26°$)

Figure 13. *Cont.*

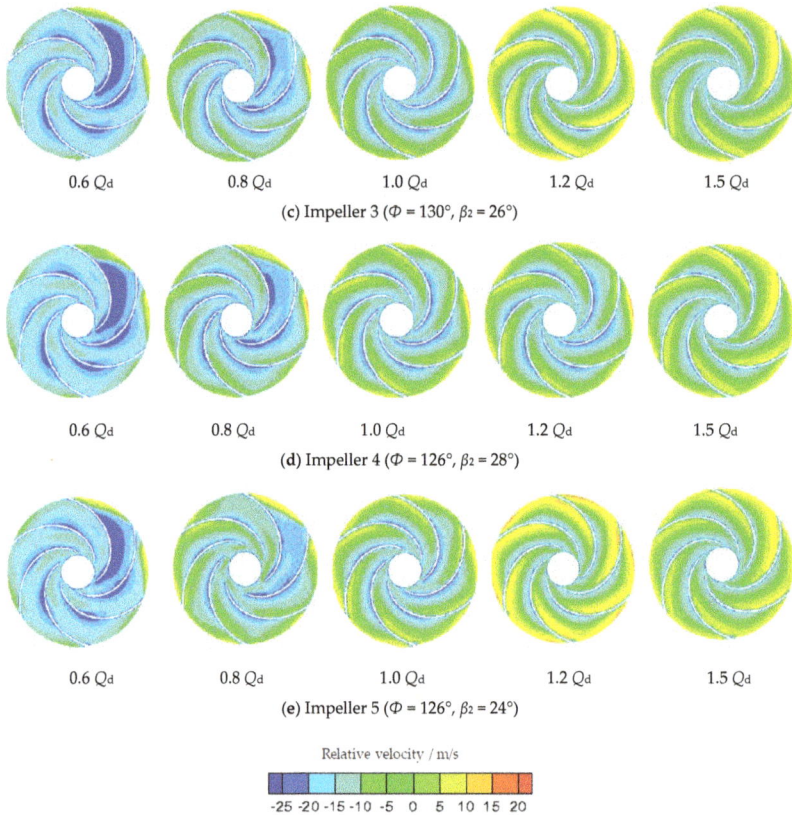

(c) Impeller 3 (Φ = 130°, β_2 = 26°)

(d) Impeller 4 (Φ = 126°, β_2 = 28°)

(e) Impeller 5 (Φ = 126°, β_2 = 24°)

Relative velocity / m/s

-25 -20 -15 -10 -5 0 5 10 15 20

Figure 13. Variation of absolute velocity in different impellers.

6.3. Variation of Streamlines

Under low flow rate conditions, streamlines in Impellers 1–4 were disorder and there were manifest vortices, especially for 0.6 Q_d condition, which could cause severe energy loss and secondary flow. For Impeller 2, the vortices were the most obvious, as shown in Figure 14b. The centrifugal pumps under these two flow rate conditions operated unstably, which could induce severe vibrations and noises [28] and the efficiency of the centrifugal pumps decreased. However, the streamlines in Impeller 5 under these two conditions were smooth. The secondary flow could be avoided successfully. Compared with low flow rate conditions, streamlines in rated and high flow rate conditions became smooth. However, streamlines of Impellers 1–4 were more disorder than that of Impeller 5. In Impellers 1, 2, and 4, the low speed region on the pressure surface was dramatic, which could let the flow in suction surface was faster. Energy loss in this region could be caused easily. Overall, the streamlines in Impeller 5 were the smoothest, as displayed in Figure 14e, which could result in lower energy loss. The flow separation and back flow could be controlled. The centrifugal pump could operate stably.

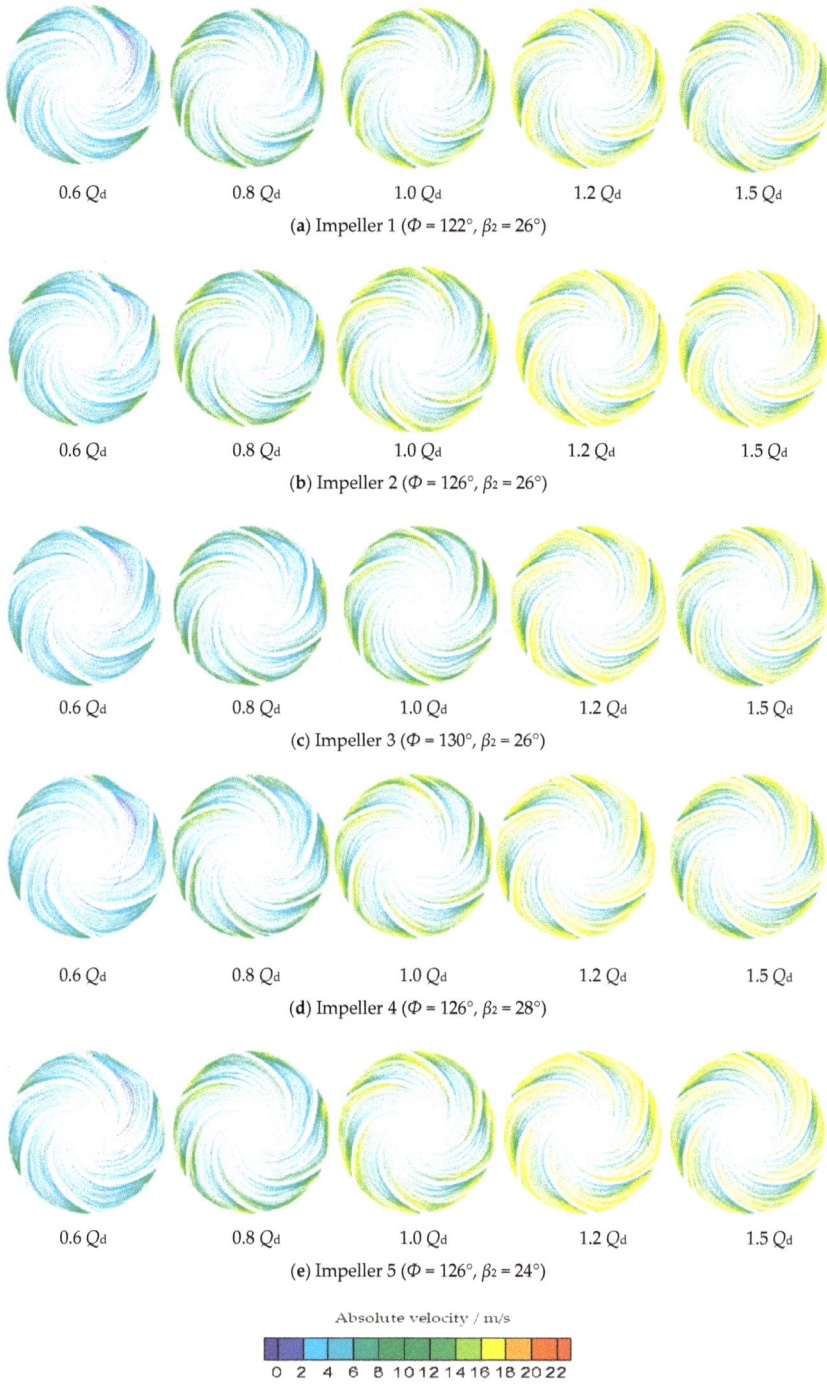

0.6 Q_d 0.8 Q_d 1.0 Q_d 1.2 Q_d 1.5 Q_d

(**a**) Impeller 1 ($\Phi = 122°$, $\beta_2 = 26°$)

0.6 Q_d 0.8 Q_d 1.0 Q_d 1.2 Q_d 1.5 Q_d

(**b**) Impeller 2 ($\Phi = 126°$, $\beta_2 = 26°$)

0.6 Q_d 0.8 Q_d 1.0 Q_d 1.2 Q_d 1.5 Q_d

(**c**) Impeller 3 ($\Phi = 130°$, $\beta_2 = 26°$)

0.6 Q_d 0.8 Q_d 1.0 Q_d 1.2 Q_d 1.5 Q_d

(**d**) Impeller 4 ($\Phi = 126°$, $\beta_2 = 28°$)

0.6 Q_d 0.8 Q_d 1.0 Q_d 1.2 Q_d 1.5 Q_d

(**e**) Impeller 5 ($\Phi = 126°$, $\beta_2 = 24°$)

Absolute velocity / m/s

0   2   4   6   8   10 12 14 16 18 20 22

Figure 14. Variation of streamlines in different impeller.

6.4. Variation of Turbulent Kinetic Energy

The distribution characteristics of turbulent kinetic energy in the middle span of the impeller were sharply different under diverse flow rate conditions, which were shown in Figure 15. The distribution scope became smaller gradually with the increase of flow rates. Under the 0.6 Q_d condition, the distribution range of turbulent kinetic energy was the largest and under the 1.5 Q_d condition, the scope was the smallest for all the impellers.

Under low flow rate conditions, the distribution scope of turbulent kinetic energy in Impeller 2 was larger than that of other impellers. Velocity vector had obvious change when water flows from the impeller to the volute. Centrifugal pump with Impeller 2 could not operate stably. Under rated flow rate conditions, the turbulent kinetic energy distribution scopes of Impellers 1, 3, and 4 were similar and they were slightly larger than that of Impeller 5. For high flow rate conditions, the distribution scope of turbulent kinetic energy was almost identical for all the impellers.

Based on the above analysis of turbulent kinetic energy and according to Equation (7), turbulent intensity l [29] of Impeller 5 was less intense than that of Impellers 1–4 overall. Thus, the energy loss in Impeller 5 was less than other impellers, which could guarantee the high transportation capacity of the centrifugal pump.

$$k = \frac{3}{2}(v_{in}l)^2 \tag{10}$$

Here, v_{in} is the inlet velocity. l is the turbulent intensity.

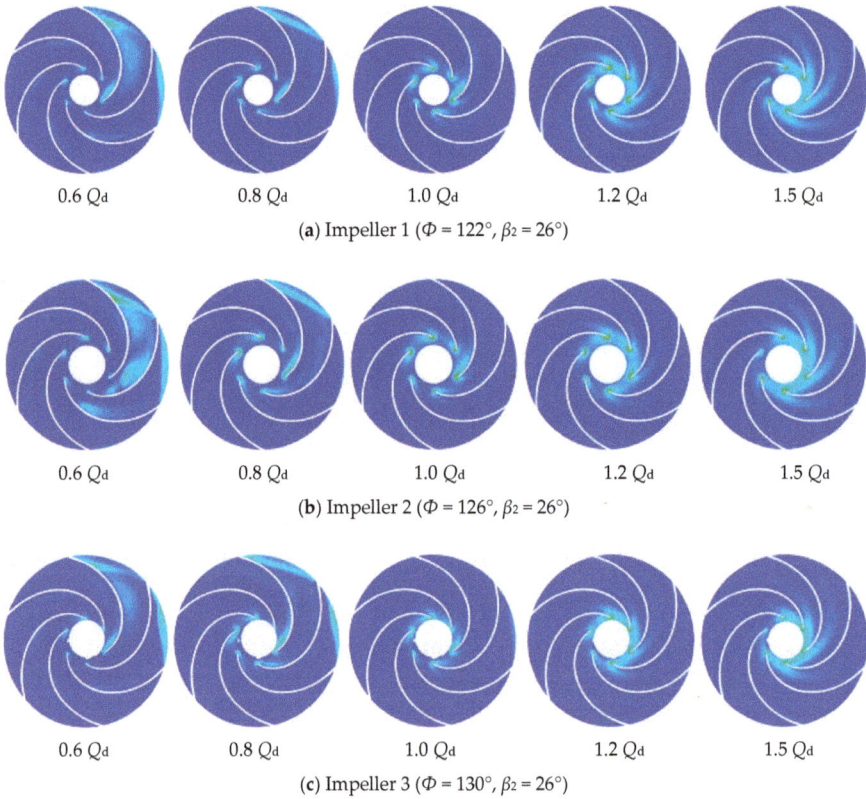

0.6 Q_d 0.8 Q_d 1.0 Q_d 1.2 Q_d 1.5 Q_d

(a) Impeller 1 ($\Phi = 122°$, $\beta_2 = 26°$)

0.6 Q_d 0.8 Q_d 1.0 Q_d 1.2 Q_d 1.5 Q_d

(b) Impeller 2 ($\Phi = 126°$, $\beta_2 = 26°$)

0.6 Q_d 0.8 Q_d 1.0 Q_d 1.2 Q_d 1.5 Q_d

(c) Impeller 3 ($\Phi = 130°$, $\beta_2 = 26°$)

Figure 15. *Cont.*

| 0.6 Q_d | 0.8 Q_d | 1.0 Q_d | 1.2 Q_d | 1.5 Q_d |

(**d**) Impeller 4 ($\Phi = 126°$, $\beta_2 = 28°$)

| 0.6 Q_d | 0.8 Q_d | 1.0 Q_d | 1.2 Q_d | 1.5 Q_d |

(**e**) Impeller 5 ($\Phi = 126°$, $\beta_2 = 24°$)

Turbulent kinetic energy / m²/s²

2 4 6 8 10 12 14 16 18 20

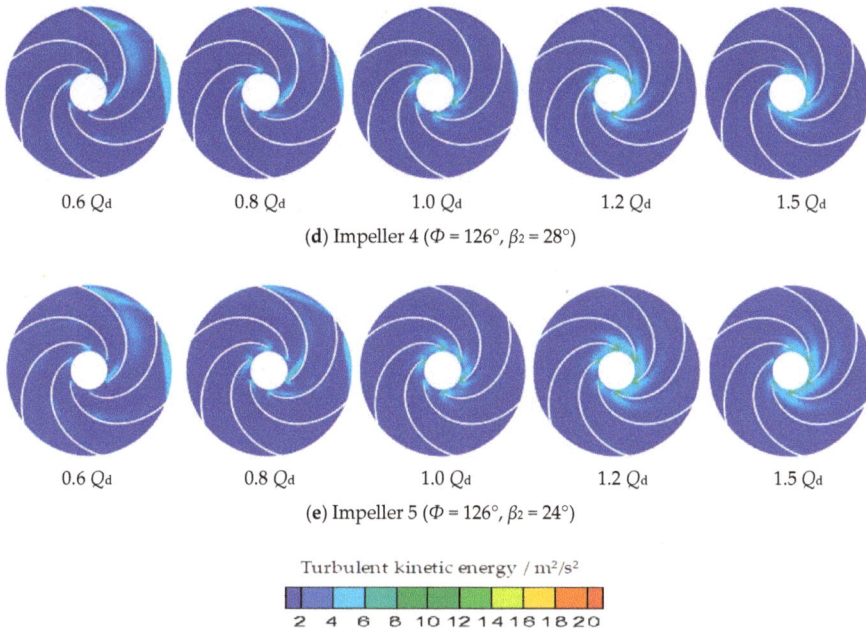

Figure 15. Variation of turbulent kinetic energy in different impellers.

6.5. Variations of Head and Efficiency

With the increase of flow rate, head decreased gradually and efficiency increased firstly then decreased, as exhibited in Figure 16. Differences of head were relatively slight. A head with $\Phi = 122°$ was higher than that of $\Phi = 126°$. The head of the centrifugal pump with $\Phi = 130°$ was the highest one. Efficiency attained the maximum under designed flow rate. Efficiency with $\Phi = 130°$ was higher than that of $\Phi = 122°$ and $126°$. Compared with Schemes 1 and 2, Scheme 3 was the optimal one.

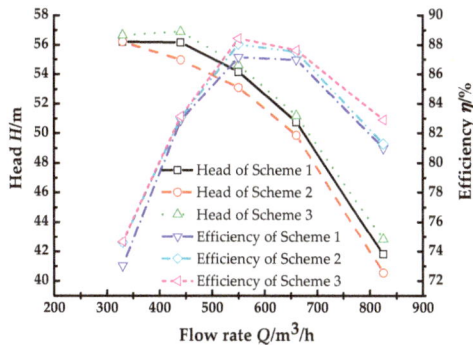

Figure 16. Hydraulic performances of the centrifugal pump with different blade wrap angles.

Schemes 2, 4, and 5 reflected the effects of blade exit angles on hydraulic performances of the centrifugal pump, as displayed in Figure 17. Head decreased constantly and efficiency increased firstly then decreased with the increase of flow rate. The efficiency attained the maximum under the rated flow rate. Head with $\beta_2 = 24°$ was slightly higher than that of $\beta_2 = 28°$. Differences of head with

$\beta_2 = 26°$ and $28°$ were sharp. The distinctions of efficiency were smaller. Efficiency of centrifugal pump with $\beta_2 = 24°$ was the highest. Compared with Schemes 2, 4, and 5, the centrifugal pump with $\beta_2 = 24°$ had the best hydraulic performances.

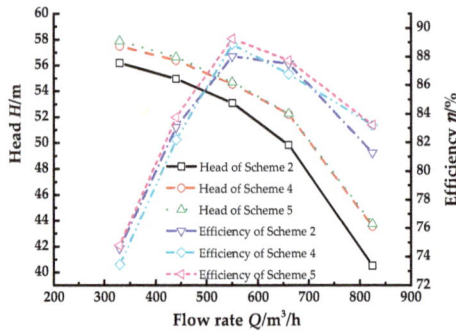

Figure 17. Hydraulic performances of the centrifugal pump with different blade exit angles.

In Scheme 3, blade wrap angle $\Phi = 130°$ was the optimal one. The corresponding blade exit angle $\beta2 = 26°$. In Scheme 5, blade exit angle $\beta2 = 24°$ was the best one. The corresponding blade wrap angle $\Phi = 126°$. Head and efficiency of Scheme 5 were higher than that of Scheme 3, as shown in Figure 18. So, Scheme 5 was selected to improve the hydraulic performances of the centrifugal pump.

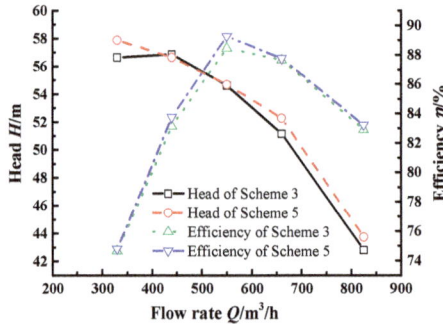

Figure 18. Hydraulic performances of Schemes 3 and 5.

7. Experimental Analysis

Via numerical simulation in Part 6, Scheme 5 was determined to be the best one. The corresponding impeller was machined. Then, the performance experiments were done in the closed performance experiment rig of Shanghai Kaiquan Pump Group to verify that head and efficiency of the centrifugal pump with the Impeller 5 showing obvious improvement.

Experiment Setup

The impeller machining process was displayed in Figure 19. At the outlet of the impeller, the allowance was 5 mm, as shown in region A. Allowance of the impeller front shroud and back shroud was 3 mm, as exhibited in region B. At the inlet of the impeller, the allowance was 3 mm too, as displayed in region C. Allowance of the hub was 3 mm, as shown in region D.

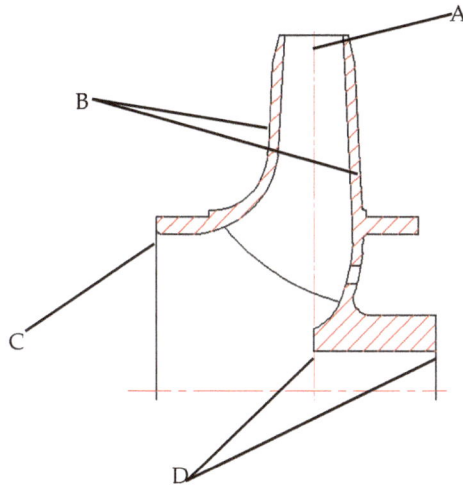

Figure 19. The machining process of the impeller.

To measure the optimized design of the impeller of Scheme 5, 3D wax pattern impeller, given in Figure 20, was machined firstly in the HRPS-V rapid prototyping machine, as shown in Figure 21. Laminated object manufacturing (LOM) in this machine—which is based on the data coming from CAD laying model, CO_2 laser beam, and hot pressing device—was employed to machine the 3D wax pattern impeller.

Figure 20. 3D wax pattern.

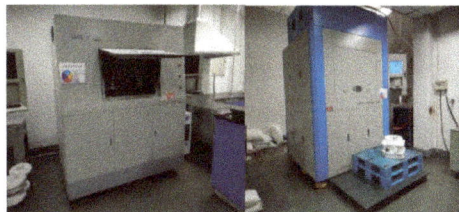

Figure 21. HRPS-V rapid prototyping machine.

Then, the casted impeller of Scheme 5 was machined, which was displayed in Figure 22. The material was 2Cr13. To guarantee that the passages of the impeller were smooth and clean, silt particles and cast joint flash in the passages were cleaned carefully.

Figure 22. The casted impeller.

The performance experiments of the centrifugal pump were done at the closed experiment rig of Shanghai Kaiquan Pump Group. The experiment rig, given in Figure 23, includes the model centrifugal pump, suction pipes, exit pipes, pressure gauges, valves, and flow meters. The type of pressure sensor is XU12087105 (1.0, Shanghai Automation Instrumentation Co. Ltd., Shanghai, China). The type of flow meter is DN300 (1.0, Tianjin Flow Meter Co. Ltd., Tianjin, China). The type of valve is ZA2.T. All testing precisions are national grade 1 (GB3216-2005 and ISO 9906-1999).

Figure 23. Performance experiment rig.

Figure 24 is the typical components of the closed performance experiment rig. It includes 16 different parts.

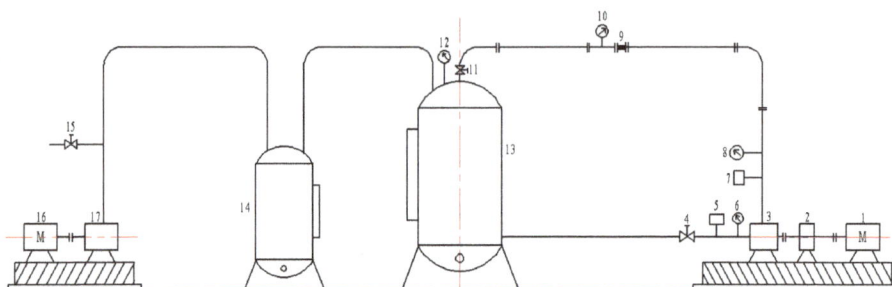

Figure 24. Components of the closed performance experiment rig. 1–electric motor; 2–torque and speed sensor; 3–centrifugal pump; 4–sluice valve; 5–inlet pressure sensor; 6–turbine flow meter; 7–outlet pressure sensor; 8–turbine flow meter; 9–expansion joint; 10–turbine flow meters; 11–regulating valve; 12–turbine flow meters; 13–cavitation tank; 14–separation tank of vapor and water; 15–ball valve; 16–electric motor; 17–vacuum pump.

Via the inlet pressure Sensor 5 and the outlet pressure Sensor 7, M_1 and M_2 were determined. According to Equations (11) and (12), the inlet pressure and the outlet pressure of the centrifugal pump could be got.

The inlet pressure is calculated by

$$\frac{p_1}{\rho g} = 102M_1 + h_1 \tag{11}$$

The outlet pressure is determined by

$$\frac{p_2}{\rho g} = 102M_2 - h_3 \tag{12}$$

where h_1 is the vertical height between the center of inlet pressure sensor with axial lead of the centrifugal pump. h_3 is the vertical height between the center of the outlet pressure sensor with the outlet pressure port.

The inlet flow rate (Q_1) and outlet flow rate (Q_2) of the centrifugal pump could be measured by turbine flow meter 6 and 8, respectively.

Thus, the inlet velocity of the centrifugal pump v_1 could be calculated by

$$v_1 = \frac{Q_1}{A_1} \tag{13}$$

The outlet velocity of the centrifugal pump v_2 could be got by

$$v_2 = \frac{Q_2}{A_2} \tag{14}$$

Here, A_1 is the inlet cross-section area. A_2 is the outlet cross-section area.

Head (H) could be calculated by Equation (15)

$$H = \left(\frac{p_2}{\rho g} - \frac{p_1}{\rho g}\right) + \left(\frac{v_2^2}{2g} - \frac{v_1^2}{2g}\right) + (Z_2 - Z_1) \tag{15}$$

where Z_2 and Z_1 are the static head.

In the performance experiment rig,

$$Z_1 = 0 \tag{16}$$

For Z_2, it could be obtained by

$$Z_2 = h_2 + h_3 \tag{17}$$

Here, h_2 is the vertical height between the center of outlet pressure sensor with axial lead of the centrifugal pump.

In the performance experiment rig,

$$h_2 = h_1 \tag{18}$$

The Equation (15) could be modified. The new form was shown in Equation (20).

$$H = 102(M_2 - M_1) + \left(\frac{v_2^2}{2g} - \frac{v_1^2}{2g}\right) \tag{19}$$

Efficiency (η) could be calculated by Equation (16)

$$\eta = \frac{\rho g Q H}{1000 P} \times 100\% \tag{20}$$

Here, *P* is the inlet power, which could be measured by torque and speed sensor 2.

Hydraulic performance curves of the improved centrifugal pump and the original centrifugal pump were shown in Figure 25. Experimental results indicated that head and efficiency of the improved centrifugal pump were higher than that of the original centrifugal pump. Under low flow rate condition (0.8 Q_d), the head increased by 3.76 m and the efficiency increased by 3.84%. With rated flow rate condition (1.0 Q_d), the head increased by 2.74 m and the efficiency increased by 5.77%, separately. For high flow rate condition (1.2 Q_d), the head and efficiency increased by 2.67 m and 5.0%, respectively. The experimental results indicated that the optimized design is successful.

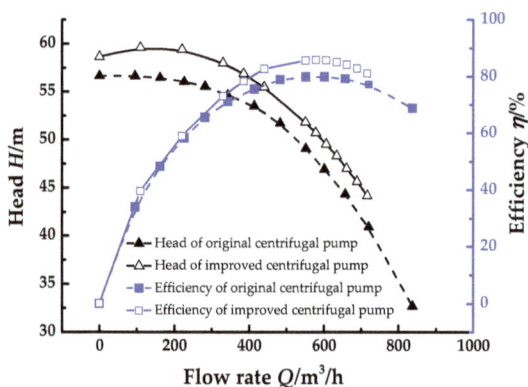

Figure 25. Hydraulic performance of improved and original centrifugal pump.

8. Conclusions

In this paper, effects of blade exit angle and blade wrap angle on the optimized design of the impeller were comprehensively investigated. Flows in the centrifugal pump with five different impellers under low flow rate, rated flow rate, and high flow rate were numerically simulated. Variations of static pressure, relative velocity, streamlines, and turbulent kinetic energy were analyzed. Head and efficiency of the centrifugal pump of different schemes were compared. The experimental head and efficiency of the centrifugal pump with the best impeller were compared with that of the original impeller. The main conclusions are as follows:

(1) The impeller with blade wrap angle 126° and blade exit angle 24° was the best one.

(2) For the best impeller, static pressure and relative velocity was the most uniform distribution. Streamlines were the smoothest and vortices did not exist. Compared with other impellers, the distribution scope of turbulent kinetic energy in the best impeller under all flow rate conditions was the smallest.

(3) For the centrifugal pump with optimized impeller, head and efficiency were higher than that of the original pump. With low flow rate (0.8 Q_d), the head and efficiency increased by 3.76 m and 3.84%. With rated flow rate, the head and efficiency increased by 2.74 m and 5.77%. With high flow rate (1.2 Q_d), the head and efficiency increased by 2.67 m and 5.0%.

Author Contributions: X.H. and Y.K. presented the optimal scheme and designed the experiments. X.H. and D.L. made the numerical simulation and performed the experiment. W.Z. analyzed the data. X.H. wrote the paper.

Acknowledgments: This research is financially supported by the National Key Basic Research Program of China (no. 2014CB239203), the National Natural Science Foundation of China (no. 51474158), and the Hubei Provincial Natural Science Foundation of China (no. 2016CFA088). We deeply acknowledge the help of Hubei Key Laboratory of Waterjet Theory and New Technology, School of Energy and Power Engineering of Lanzhou University of Technology, and Shanghai Kaiquan Group.

Conflicts of Interest: The authors declare no conflict of interest.

References

1. Yedidiah, S. *Centrifugal Pump User's Guidebook: Problems and Solutions*; Springer Group: Berlin, Germany, 1996.
2. Gülich, J.F. *Centrifugal Pumps*; Springer Group: Berlin, Germany, 2008.
3. Karassik, I.J. *Centrifugal Pump Clinic*; Taylor Francis Inc.: Oxford, UK, 1989.
4. Tuzson, J. *Centrifugal Pump Design*; Wiley-Interscience: Washington, DC, USA, 2000.
5. Grist, E. *Cavitation and the Centrifugal Pump: A Guide for Pump Users*; Taylor Francis Inc.: Oxford, UK, 1998.
6. Lobanoff, V.S.; Ross, R.R. *Centrifugal Pump: Design and Application*; Gulf Professional Publishing: Houston, TX, USA, 1992.
7. Girdhar, P.; Moniz, O. *Practical Centrifugal Pumps. Design, Operation, Maintenance*; Newnes: Amsterdam, The Netherlands, 2005.
8. Wu, Z.H. Three-dimensional turbomachine flow equations expressed with respect to non-orthogonal curvilinear coordinates and non-orthogonal velocity components and methods of solution. *J. Mech. Eng.* **1979**, *15*, 1–23.
9. Zhang, Z.H. Streamline similarity method for flow distribution and shock losses at the impeller inlet of the centrifugal pump. *J. Hydrodyn.* **2018**, *30*, 140–152. [CrossRef]
10. Pei, J.; Wang, W.J.; Yuan, S.Q.; Zhang, J.F. Optimization on the impeller of a low-specific-speed centrifugal pump for hydraulic performance improvement. *Chin. J. Mech. Eng.* **2017**, *29*, 992–1002. [CrossRef]
11. Tan, L.; Zhu, B.S.; Cao, S.L.; Bing, H.; Wang, Y.M. Influence of blade wrap angel on centrifugal pump performance by numerical and experimental study. *Chin. J. Mech. Eng.* **2014**, *27*, 171–177. [CrossRef]
12. Kim, J.H.; Lee, H.C.; Kim, J.H.; Kim, S.; Yoon, J.Y.; Choi, Y.S. Design techniques to improve the performance of a centrifugal pump using CFD. *J. Mech. Sci. Technol.* **2015**, *29*, 215–225. [CrossRef]
13. Stepanoff, A.J. *Centrifugal and Axial Pump: Theory, Design and Application*; John Wiley Sons Inc.: Hoboken, NJ, USA, 1986.
14. Yang, A.L.; Lang, D.P.; Li, G.P.; Chen, E.Y.; Dai, R. Numerical research about influence of blade outlet angel on flow-induced noise and vibration for centrifugal pump. *Adv. Mech. Eng.* **2015**, *6*, 1–11.
15. Heo, M.W.; Kim, J.H.; Seo, T.W.; Kim, K.Y. Aerodynamic and aeroacoustic optimization for design of a forward-curved blades centrifugal fan. *Proc. Inst. Mech. Eng. Part A J. Power Energy* **2017**, *230*, 154–174. [CrossRef]
16. Guan, X.F. *Modern Pumps Theory and Design*; China Astronautic Publishing House: Beijing, China, 2011. (In Chinese)
17. Bai, Y.X.; Kong, F.Y.; Yang, S.S.; Chen, K.; Dai, T. Effect of blade wrap angel in hydraulic turbine with forward-curved blades. *Int. J. Hydrogen Energy* **2017**, *42*, 18709–18717. [CrossRef]
18. Wang, W.J.; Pei, J.; Yuan, S.Q.; Zhang, J.F.; Yuan, J.P.; Xu, C.Z. Application of different surrogate models on the optimization of centrifugal pump. *J. Mech. Sci. Technol.* **2016**, *30*, 567–574. [CrossRef]
19. Zhou, L.; Shi, W.D.; Wu, S.Q. Performance optimization in a centrifugal pump impeller by orthogonal experiment and numerical simulation. *Adv. Mech. Eng.* **2013**, *6*, 1–11. [CrossRef]
20. Nelik, L. *Centrifugal and Rotary Pumps: Fundamentals with Applications*; CRC Press: London, UK, 1999.
21. Li, W.G.; Su, F.Z.; Ye, Z.M.; Xia, D.L. Experiment on effect of blade pattern on performance of centrifugal oil pumps. *Chin. J. Appl. Mech.* **2002**, *19*, 31–34. [CrossRef]
22. Lohner, R. *Applied Computational Fluid Dynamics Techniques: An Introduction Based on Finite Element Methods*; John Wiley & Sons: Hoboken, NJ, USA, 2008.
23. Yakhot, V.; Orzag, S.A. Renormalization group analysis of turbulence: Basic theory. *J. Sci. Comput.* **1986**, *1*, 3–51. [CrossRef]
24. Zhou, L.; Shi, W.D.; Li, W.; Agarwal, R. Numerical and experimental study of axial force and hydraulic performance in a deep-well centrifugal pump with different impeller rear shroud radius. *J. Fluid Eng.* **2013**, *135*, 749–760. [CrossRef]
25. Posa, A.; Lippolis, A. A LES investigation of off-design performance of a centrifugal pump with variable-geometry diffuser. *Int. J. Heat Fluid Flow* **2018**, *70*, 299–314. [CrossRef]
26. Chen, H.X.; He, J.W.; Liu, C. Design and experiment of the centrifugal pump impellers with twisted inlet vice blades. *J. Hydrodyn.* **2017**, *29*, 1085–1088. [CrossRef]
27. Yang, S.L.; Kong, F.Y.; Chen, H.; Su, X.H. Effects of blade wrap angel influencing a pump as turbine. *J. Fluid Eng.* **2012**, *134*, 1–8. [CrossRef]

28. Cheah, K.W.; Lee, T.S.; Winoto, S.H.; Zhao, Z.M. Numerical flow simulation in a centrifugal pump at design and off-design conditions. *Int. J. Rotating Mach.* **2007**, *2*, 1–9. [CrossRef]

29. Ferziger, J.H.; Peric, M. *Computational Methods for Fluid Dynamics*; Springer Group: Berlin, Germany, 2002.

energies

MDPI

Article

Numerical Study on the Transient Thermal Performance of a Two-Phase Closed Thermosyphon

Zhongchao Zhao *, Yong Zhang, Yanrui Zhang, Yimeng Zhou and Hao Hu

School of Energy and Power, Jiangsu University of Science and Technology, Zhenjiang 212000, China; 18896658850@163.com (Y.Z.); zyr1056187930@163.com (Y.Z.); 18851406053@163.com (Y.Z.); 00001111zy@sina.com (H.H.)

* Correspondence: zhongchaozhao@just.edu.cn; Tel.: +86-0511-8449-3050

Received: 5 May 2018; Accepted: 28 May 2018; Published: 3 June 2018

check for updates

Abstract: The transient thermal performance of phase change and heat and mass transfer in a two-phase closed thermosyphon are studied with computational fluid dynamics (CFD). A CFD model based on the volume of fluid technique is built. Deionized water is specified as the working fluid of this thermosyphon. The CFD model reproduces evaporation and condensation in the thermosyphon at different heating inputs. The average wall temperatures are also analyzed. Variations of average wall temperatures indicate that this thermosyphon reaches a steady state after 19 s, and starts to work in advance when the heating input increases. Moreover, thermal resistance is decreased until a minimum (0.552 K/W) by increasing the heating input, and the effective thermal conductivity is elevated to a maximum (2.07×106 W/m·K).

Keywords: thermosyphon; volume of fluid; multiphase flow; evaporation and condensation

1. Introduction

A two-phase closed thermosyphon, like a wicked heat pipe, transmits heat by the evaporation and condensation of a working fluid that is circulating in a sealed container. However, the thermosyphon relies on the force of gravity to return the working fluid from a condenser to an evaporator rather than relying on the capillary forces produced by the wick in the wicked heat pipe [1]. As a kind of high-efficiency heat transfer component, a thermosyphon, whose internal complex heat and mass transfer is the focus of this research about the heat transfer mechanism of a heat pipe, has already been widely used in the field of heat exchange. Thermosyphon technology is playing an important role in many industrial applications, particularly in the heat transfer of heat exchangers and in energy savings in applications. For the advantages of its light weight, wide operating temperature range, compact structure, flexibility, high capacity of heat transfer, and great isothermal performance, the thermosyphon has been widely used in many cooling fields that contain electronic components and products [2–4].

A thermosyphon consists of a condenser, an adiabatic section, and an evaporator section. The evaporator section is heated by a hot source, and the condenser section is cooled by a cold source. The adiabatic section placed between the condenser and evaporator sections is surrounded with thermal insulation layers. Heat is absorbed by the evaporator section with a liquid pool, which is situated at the bottom of the thermosyphon, following which liquid that absorbs the latent heat of evaporation is turned into vapor. Afterwards, vapor rises up to the condenser section at the top of the thermosyphon, where it is condensed and gives up latent heat. The condensed working fluid, which forms a liquid film along the walls, moves back to the evaporator under gravity.

A great variety of experiments have been done to understand the thermal characteristics of the thermosyphon [5–11], although with an unknown internal heat and mass transfer mechanism.

Annamalai and Ramalingam [12] built a model for a wicked heat pipe by using ANSYS CFX and performing an experimental study. They assumed that the section inside the heat pipe was filled by the vapor phase, and the liquid phase only occurred in the wick section. The wall temperature of the condenser and evaporator and the vapor temperature were compared. Legierski et al. [13] provided a model for simulating the temperature along the heat pipe walls at different moments and the vapor velocity inside the heat pipe. They considered the heat pipe as an open system, wherein the evaporator was treated as the inlet of mass flow and the condenser was treated as the outlet. The start-up time of the heat pipe was 20~30 s, and the evaporation coefficient of water was 1.58×10^{-3}. The range of effective thermal conductivity was also provided.

De Schepper et al. [14] employed the user-defined function (UDF) and volume of fluid (VOF) techniques to calculate the phase-change of a hydrocarbon feedstock. They proposed that UDFs were specified as source terms, which were used to simulate heating and boiling in the convection section of a steam cracker. Lin et al. [15] carried out a computational fluid dynamics (CFD) analysis of miniature oscillating heat pipes using VOF and mixture methods. Alizadehdakhel et al. [16] performed an experimental study under different operating conditions in a thermosyphon. They also created a CFD model to compare the predicted temperature profile with experimental results that employed the VOF model.

Fadhl et al. [17] carried out an experimental study and CFD analysis of a two-phase closed thermosyphon using the VOF technique. The experimental investigation was performed with the input heat powers of 100.41, 172.87, 225.25, 275.6, 299.52, and 376.14 W. The temperature profile predicted by CFD was consistent with the experimental data at the same input heat powers. They then [18] established a two-dimensional model with three working fluids, i.e., distilled water, R134a, and R404a. The refrigerants (R134a and R404a) and distilled water showed different boiling phenomena in the thermosyphon.

Most currently available studies only focused on the steady state thermal performance of the thermosyphon, but numerical investigation of the thermal performance during start-up remains limited. Therefore, the purpose of this study is to simulate a vapor–liquid two-phase flow, evaporation and condensation, and other complex physical processes during the operation of a closed thermosyphon. The VOF technique was employed to develop a CFD model to track the liquid–vapor interface. In order to calculate the phase change, the mass and energy sources compiled by UDFs were used to complete the FLUENT code.

2. CFD Modeling

The objective of this study was to analyze the mechanism of phase change and the thermal performance of working fluids in a two-phase closed thermosyphon. There are three main methods for the numerical calculations of Euler–Euler multiphase flows, which are the VOF method, the mixture method, and the Eulerian method. The mixture method and the Eulerian method are applicable when the volume fractions of dispersed-phase exceed 10%, wherein the phase is mixed or separated in the flows. As the phase is immiscible with multiphase flows in the thermosyphon, the VOF method was adopted as it can simulate either stratified flow or free surface flow.

2.1. VOF Model

A single set of Navier–Stokes equations in the VOF model were applied to immiscible phases, and the volume fraction of each phase was recorded into each cell throughout the computation domain. Thus, the summation of volume fraction of all phases is unity:

$$\alpha_l + \alpha_v = 1 \tag{1}$$

The cell is occupied by liquid phase if $\alpha_l = 1$ and by vapor phase if $\alpha_v = 0$. The relationship of $0 < \alpha_l < 1$ represents that the cell is at the interface between liquid and vapor phases. The continuity equations for the VOF model have the following forms:

$$\nabla \cdot \left(\alpha_l \rho_l \vec{u}\right) = -\frac{\partial}{\partial t}(\alpha_l \rho_l) + M_l \tag{2}$$

$$\nabla \cdot \left(\alpha_v \rho_v \vec{u}\right) = -\frac{\partial}{\partial t}(\alpha_v \rho_v) + M_v \tag{3}$$

where ρ_l is the density of liquid phase and ρ_v is the density of vapor phase.

The continuity equations treat velocity u as the mass-averaged velocity. Meanwhile, the rates of mass transfer passing the two-phase interface are represented as M_l and M_v.

Considering the effect of gravity and volumetric surface tension, the momentum equation for the VOF model is described as follows:

$$\frac{\partial}{\partial t}\left(\rho \vec{u}\right) + \nabla \cdot \left(\rho \vec{u}\vec{u}\right) = \rho \vec{g} - \nabla p + \nabla \cdot \left[\mu\left(\nabla \vec{u} + \nabla \vec{u}^T\right) - \frac{2}{3}\mu \nabla \cdot u I\right] + F_{CSF} \tag{4}$$

where g represents the acceleration of gravity, p represents the pressure, and I represents the unit tensor.

In Equation (4), the volumetric surface tension force is calculated by the continuum surface force (CSF) model proposed by Brackbill et al. [19]. Therefore, the surface tension, as a source term, has the following form:

$$F_{CSF} = 2\sigma_{vl}\frac{\alpha_l \rho_l k_v \nabla \alpha_v + \alpha_v \rho_v k_l \nabla \alpha_l}{(\rho_l + \rho_v)} \tag{5}$$

where σ_{vl} is the surface tension coefficient and k is the surface curvature.

In the momentum equation, density and viscosity rely on the volume fraction of the phases. Therefore, density ρ and the dynamic viscosity μ are expressed as follows:

$$\rho = a_l \rho_l + a_v \rho_v \tag{6}$$

$$\mu = a_l \mu_l + a_v \mu_v \tag{7}$$

where μ_l and μ_v are the dynamic viscosities of liquid and vapor, respectively.

The energy equation for the VOF model has the following form:

$$\frac{\partial}{\partial t}(\rho e) + \nabla \cdot \left(\vec{v}(\rho e + p)\right) = \nabla \cdot (k\nabla T) + E \tag{8}$$

where $k = \alpha_l k_l + \alpha_v k_v$. The energy source term used to compute the rates of heat transfer through the interface is represented as E.

In the VOF model, T is the mixture temperature rather than the temperature of a specific phase for the energy equation. Meanwhile, the internal energy e treated as a mass-averaged variable is given by the following:

$$e = \frac{\alpha_l \rho_l e_l + \alpha_v \rho_v e_v}{\alpha_l \rho_l + \alpha_v \rho_v} \tag{9}$$

where e_l and e_v are calculated on the basis of the specific heat c_p of the phase and the mixture temperature:

$$e_l = c_{p,l}(T - T_{sat}) \tag{10}$$

$$e_v = c_{p,v}(T - T_{sat}) \tag{11}$$

2.2. Mass and Heat Transfer Model

FLUENT 6.3 [20] was used to solve the governing equations of the VOF model. UDF was linked to the governing equations, which simulated the evaporation and condensation in the thermosyphon. In UDF, the mass and heat transfer rates are defined by source terms proposed by De Schepper et al. [14]. Therefore, all source terms are required to calculate the mass and heat transfer rates in the following forms:

For mass source terms in the evaporation process

$$M_l = -C\rho_l \alpha_l \frac{T_{mix} - T_{sat}}{T_{sat}} \tag{12}$$

$$M_v = -(S_{Ml}) \tag{13}$$

For mass source terms in the condensation process:

$$M_l = C\rho_v \alpha_v \frac{T_{sat} - T_{mix}}{T_{sat}} \tag{14}$$

$$M_v = -(S_{Ml}) \tag{15}$$

For energy source terms in the evaporation and condensation processes, respectively:

$$E_e = \left(-C\rho_l \alpha_l \frac{T_{mix} - T_{sat}}{T_{sat}}\right) \cdot LH \tag{16}$$

$$E_c = \left(C\rho_v \alpha_v \frac{T_{sat} - T_{mix}}{T_{sat}}\right) \cdot LH \tag{17}$$

where T_{mix} is the mixture temperature, T_{sat} is the saturation temperature, and LH is the latent heat. Coefficient C, used to compute the condensation/evaporation rate, is normally specified as 0.1 [17].

3. Geometry and Boundary Condition

A two-dimensional numerical model was built for simulating the two-phase flow, taking the axisymmetric structure of a thermosyphon into consideration. The geometry of the computational domain (Figure 1) is divided into the condenser, adiabatic, and evaporator sections. Furthermore, the lengths of evaporator and condenser sections are 100 mm, while the adiabatic section is fixed at 50 mm. The inner diameter and wall thickness are 8.32 and 0.6 mm, respectively. The grids were created with GAMBIT 2.4.6, and the type of grid was Quad. The grid independence was tested by different mesh sizes, and the average temperature of the evaporator (Te.av) and condenser (Tc.av) sections for different mesh sizes were monitored. For the water-charged thermosyphon and the heating power of 40 W, it was found that both Te.av and Tc.av were almost the same for a different number of grids, such as 156,900, 80,000, and 58,045.

Thus, the grid quantity of 80,000 was selected for the simulation analysis, and it was adopted in order to ensure the grid independence. Eight layers of cells were meshed, giving the initial size and growth factor of 0.035 and 1.2, respectively. A very thin liquid film was developed near the inner walls of the thermosyphon, and the boundary layer technique was applied on the left and right inner walls. In fluid dynamics, there is zero velocity between the fluid and the solid boundary, relatively. Therefore, a non-slip boundary condition was applied on the inner walls. A constant heat flux was specified as the boundary condition for the evaporator section, and a convection heat transfer coefficient was specified as the boundary condition for the condenser section. The heat flux, as the boundary condition for the adiabatic section, was zero, considering this section was insulated. Furthermore, water was adopted as the working fluid and fixed at 60% of the evaporator volume, and the density depending on the temperature in this model is defined by the steam table in the following form:

$$\rho_l = 249.46 + 6.625 \times T - 0.0184 \times T^2 + 1.532 \times 10^{-5} \times T^3 \tag{18}$$

where T represents the mixture temperature.

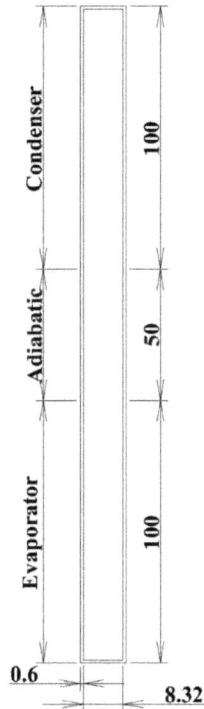

Figure 1. Model geometry and dimensions.

4. CFD Solution Method and Model Validation

Taking the start-up and steady-state operation into consideration, the whole simulation process of the thermosyphon is transient with a time step of 0.0001 s. In order to solve the pressure–velocity coupling, the SIMPLE algorithm was selected from the segregated algorithms provided by FLUENT. The first order upwind was adopted to discretize the momentum and energy equations, which reduced the difficulty of convergence. Meanwhile, Geo-Reconstruct for the volume fraction combined with PRESTO for the pressure was included in this model. In the present simulation, the scaled residual of mass and velocity components should be 10^{-4} for the convergence of numerical computation. Water vapor was taken as the primary (vapor) phase, and the saturation temperature was 308.55 K.

For validating the reliability of the numerical model and method, the experimental data should be compared with the results of the simulation. The thermosyphon was made of copper with the internal and external diameters of 8.32 mm and 9.52 mm, respectively (Figure 2a). The lengths of evaporator and condenser sections were designed to be 100 mm, and the length of the adiabatic section was designed to be 50 mm. The whole experimental apparatus consisted of a thermosyphon, heating system, cooling system, vacuum pumping system, and data acquisition system (Figure 2b). The filling ratio, which means the ratio of volume of liquid to that of thermosyphon, was 24%, and the working fluid was 3.26 g deionized water. The experiments were performed at the heating input powers of 40 W, 60 W, and 80 W. Three of the thermocouples were sited on the evaporator section, one on the adiabatic section, and two on the condenser section (one at the inlet and one at the outlet of the cooling water jacket, respectively).

Figure 3 shows the comparisons of the simulation and experimental results of the wall temperature of the evaporator and condenser during the start-up process at the heating input power of 40 W. The experimental data showed the same trend as the simulation results of the VOF model, and the start-up time is nearly equal. The maximum error of temperature difference of the evaporator and condenser between the simulation and the experiment are 0.53% and 1.02%, respectively. Considering the influence of the experimental environment, the errors are acceptable. Therefore, the numerical data were in good fit to the experimental data, and the numerical model and method were reliable.

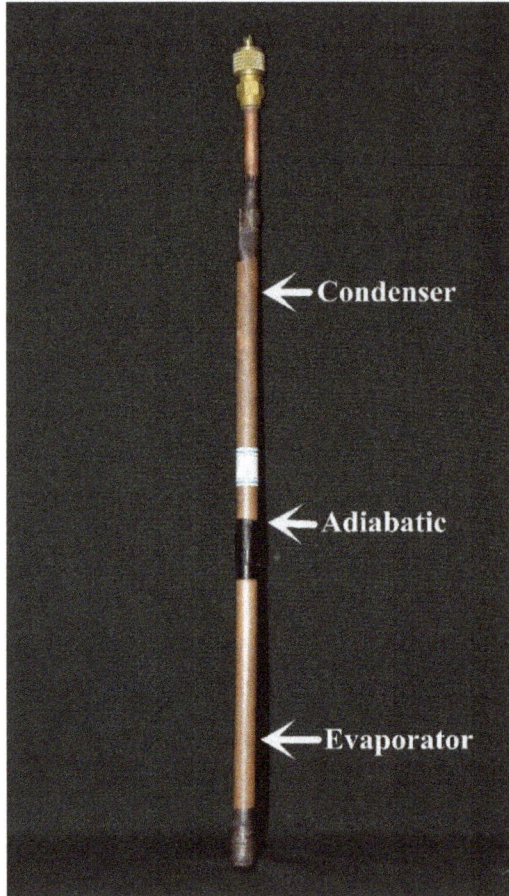

(a)

Figure 2. *Cont.*

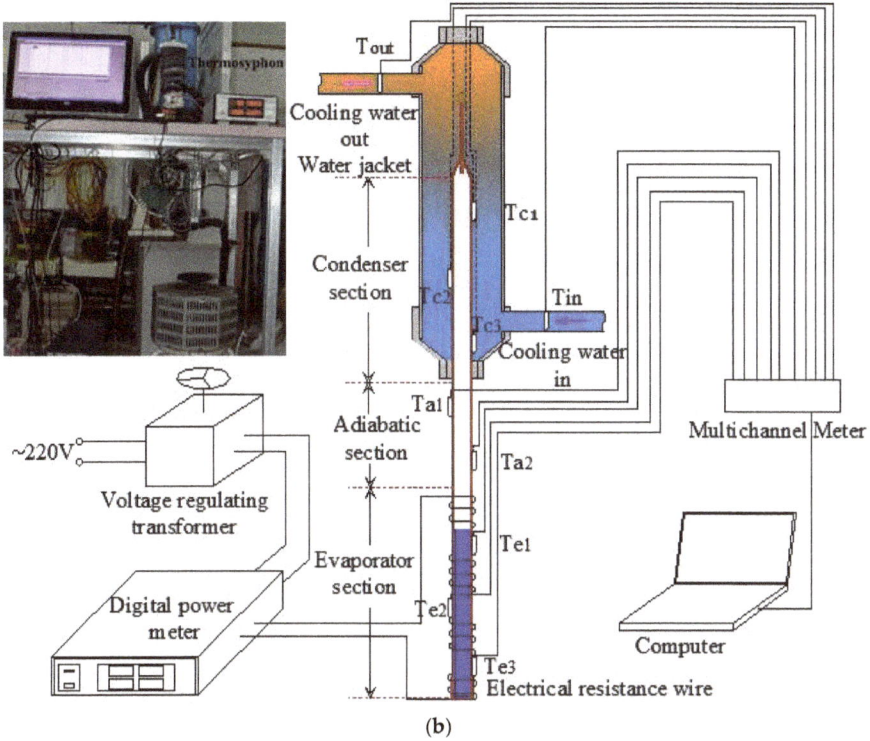

Figure 2. (a) The real thermosyphon and (b) Schematic diagram of experimental apparatus.

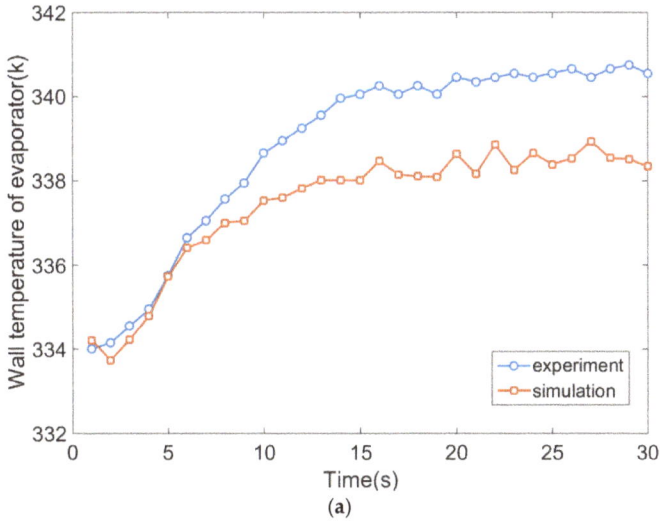

Figure 3. *Cont.*

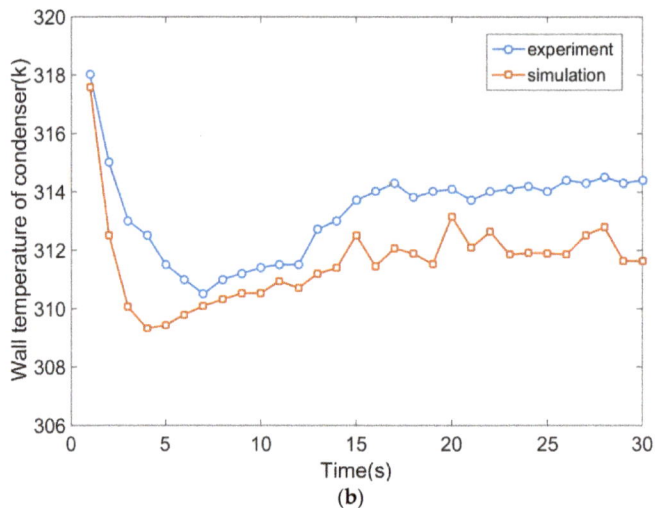

Figure 3. Start-up characteristics comparison of the simulation and experimental wall temperature of (**a**) Evaporator and (**b**) Condenser at the heating input power of 40 W.

5. Results and Discussion

The numerical simulation was performed to observe the evaporation and condensation phenomenon in the thermosyphon. Furthermore, the wall temperature profile, the thermal resistance, and the effective thermal conductivity were obtained during the start-up of the thermosyphon under heating inputs of 40 W, 60 W, and 80 W.

5.1. The Evaporation and Condensation Process of the Numerical Simulation of the Thermosyphon

Pool boiling and vapor-liquid distribution in the evaporator section and the liquid film in the lower region of the condenser section are visualized in Figures 4 and 5, respectively. The blue color represents that the liquid volume fraction is 1, indicating that only liquid exists. As evidenced by the liquid volume fraction (value: 0) represented by the red color, only vapor exists. The evaporator section is represented by "We" in Figure 4, 60% of which is occupied by the liquid pool at 0.1 s.

The heating power transferred the heat into the liquid pool through the walls of the evaporator section. Evaporation occurred at the nucleation site, where liquid reached the saturation temperature at 0.3 s. Vapor bubbles formed and departed from the inner walls of the evaporator section at 0.5 s. As the heating input increased, more vapor bubbles formed and then started to grow upwards and to coalesce, at 1 s and 1.5 s respectively. Finally, the vapor bubbles rose up to the interface between the liquid phase and vapor phase, where they broke up and passed the vapor phase. Moreover, the quantities, size, and shape of the vapor bubbles changed along with the heating input.

Conversely, saturated vapor condensed when it rose up to the condenser section through the adiabatic section. The adiabatic and condenser sections are represented by "Wa" and "Wc" in Figure 5. Condensed liquid film formed on the inner walls of the condenser section, where heat was transferred to the external environment. At the beginning, the film was discontinuous at 5 s, 6 s, 7 s, and 8 s, and the heating input was 40 W (Figure 5a). Based on continuous transportation of vapor from the evaporator section, more liquid was condensed on the inner walls of the condenser section. Eventually, at the heating input of 40 W, a continuous liquid film formed (visualized in Figure 5a) at 9 s and 10 s, and it returned to the liquid pool of the evaporator section through the adiabatic section under gravity.

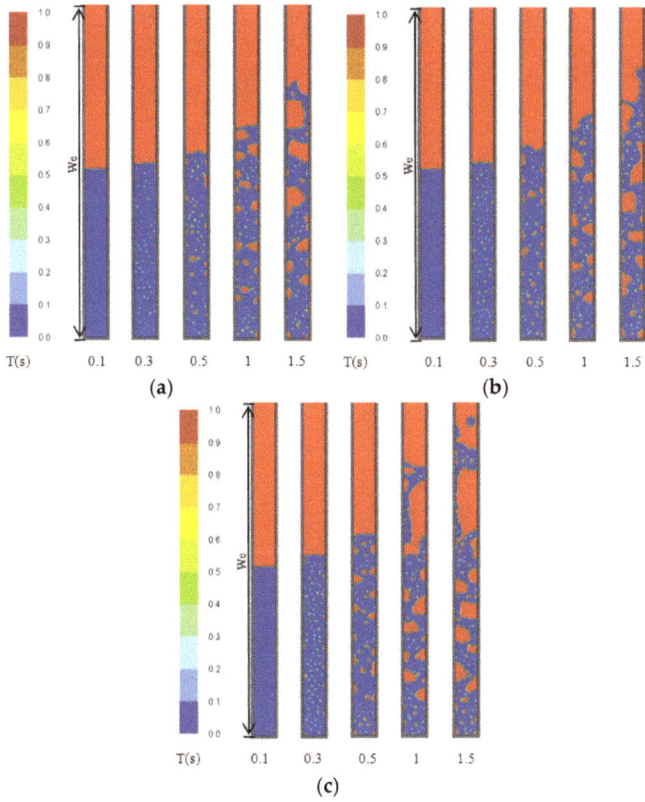

Figure 4. Vapor–liquid distribution of the evaporator section at different heating input powers of (**a**) Q = 40 W, (**b**) Q = 60 W, and (**c**) Q = 80 W.

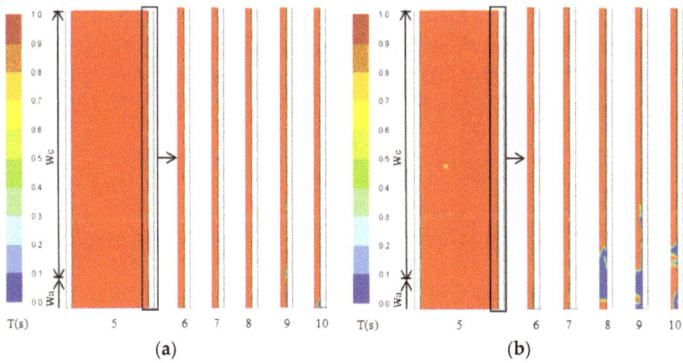

Figure 5. *Cont.*

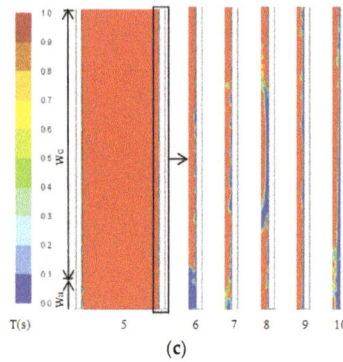

(c)

Figure 5. Liquid film in the lower region of condenser section at different heating input powers of (a) Q = 40 W, (b) Q = 60 W, and (c) Q = 80 W.

Apparently, the time required by the formation of a continuous liquid film is shortened by increasing the heating input power. Meanwhile, the thickness of the liquid film is increased by raising the heating input power at 10 s. More vapor appears and reaches the condenser section with rising heating input power, which increases the condensed liquid and accelerates the formation of a continuous liquid film. The simulation results prove that this numerical model can reproduce the difference in pool-boiling and film-wise condensation between different heating inputs.

5.2. The Wall Temperature Profile of the Numerical Simulation of the Thermosyphon

The average wall temperatures, versus time, for evaporator, adiabatic, and condenser sections are specified in Figure 6. With a rising power input, the average wall temperature of the evaporator section increased between 0 s and 19 s (Figure 6a). By the heat dissipation of the condenser section, the wall temperature of the evaporator section was relatively constant after 19 s, indicating that the thermosyphon needs 19 s to reach stable operation. The wall temperature of the evaporator also increased by elevating the heating input. Meanwhile, at the heating input of 80 W, its wall temperature volatility was also increased. Probably, geyser boiling, which was caused by the larger bubbles in Figure 4c, reduced the stability of the thermosyphon. Compared to the evaporator section, the average wall temperature of the adiabatic section barely varies (Figure 6b). Therefore, the center temperature of the adiabatic section can be specified as the saturation temperature of the thermosyphon.

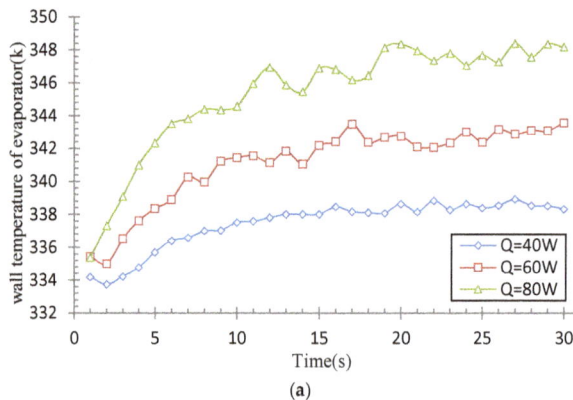

(a)

Figure 6. *Cont.*

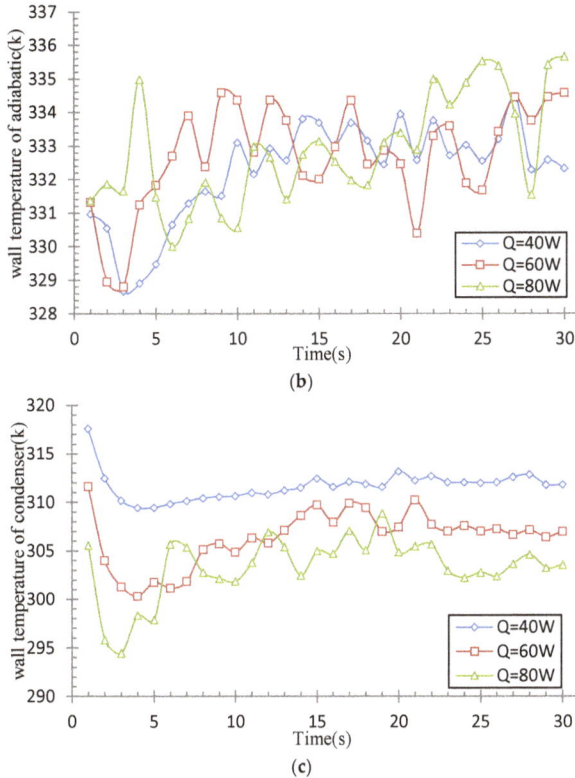

Figure 6. Average wall temperature profiles for (**a**) Evaporator, (**b**) Adiabatic, and (**c**) Condenser sections at different heating input powers.

As shown in Figure 6c, the average wall temperature of the condenser section temporarily decreased from 0 s to 4 s at the heating input of 40 W and 60 W, because a constant heat flux boundary condition was applied on the walls of the condenser section at the beginning of the numerical simulation. The heat of the vapor phase region was transferred to the external environment while that generated by boiling had not yet been transferred to the condenser section. The average wall temperature of the condenser section began to increase when the heat generated by boiling started after 4 s at the heating inputs of 40 W and 60 W. Therefore, this thermosyphon started to work at 4 s at the heating inputs of 40 W and 60 W. In addition, the condenser section started to increase at 3 s at the heating input of 80 W, suggesting that the thermosyphon can work in advance by increasing the heating input.

5.3. The Thermal Resistance of the Numerical Simulation of the Thermosyphon

The thermal resistance of the whole thermosyphon can be defined as follows:

$$R_{ave} = \frac{(T_e - T_c)}{Q_e} \tag{19}$$

where T_e and T_c are the average wall temperatures of the evaporator and the condenser, respectively, and Q_e represents the power input of the evaporator. Based on Figure 7, the thermal resistance increases with extended time and then gradually levels off. In addition, the minimum of thermal resistance is 0.552 K/W, and its decrease is slowed down with increasing heating input (Table 1). r_1 and r_2 have the following forms:

$$r_1 = \frac{R_{40} - R_{60}}{R_{40}} \times 100\% \tag{20}$$

$$r_2 = \frac{R_{60} - R_{80}}{R_{60}} \times 100\% \tag{21}$$

where R_{40}, R_{60}, and R_{80} represent the thermal resistances at the heating inputs of 40 W, 60 W, and 80 W, respectively. The thermosyphon thermal resistances at various power inputs have been compared in this paper; the thermosyphon thermal resistance gradually decreased by increasing the power inputs.

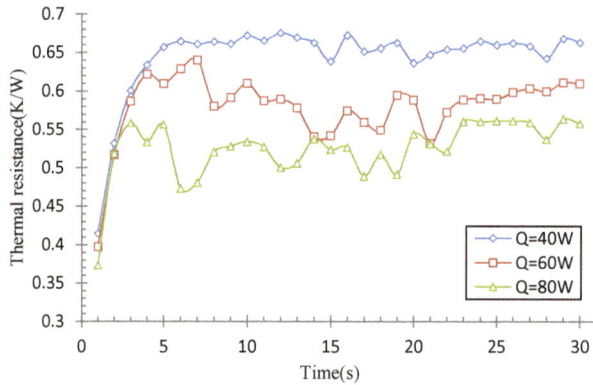

Figure 7. Thermal resistance variations of the thermosyphon at different heating input powers.

Table 1. Thermal resistances of heat pipes at different input powers.

	40 W	60 W	80 W	r_1 (%)	r_2 (%)
R (K/W)	0.656	0.582	0.552	11.3	5.15

5.4. The Effective Thermal Conductivity of the Numerical Simulation of the Thermosyphon

To evaluate the thermal performance of the thermosyphon filled with working fluid, the heat transfer capability is treated as a solid cylinder with equal size. The thermal conductivity of this solid cylinder is specified as the effective one of the thermosyphon with the following correlation:

$$\lambda_{eff} = \frac{4L^2}{d_o^2}[1/(\frac{\ln(d_o/d_i)}{2(L_e/L)} + \frac{\lambda_w}{h_e d_o(L_e/L)} + \frac{\lambda_w}{h_c d_o(L_c/L)} + \frac{\ln(d_o/d_i)}{2(L_c/L)})]\lambda_w \tag{22}$$

where L, L_e, and L_c are the lengths of the total thermosyphon, evaporator, and condenser respectively; d_o and d_i are the outer diameter and inner diameter respectively; and λ_w is the thermal conductivity of the wall. h_e and h_c represent the heat transfer coefficients of the evaporator and the condenser with the following forms, respectively:

$$h_e = \frac{Q}{\pi d_o L_e(T_e - T_s)} \tag{23}$$

$$h_c = \frac{Q}{\pi d_o L_c(T_s - T_c)} \tag{24}$$

where T_e and T_c are the average wall temperatures of the evaporator and the condenser, respectively, and Q represents the power input.

The effective thermal conductivity, versus time, is displayed in Figure 8. Apparently, it is highest initially, and it then decreases with prolonged time. Eventually, the effective thermal conductivity hardly fluctuates. As shown in Table 2, with r_3 and r_4 calculated by (17-1) and (17-2), the effective thermal conductivity increases to a maximum (2.07×106 W/m·K^{-1}) with the rising of the heat flux, but the increment amplitude changes little.

$$r_3 = \frac{\lambda_{eff60} - \lambda_{eff40}}{\lambda_{eff40}} \times 100\% \tag{25}$$

$$r_4 = \frac{\lambda_{eff80} - \lambda_{eff60}}{\lambda_{eff60}} \times 100\% \tag{26}$$

where λ_{eff40}, λ_{eff60}, and λ_{eff80} represent the thermal resistances at the heating inputs of 40 W, 60 W, and 80 W, respectively.

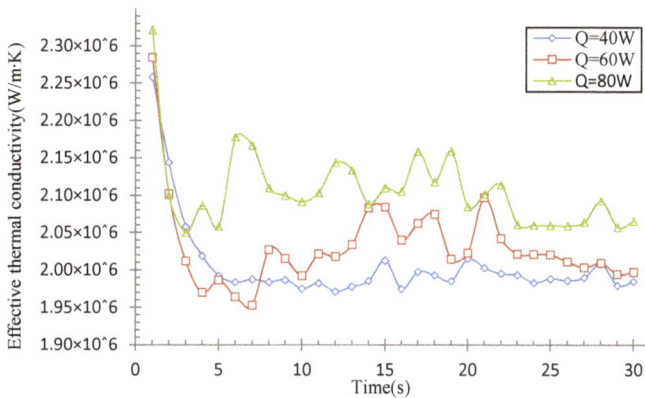

Figure 8. Effective thermal conductivity variations of the thermosyphon at different heating input power.

Table 2. Effective thermal conductivities of heat pipes at different input powers.

	40 W	60 W	80 W	r_3 (%)	r_4 (%)
λ_{eff} (W/m · k)	1.993×10^6	2.032×10^6	2.072×10^6	1.96	1.97

6. Conclusions

A transient two-dimensional model was built with the VOF technique to simulate the complex phenomenon in a thermosyphon. In order to calculate the heat and mass transfer, UDFs were associated with the governing equations of FLUENT by source terms. The CFD results confirmed that this model managed to reproduce pool boiling in the evaporator section and the formation of liquid film in the condenser section during the start-up of thermosyphon. Moreover, the simulation results showed that the quantities, size, shape, and position of vapor bubbles produced by boiling in the evaporator changed along with the heating power at the same time. With rising heating power, the time required by the formation of continuous liquid film was shortened, and the thickness also increased.

The average wall temperatures for condenser, adiabatic, and evaporator sections were investigated at different heating inputs during the operation of the thermosyphon. The prediction results showed that the thermosyphon reached a steady state after 19 s. Variations of the condenser wall temperature suggested that the heat pipes filled with pure water started to work in advance at appropriate heating power, which was around 3 s in the fastest case. We can increase heating power properly to reduce the start-up time and increase the effective thermal conductivity of the thermosyphon.

The thermal performance of the thermosyphon was evaluated by studying the thermal resistance and effective thermal conductivity. With increasing heating power, the thermal resistance decreased to a minimum value (0.552 K/W), whereas the effective thermal conductivity rose to a maximum (2.07×106 W/m·K).

Author Contributions: All authors contributed to the paper. Z.Z. designed the research, conceived the idea of the study, and wrote this paper; Y.Z. (Yong Zhang) and Y.Z. (Yanrui Zhang) conducted the numerical simulation and data analyses; Y.Z. (Yimeng Zhou) and H.H. discussed experimental results; all authors performed the experiment.

Acknowledgments: The authors gratefully acknowledge the support of the Jiangsu Marine and Fishery Science and Technology Innovation and Extension project (HY2017-8) and Zhenjiang funds for the Key Research and Development project (GY2016002-1).

Conflicts of Interest: The authors declare no potential conflicts of interest with respect to the research, authorship, and/or publication of this article.

Nomenclature

C	Constant of evaporation/condensation
d	Diameter (m)
E	Energy source term (W/m^3)
e	Internal energy (J/kg)
F_{CSF}	Volumetric surface tension force (N)
g	Acceleration of gravity (m/s^2)
I	Unit tensor
k	Surface curvature (1/m)
LH	Latent heat (J/kg)
L	Length (m)
M	Mass transfer rate (kg/s)
p	Pressure (Pa)
T	Temperature (K)
Greek symbols	
α	Volume fraction
λ	Thermal conductivity (W/m K)
μ	Dynamic viscosity (Pa s)
ρ	Density (kg/m^3)
σ_{vl}	Surface tension coefficient (N/m)
Subscript	
CSF	Continuum surface force
c	Condenser
e	Evaporator
eff	Effective
i	Inner
l	Liquid phase
mix	Mixture
o	Out
sat	Saturation
v	Vapor phase
W	Wall

References

1. Faghri, A. Heat Pipe Science and Technology. *Fuel Energy Abstr.* **1995**, *36*, 285. [CrossRef]
2. Weng, Y.C.; Cho, H.P.; Chang, C.C.; Chen, S.L. Heat pipe with PCM for electronic cooling. *Appl. Energy* **2011**, *88*, 1825–1833. [CrossRef]

3. Vasiliev, L.; Lossouarn, D.; Romestant, C.; Alexandre, A.; Bertin, Y.; Piatsiushyk, Y.; Romanenkov, V. Loop heat pipe for cooling of high-power electronic components. *Int. J. Heat Mass Transf.* **2008**, *52*, 301–308. [CrossRef]
4. Zhang, M.; Liu, Z.; Ma, G.; Cheng, S. Numerical simulation and experimental verification of a flat two-phase thermosyphon. *Energy Convers. Manag.* **2009**, *50*, 1095–1100. [CrossRef]
5. Farsi, H.; Joly, J.L.; Miscevic, M.; Platel, V.; Mazet, N. An experimental and theoretical investigation of the transient behavior of a two-phase closed thermosyphon. *Appl. Therm. Eng.* **2003**, *23*, 1895–1912. [CrossRef]
6. Zhang, P.L.; Wang, B.L.; Shi, W.X.; Li, X.T. Experimental investigation on two-phase thermosyphon loop with partially liquid-filled downcomer. *Appl. Energy* **2015**, *160*, 10–17. [CrossRef]
7. Tong, Z.; Liu, X.H.; Jiang, Y. Three typical operating states of an R744 two-phase thermosyphon loop. *Appl. Energy* **2017**, *206*, 181–192. [CrossRef]
8. Tong, Z.; Liu, X.H.; Jiang, Y. Experimental study of the self-regulating performance of an R744 two-phase thermosyphon loop. *Appl. Energy* **2017**, *186*, 1–12. [CrossRef]
9. Naresh, Y.; Balaji, C. Experimental investigations of heat transfer from an internally finned two phase closed thermosyphon. *Appl. Therm. Eng.* **2017**, *112*, 1658–1666. [CrossRef]
10. Jafari, D.; Di Marco, P.; Filippeschi, S.; Franco, A. An experimental investigation on the evaporation and condensation heat transfer of two-phase closed thermosyphons. *Exp. Therm. Fluid.* **2017**, *88*, 111–123. [CrossRef]
11. Chen, L.; Deng, B.L.; Zhang, X.R. Experimental investigation of CO_2 thermosyphon flow and heat transfer in the supercritical region. *Int. J. Heat Mass Transf.* **2013**, *64*, 202–211. [CrossRef]
12. Annamalai, A.S.; Ramalingam, V. Experimental investigation and computational fluid dynamics analysis of air cooled condenser heat pipe. *Therm. Sci.* **2011**, *15*, 759–772. [CrossRef]
13. Legierski, J.; Wiecek, B.; de Mey, G. Measurements and simulations of transient characteristics of heat pipes. *Microelectron. Reliab.* **2006**, *46*, 109–115. [CrossRef]
14. De Schepper, S.C.K.; Heynderickx, G.J.; Marin, G.B. Modeling the evaporation of a hydrocarbon feedstock in the convection section of a steam cracker. *Comput. Chem. Eng.* **2009**, *33*, 122–132. [CrossRef]
15. Lin, Z.; Wang, S.; Shirakashi, R.; Winston Zhang, L. Simulation of a miniature oscillating heat pipe in bottom heating mode using CFD with unsteady modeling. *Int. J. Heat Mass Transf.* **2013**, *57*, 642–656. [CrossRef]
16. Alizadehdakhel, A.; Rahimi, M.; Alsairafi, A. CFD modeling of flow and heat transfer in a thermosyphon. *Int. Commun. Heat Mass* **2010**, *37*, 312–318. [CrossRef]
17. Fadhl, B.; Wrobel, L.; Jouhara, H. Numerical modeling of the temperature distribution in a two-phase closed thermosyphon. *Appl. Therm. Eng.* **2013**, *60*, 122–131. [CrossRef]
18. Jouhara, H.; Fadhl, B.; Wrobel, L. CFD modeling of a two-phase closed thermosyphon charged with R134a and R404a. *Appl. Therm. Eng.* **2015**, *78*, 482–490. [CrossRef]
19. Brackbill, J.; Kothe, D.; Ca, Z. A continuum method for modeling surface tension. *J. Comput. Phys.* **1995**, *100*, 335–354. [CrossRef]
20. Li, J.L.; Li, C.X.; Hu, R.X. *Master Fluent 6.3 Flow Field Analysis*, 1st ed.; Chemical Industry Press: Beijing, China, 2009; pp. 16–55.

energies

MDPI

Article

An Experimental Study on the Radiation Noise Characteristics of a Centrifugal Pump with Various Working Conditions

Chang Guo [1], Ming Gao [1,*], Dongyue Lu [2] and Kun Wang [1]

[1]   School of Energy and Power Engineering, Shandong University, Jinan 250061, China;
       gg3263@163.com (C.G.); sduwang1993@163.com (K.W.)
[2]   State Nuclear Power Engineering Company, Shanghai 200233, China; ldy11energy@163.com
*   Correspondence: gm@sdu.edu.cn; Tel.: +86-531-8839-9008

Received: 8 November 2017; Accepted: 12 December 2017; Published: 15 December 2017

Abstract: To investigate the radiation noise characteristics of a centrifugal pump under various working conditions, a noise measurement system is established; afterwards, the distribution of different points and intervals, as well as the overall level of noise, are studied. The total sound pressure level distribution for different points manifests the dipole and asymmetric directivity characteristics. Additionally, the acoustic energy is introduced to compare the noise of different intervals to reveal the asymmetric characteristics, and it is found that variation in working conditions has little impact on the acoustic energy distribution, and the ratio of the acoustic energy in the direction facing the tongue, as well as that in the direction against the tongue, to total acoustic energy fluctuate around 0.410 and 0.160, respectively, under various working conditions. Also, the A-weighted average sound pressure level (L_{pA}) is applied to describe the overall level of noise, and L_{pA} increases gradually with the growth of rotational speed, but the growth slope decreases. While in the operation of throttling regulation, L_{pA} shows the trend that first increases, then remains stable, and increases again with the growth of flow rate. This study could provide guidance for optimizing the operating conditions and noise control of centrifugal pumps.

Keywords: centrifugal pump; radiation noise; distribution characteristic; acoustic energy; experimental research

1. Introduction

Widespread concerns about environmental protection are accelerating noise abatement investment, accounting for 15–20% of the environmental protection investment [1]. Centrifugal pumps, widely applied in many fields of national economy [2,3], radiate a lot of noise during operation. The unexpected noise could affect the flow performance and deteriorate the working environment. To establish theoretical basis of radiation noise control technology for centrifugal pumps, this study reveals the changing rules of radiation noise under various working conditions.

Experimental studies could provide the most reliable results for scholars to detect the noise from centrifugal pumps. Choi et al. [4] conducted experiments and revealed that the main cause of radiation noise generated by an impeller without volute was the pressure fluctuation on the blade surface. Chu et al. [5,6] took the volute into consideration and attributed the primary noise to the interaction between the non-uniform outflux from the impeller and tongue [5,6], and further research showed that a slight increase of the gap between impeller and tongue would reduce the noise significantly [7]. Parrondo et al. [8] established an acoustic model to characterize the internal sound field at low frequency range, and concluded that the internal sound field could be characterized by a dipole-like source near the tongue. Cai et al. [9] measured the pressure fluctuation near the wall of the tongue under various rotational speed conditions, and found that the pressure fluctuation intensity increased more rapidly than that of rotational speed. The four-port model was introduced for research on internal sound fields in pipes, making it convenient to change the rules of noise in pipes under various working conditions [10–12]. Based on LabVIEW (LabVIEW8.6, National Instruments, Austin, Texas, USA), Yuan et al. [13] designed a measurement system for internal noise analysis of centrifugal pumps with synchronous measurement for noise signal, pressure and flow rate, and laid a foundation for the follow-up study about the influence of different structures [14–16] on sound pressure levels (SPL) at different frequencies. To characterize far field radiation noise, Ye et al. [17] measured the noise amplitude outside a centrifugal pump under various flow rate conditions, and found that noise amplitude reached maximum at the highest efficiency point.

Currently, as a useful research tool, numerical simulation compensated for weaknesses in the experiment. A hybrid method combining computational fluid dynamics (CFD) with Lighthill acoustic analogy was widely used to elucidate the acoustic generation [18,19]. Langthjem and Olhoff [20,21] applied the hybrid method for noise calculation in a two-dimensional centrifugal pump, and concluded that the main cause of noise was the dipole source. The dipole source was defined as unsteady fluid force acting on the wall surface, including the impeller and volute source, in centrifugal pumps. Huang et al. [22], Ma et al. [23] and Liu et al. [24] considered the two dipole sources to be the noise source for acoustic calculation, and compared the SPL at different orders of blade passing frequency (BPF) with different structures, and this research provided guidance for structural optimization of centrifugal pumps. In addition, Gao et al. [25] discovered that the radiation noise generated by an impeller demonstrated diploe directivity, while the volute-generated noise appeared to have asymmetric directivity, i.e., the noise level in the direction facing the tongue was higher than that of the direction against the tongue. Dong et al. [26] pointed out that the volute-generated noise increased monotonously and nonlinearly with the increase of rotational speed.

Obviously, the literature shows that previous experimental research mainly focused on the internal noise characteristics in pipes, while merely experimental research focused on far field radiation noise. During the numerical simulation, the previous research mainly considered the influence of the impeller dipole source and volute dipole source on the radiation noise separately, and most are the SPL at different orders of BPF. Actually, the radiation noise results from the interaction of the two dipole sources when pumps are running, so the interaction influence should be considered to obtain more accurate results. In addition, it is necessary to calculate the total sound pressure level (TSPL) of different monitoring points, calculated by the superposition of SPL at different characteristic frequencies, to acquire the directivity characteristic. Noise propagation is the spread of acoustic energy in the medium, so the acoustic energy changes with noise propagation process and the directivity characteristic can be detected. The overall level of radiation noise should also be conducted to evaluate the noise intensity under specific working conditions. In general, the pumps are always working at various conditions with various working demands, and the radiation noise changes accordingly, and it is also indispensable to carry out the various working condition study for radiation noise to figure out these changing rules.

In this paper, a centrifugal pump radiation noise measurement system is established. Then, the TSPL and acoustic energy distribution, as well as the overall level of radiation noise, are analyzed, which may provide guidance for optimizing the working conditions and noise control of centrifugal pumps.

2. Experimental Facility and Procedure

2.1. Parameters of the Test Pump

In this study, IS-80-50-250 (IS-80-50-250, Shanghai Pump Manufacture Co., Ltd, Shanghai, China) is chosen as the test pump, and the prototype figure is shown in Figure 1. Water at normal temperature is used as medium, and the geometric and performance parameters of the test pump are listed in Table 1.

Figure 1. Prototype figure of the test pump.

Table 1. Geometric and performance parameters of the test pump.

Parameter	Value
Inlet diameter, mm	80
Impeller diameter, mm	250
Outlet diameter, mm	50
Nominal flow rate, m^3/h	50
Best efficiency	63% (50 m^3/h)
Design head, m	80
Nominal rotational speed, rpm	2900
Blade number	6

2.2. Radiation Noise Measurement System

The experimental apparatuses used in the system are shown schematically in Figure 2, which include a soundproof room, water circulation system, circuit control system, and data acquisition and storage system. In the measurement system, to reduce the influence of the surrounding environment and motor operation on the measurement results, the centrifugal pump and motor are insulated by a soundproof room; the interior and exterior walls of the soundproof room, along with the motor, are surrounded by soundproof cotton. The pump is driven by the YVF2180L-2 type three-phase asynchronous motor (YVF2180L-2, Shanghai Nama Electric Co., Ltd, Shanghai, China), and the rotational speed is regulated by the Y0300G3 type frequency converter (Y0300G3, Shanghai Nama Electric Co., Ltd, Shanghai, China). The flow rate, inlet pressure, outlet pressure, as well as the radiation noise level are measured and recorded by corresponding instrument listed in Table 2.

During the operation, when the operation system reaches stable, the flow rate, pressure, radiation noise level are measured sequentially and stored in the computer terminal. After that, the rotational speed and flow rate are adjusted via frequency converter and artificial regulation, respectively, then these parameters under various working conditions are recorded for analysis. Figure 3 shows the structure diagram of the measurement system.

Table 2. Measurement characteristic of instruments.

Instruments	Type	Application	Measuring Range	Accuracy or Sensitivity
Flow meter	SLDG-800 (SLDG-800, Nanjing Shunlaida Measurement and Control Equipment Co., Ltd., Nanjing, Jiangsu, China)	Measuring flow rate	0–100 m³/h	0.2% (accuracy)
Pressure transducer	MIK-300 (MIK-300, Hangzhou Meikong Automation Technology Co., Ltd., Hangzhou, China)	Measuring inlet pressure	−100–0 kPa (inlet pipe)	0.5% (accuracy)
		Measuring outlet pressure	0–1600 kPa (outlet pipe)	0.5% (accuracy)
Pressure recorder	RX-200D (RX-200D, Hangzhou Meikong Automation Technology Co., Ltd., Hangzhou, China)	Recording pressure	/	/
Microphone	AWA14423L (AWA14423L, Hangzhou Aihua Instruments Co., Ltd., Hangzhou, China)	Measuring radiation noise	10–20 kHz	50 mV/Pa (sensitivity)
Two channel signal analyzer	AWA6290M+ (AWA6290M+, Hangzhou Aihua Instruments Co., Ltd, Hangzhou China)	Recording radiation noise	/	/

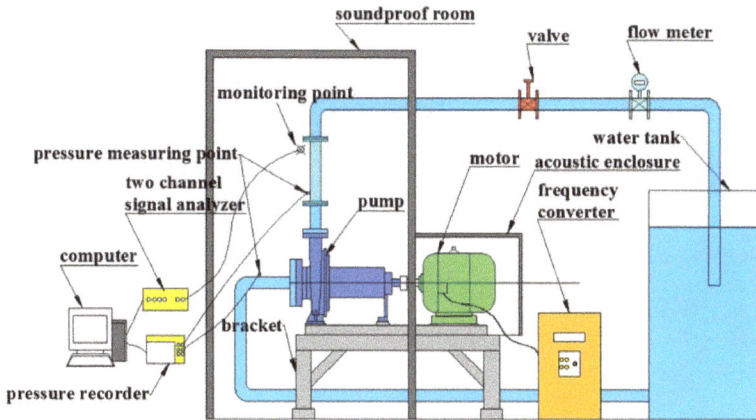

Figure 2. Layout and instrumentation of measurement system.

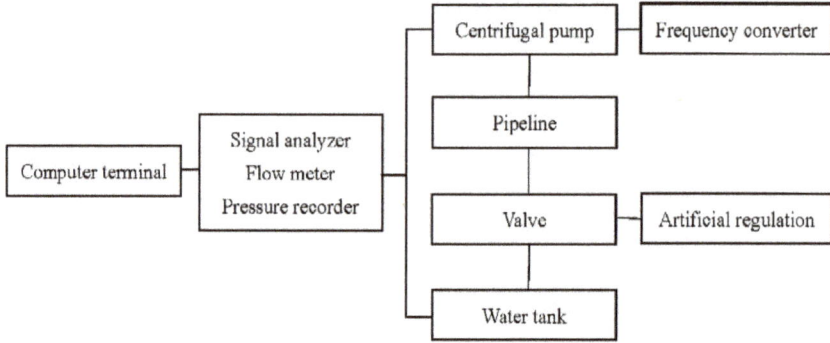

Figure 3. Structure diagram of measurement system.

2.3. Arrangement of the Monitoring Points

To acquire the distribution characteristic of radiation noise, 16 monitoring points are arranged on the measurement surface around the pump. As shown in Figure 4, the monitoring points are 1000 mm away from the center of impeller, and arranged evenly in circumferential directions [27]. During the measurement process, SPL characteristic of every point is measured sequentially by a microphone. Here, SPL is defined as:

$$\text{SPL} = 20 \log \frac{P_e}{P_{ref}} \tag{1}$$

$$P_e = \sqrt{\frac{1}{T} \int_0^T p'^2 dt} \tag{2}$$

$$T = \frac{t}{l} \tag{3}$$

where P_{ref} is the reference acoustic pressure (2×10^{-5} Pa in air), P_e is the effective value of acoustic pressure, p' is instantaneous acoustic pressure, t is the time of one revolution of the impeller, and l is the number of blades ($l = 6$ in this paper).

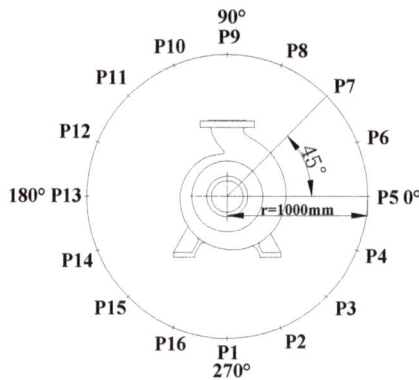

Figure 4. Arrangement of monitoring points in circumferential direction.

To reveal the noise intensity of different monitoring points and the directivity characteristic, it is necessary to derive a temporal intensity profile involving a superposition of SPL at each Fourier frequency. Thus TSPL is introduced and given by:

$$\text{TPSL} = 10\lg\sum_{i=1}^{n} 10^{\text{SPL}_i/10} \tag{4}$$

where n represents the number of frequencies. However, TSPL only represents the noise level of a certain monitoring point, and it can't be superimposed by arithmetic to describe the noise level of an interval. Therefore, the acoustic energy is more suitable for comparing the noise level of different monitoring intervals since it can be superimposed by arithmetic, moreover, the propagation of sound is essentially the propagation of energy. Briefly, the application of acoustic energy could reveal the noise level relationship between different intervals intuitively, so the average acoustic energy density [28] is analyzed, and it is defined as:

$$\varepsilon = \frac{P_e^2}{\rho c^2} \tag{5}$$

where ε, ρ and c represent the acoustic energy density, medium density (1.29 kg/m^3 in air) and the sound speed in medium (343 m/s in air), respectively.

To further evaluate the overall level of radiation noise on measurement surface, the average A-weighted sound pressure level (L_{pA}) of the measurement surface is introduced and expressed as:

$$L_{pA} = 10\lg\left(\frac{1}{m}\sum_{i=1}^{m} 10^{\text{TSPL}_i/10}\right) \tag{6}$$

where m represents the number of monitoring points, and the L_{pA} of 16 monitoring points on the measurement surface around the pump is calculated for analysis.

3. Radiation Noise Characteristic under Various Rotational Speed Conditions

The variable speed regulation has no throttling loss, and is an ideal adjustment method. In this section, on the basis of keeping the flow valve installed downstream on the outlet pipe opened fully and unchanged, then the radiation noise characteristic under various rotational speed conditions are studied. According to the similarity law, the relationship between flow rate and rotational speed is defined as:

$$\frac{Q_1}{Q_2} = \frac{n_1}{n_2} \tag{7}$$

where Q and n represent the flow rate and rotational speed, while the subscript 1 and 2 represent two different working conditions. It can be found that the flow rate is proportional to the rotational speed, so the flow rate of the centrifugal pump increases with the rise in rotational speed. To ensure the safety of the running system, seven different rotational speeds that are less than or equal to the nominal rotational speed are considered. The rotational speeds and corresponding flow rates are shown in Table 3.

Table 3. Rotational speeds and corresponding flow rates.

Rotational Speed, rpm	Flow Rate, m^3/h
1700	56.9
1900	64.8
2100	69.6
2300	74
2500	78.2
2700	82.1
2900	86

3.1. Directivity Characteristic of Radiation Noise under Various Rotational Speed Conditions

By measuring the SPL characteristic and calculating the TSPLs of 16 different points in the circumferential direction, the directivity of radiation noise is obtained under various rotational speed conditions.

As shown in Figure 5, it can be found that the TSPL increases from 80 to 100 dB with the increase of rotational speed. As a result of the symmetric characteristic of impeller, the directivity profile diagram of TSPL demonstrates the dipole symmetry, and one sees that the two TSPL valleys appear at 0° and 180°. In addition, apparent asymmetric characteristics also can be found due to the asymmetric structure of the volute, more concretely, the noise level in the direction facing the tongue (in the interval from 90° to 157.5°) is higher than that in the direction against the tongue (in the interval from 292.5° to 0°), which means that the radiation noise is the interaction result of impeller and volute dipole source.

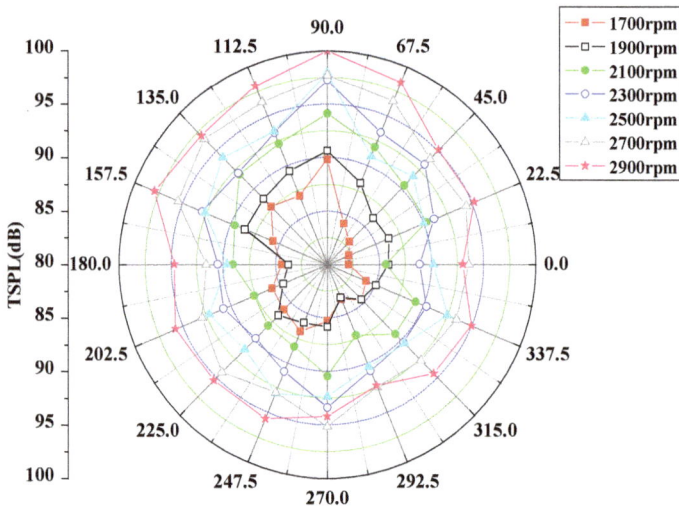

Figure 5. Directivity characteristic of radiation noise under various rotational speed conditions.

To further reveal the asymmetric characteristics under various rotational speed conditions quantitatively, the ratio of the acoustic energy in the interval from 90° to 157.5° (ε_1), as well as that in the interval from 292.5° to 0° (ε_2) to the total acoustic energy (ε_t) are calculated. As shown in Figure 6, the values of $\varepsilon_1/\varepsilon_t$ and $\varepsilon_2/\varepsilon_t$ change little with the change of rotational speed, and the value of $\varepsilon_1/\varepsilon_t$ fluctuates around 0.413, while the value of $\varepsilon_2/\varepsilon_t$ fluctuates near 0.157. This proves that the change of rotational speed would affect the acoustic energy, but it has little impact on the acoustic energy distribution. According to the past research [5–7], the volute tongue is the major noise source, and the measurement interval in the direction facing the tongue is closer to the tongue than other intervals. On the other hand, acoustic energy weakens gradually as the distance from the sound source increases, so the acoustic energy has less attenuation in the direction facing the tongue than that in the direction against the tongue, causing a much higher ratio of acoustic energy in the direction facing the tongue.

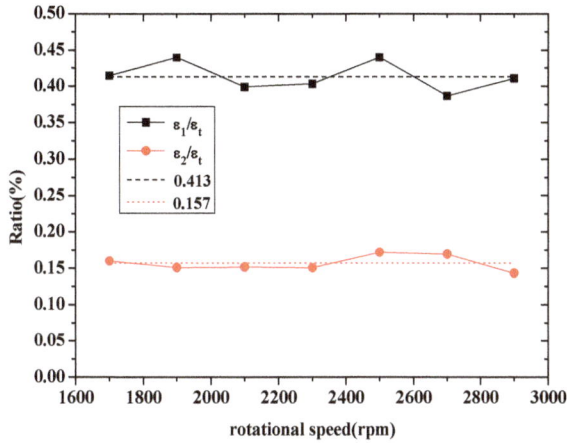

Figure 6. The acoustic energy changing curves with rotational speed.

3.2. Changing Rules of Radiation Noise under Various Rotational Speed Conditions

To analyze the noise level changing rules of different points under various rotational speed conditions, P1 (270°, in the direction against the outlet), P5 (0°, the valley point), P9 (90°, in the direction facing the outlet), P11 (135°, in the direction facing the tongue) and P13 (180°, the valley point) are selected. As shown in Figure 7, it can be more visually seen that the TSPLs of different monitoring points rise with the rotational speed increasing. In general, the TSPLs of P5 and P13 are lower than the others because the two points are located at the valley of dipole characteristic, meanwhile, TSPL of P1 lies between P5 and P11. As a result of the influence of volute tongue noise source, P9 and P11 have a higher noise level, and P9 is located in outlet direction, which is more affected by the internal flow noise in outlet pipe, so P9 is the highest noise level point.

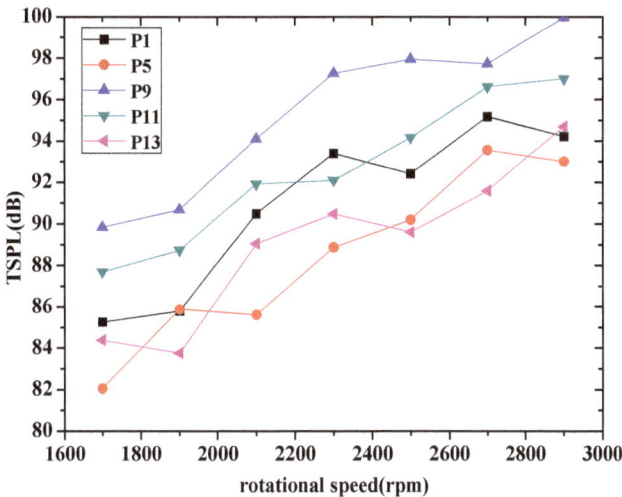

Figure 7. TSPL changing curves with rotational speed.

To evaluate the overall level of radiation noise, the L_{pA} of the 16 monitoring points is compared in Figure 8 under various rotational speed conditions. It can be observed that L_{pA} climbs gradually with the increase of rotational speed, and reaches to maximum at 2900 rpm. When rotational speed increases from 1700 to 2900 rpm, the L_{pA} increases by 12.40%. However, the growth slope decreases gradually with the increase of rotational speed generally, in particular, the L_{pA} grows rapidly in the interval between 1700 and 2300 rpm, with an average increase of 1.12 dB per 100 rpm, while in the range from 2300 to 2900 rpm, the growth rate of L_{pA} slows down, with an average increase of 0.65 dB per 100 rpm, which is less than that in the range from 1700 to 2300 rpm. It could be explained by that with the increase of rotational speed, the pressure fluctuation on wall surface also increases, but the increment rate of the variance of pressure fluctuation decreases [29], and causes a slow increase in acoustic pressure around the pump, and as a result, the growth slope of noise level decreases steadily.

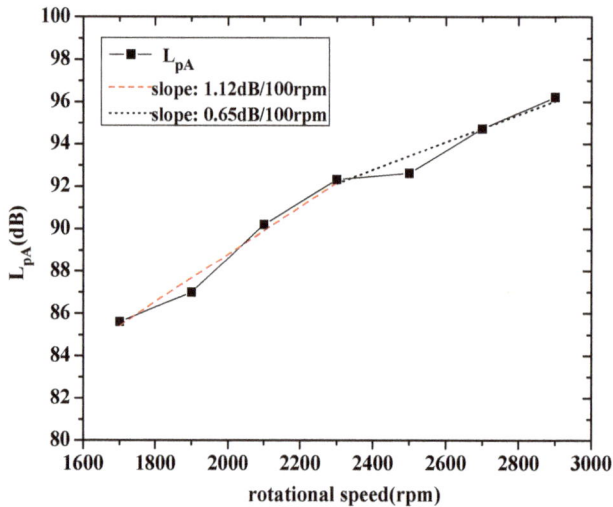

Figure 8. L_{pA} changing curve with rotational speed.

4. Radiation Noise Characteristics under Various Flow Rate Conditions

Compared with the variable speed regulation, the throttling regulation has the advantages of easy operation and low cost. Considering the two factors mentioned, throttling regulation is widely adopted at some occasions. In this section, the throttling regulation is realized by adjusting the valve opening on outlet pipe, then five different flow rates, i.e., 37.5, 50 (nominal flow rate), 62.5, 75 and 86 m³/h (valve open fully), are studied to explore the radiation noise characteristic under various flow rate conditions when rotational speed is set as 2900 rpm.

4.1. Directivity Characteristic of Radiation Noise under Various Flow Rate Conditions

By calculating the TSPLs of 16 different points in the circumferential direction, then the directivity characteristic of radiation noise is analyzed under various flow rate conditions.

It is apparent in Figure 9 that TSPL changes between 86 dB and 100 dB, not only the dipole characteristic, but the asymmetric characteristic are presented in circumferential direction. And then the ratio of the acoustic energy in the interval from 90° to 157.5° (ε_1), as well as that in the interval from 292.5° to 0° (ε_2) to the total acoustic energy (ε_t) are compared under various flow rate conditions. As shown in Figure 10, the values of $\varepsilon_1/\varepsilon_t$ and $\varepsilon_2/\varepsilon_t$ show the fluctuation trend in the vicinity of 0.410 and 0.166, respectively. And these results are closer to those under various rotational speed conditions. It also can be found that the deviation between the value of $\varepsilon_1/\varepsilon_t$ and average value (0.410) is larger at 50 m^3/h. It could be explained by that in the measurement process, the measurement results are affected by the severe disturbance of the surrounding environment, but it does not affect the conclusion that the change of flow rate also has little impact on the acoustic energy distribution.

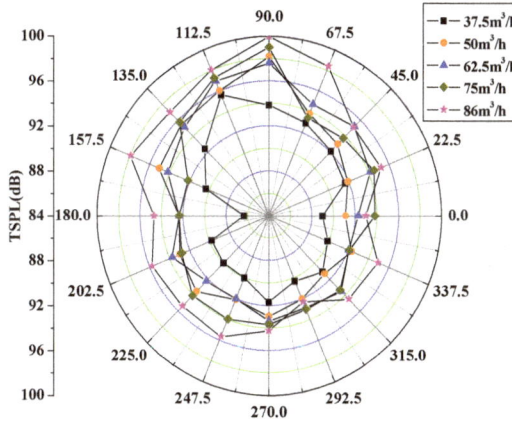

Figure 9. Directivity characteristics of radiation noise under various flow rate conditions (2900 rpm).

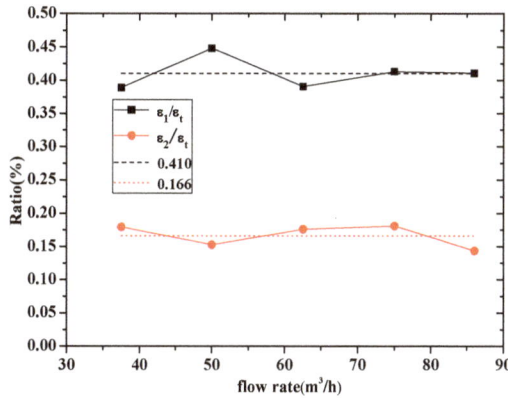

Figure 10. The acoustic energy changing curves with flow rate (2900 rpm).

4.2. Changing Rules of Radiation Noise under Various Flow Rate Conditions

In Figure 11, TSPLs of P1, P5, P9, P11 and P13 are also analyzed to reveal the noise level changing rules under various flow rate conditions. It is revealed that TSPLs of different points show similarly changing rules, specifically, the TSPL increases rapidly in small flow range from 37.5 to 50 m^3/h, then basically levels off in the interval from 50 to 75 m^3/h, and continues to increase when flow rate is higher than 75 m^3/h. What's more, the TSPL distribution characteristic of different points under various flow rate conditions also shows that the noise level at P5 and P13 are lower than others, while TSPLs of P9 and P11 are higher than others, and P9 is highest noise level point, which coincides with the distribution under various rotational speed conditions.

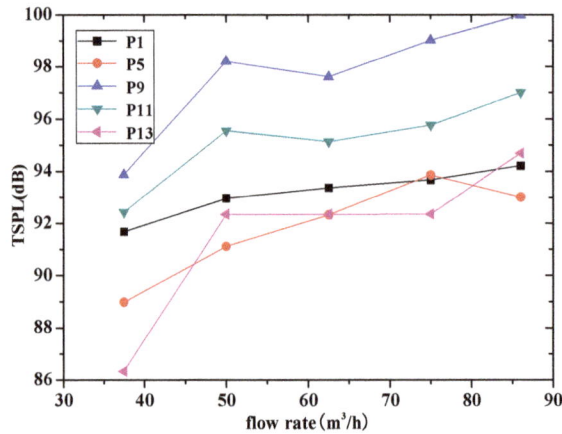

Figure 11. TSPL changing curves with flow rate (2900 rpm).

Furthermore, L_{pA} of 16 monitoring points is also studied under various flow rate conditions. It can be found in Figure 12 that the L_{pA} also increases sharply initially at small flow rate, then maintains stable, and reaches to maximum value at large flow rate interval. Additionally, L_{pA} changes a little with the increase of flow rate, which increases by 5.10% when flow rate grows from 37.5 to 86 m^3/h.

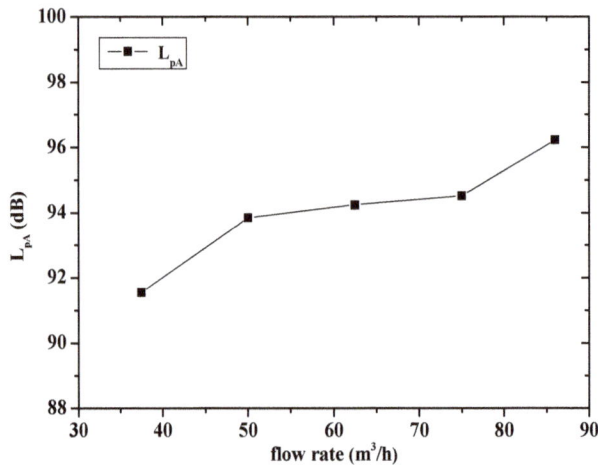

Figure 12. L_{pA} changing curve with flow rate (2900 rpm).

To explore the reasons for the changing rules of L_{pA} under various flow rate conditions, the head and efficiency are calculated, respectively. The changing rules of the two parameters are shown in Figure 13, it can be found that with the increase of flow rate, the head decreases gradually. In addition, the efficiency shows the tendency that increases from 37.5 to 50 m^3/h, and reaches to the maximum at 50 m^3/h, then remains essentially unchanged in the range from 50 to 75 m^3/h, which is the best efficiency range, and it keeps decreasing subsequently with the flow rate continues to grow. At low flow rate conditions, the pressure fluctuation inside the pump is low and causes low noise levels. With the increase of flow rate, the pressure fluctuation inside the pump increases accordingly, while in the best efficiency range, i.e., in the range from 50 to 75 m^3/h, the pressure becomes stable [17], and leads to little change for the noise level. However, as the flow rate continues to increase, the head and efficiency decrease dramatically, which may be caused by the occurrence of cavitation inside the pump, and contribute to the dramatic increase of L_{pA} when flow rate is greater than 75 m^3/h.

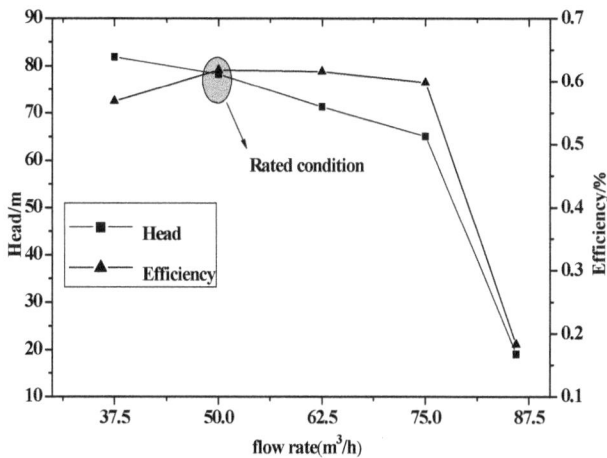

Figure 13. Head and efficiency changing curve with flow rate (2900 rpm).

5. Conclusions

In this paper, a centrifugal pump radiation noise measurement system is established, then the directivity distribution characteristics and overall level of radiation noise are analyzed under various working conditions. The main conclusions are drawn as follows:

(a) Under various working conditions, the total sound pressure level distribution for different monitoring points is dipolar; specifically, the two valley values appear at 0° (in the direction against the tongue) and 180°, and the minimum valley values are presented at the minimum rotational speed and minimum flow rate condition. Additionally, asymmetry is also validated, i.e., the noise level in the direction facing the volute tongue is higher than that in the direction against the tongue, and the monitoring point in outlet direction is the highest noise level point.

(b) The change in working conditions has little impact on the acoustic energy distribution of different intervals, and the ratios of the acoustic energy in the direction facing the tongue (ε_1), as well as that in the direction against the tongue (ε_2), to the total acoustic energy (ε_t) fluctuate around 0.410 and 0.160, respectively.

(c) In the operation of variable speed regulation, the average A-weighted sound pressure level (L_{pA}) increases gradually with the increasing of rotational speed, and it increases by 12.40% when rotational speed increases from 1700 to 2900 rpm, but the growth slope decreases gradually with the rise of pump rotational speed. While in the operation of throttling regulation, L_{pA} first increases, then remains stable, and continues to increase with the increase in flow rate, and it increases by 5.10% when flow rate grows from 37.5 to 86 m^3/h.

Acknowledgments: This paper is supported by National Natural Science Foundation of China (No. 51776111), Shandong Province Natural Science Foundation (No. ZR2016EEM35), and National Development and Reform Commission Foundation (No. 2013-1819).

Author Contributions: Chang Guo, Dongyue Lu designed the study, conducted the experiment and collected the experimental data; Kun Wang analyzed the experimental data; Chang Guo wrote the manuscript; Ming Gao reviewed and edited the manuscript.

Conflicts of Interest: The authors declare no conflict of interest.

References

1. Mao, D.X.; Hong, Z.H. *Environmental Noise Control Engineering*; Higher Education Press: Beijing, China, 2002.
2. Shah, S.R.; Jain, S.V.; Patel, R.N.; Lakhera, V.J. CFD for centrifugal pumps: A review of the state-of-the-art. *Procedia Eng.* **2013**, *51*, 715–720. [CrossRef]
3. Wu, D.F.; Liu, Y.S.; Li, D.L.; Zhao, X.F.; Li, C. Effect of materials on the noise of a water hydraulic pump used in submersible. *Ocean Eng.* **2017**, *131*, 107–113. [CrossRef]
4. Choi, J.S.; Mclaughlin, D.K.; Thompson, D.E. Experiments on the unsteady flow field and noise generation in a centrifugal pump impeller. *J. Sound Vib.* **2003**, *263*, 493–514. [CrossRef]
5. Chu, S.; Dong, R.; Katz, J. Relationship between unsteady flow, pressure fluctuation, and noise in a centrifugal pump—Part A: Use of PDV data to compute the pressure field. *J. Fluid Eng.* **1995**, *117*, 24–29. [CrossRef]
6. Chu, S.; Dong, R.; Katz, J. Relationship between unsteady flow, pressure fluctuation, and noise in a centrifugal pump—Part B: Effects of blade-tongue interactions. *J. Fluid Eng.* **1995**, *117*, 30–35. [CrossRef]
7. Dong, R.; Chu, S.; Katz, J. Effect of Modification to Tongue and Impeller Geometry on Unsteady Flow, Pressure Fluctuations, and Noise in a Centrifugal Pump. *J. Turbomach.* **1997**, *119*, 506–515. [CrossRef]
8. Parrondo, J.; Pérez, J.; Barrio, R.; González, J. A simple acoustic model to characterize the internal low frequency sound field in centrifugal pumps. *Appl. Acoust.* **2011**, *72*, 59–64. [CrossRef]
9. Cai, J.C.; Pan, J.; Andrew, G. Experimental study of the pressure fluctuation around the volute tongue of a centrifugal pump at variable rotating speed. *Fluid Mach.* **2015**, *43*, 13–16.
10. Rzentkowski, G.; Zbroja, S. Experimental characterization of centrifugal pumps as an acoustic source at the blade-passing frequency. *J. Fluid Struct.* **2000**, *14*, 529–558. [CrossRef]
11. Wang, Q.Y. Numerical and Experimental Research on the Flow-Induced Noise of Centrifugal Pump in the Seawater Pipe. Master's Thesis, Harbin Engineering University, Harbin, China, 2010.
12. Si, Q.R.; Yuan, S.Q.; Yuan, J.P. Experimental study on the influence of impeller-tongue gap on the performance and flow-induced noise characteristics of centrifugal pumps. *J. Vib. Shock* **2016**, *35*, 164–168.
13. Yuan, S.Q.; Yang, Y.; Yuan, J.P.; Luo, Y. Measurement system design of flow-induced noise in centrifugal pumps. *Drain. Irrig. Mach.* **2009**, *27*, 10–14.
14. Liu, H.L.; Wang, Y.; Yuan, S.Q.; Tan, M.G. Effects of impeller outlet width on the vibration and noise from centrifugal pumps induced by flow. *J. Huazhong Univ. Sci. Technol.* **2012**, *40*, 123–127.
15. Tan, M.G.; Wang, Y.; Liu, H.L.; Wu, X.F.; Wang, W. Effects of number of blades on flow induced noise vibration and noise of centrifugal pumps. *J. Drain. Irrig. Mach. Eng.* **2012**, *30*, 131–135.
16. Wang, Y.; Liu, H.L.; Liu, D.X.; Wang, J.; Wu, X.F. Effects of vane wrap angle on flow induced vibration and noise of centrifugal pumps. *Trans. Chin. Soc. Agric. Eng.* **2013**, *29*, 72–77.
17. Ye, X.M.; Pei, J.J.; Li, C.X.; Liu, Z. Experimental Study on Nosie Characteristics of Centrifugal Pump Based on Near-Field Acoustic Pressure Method. *Chin. J. Power Eng.* **2013**, *33*, 375–380.
18. Wang, M.; Freund, J.B.; Lele, S.K. Computational prediction of flow-generated sound. *Annu. Rev. Fluid Mech.* **2015**, *38*, 483–512. [CrossRef]
19. Liu, H.L.; Dai, H.W.; Ding, J.; Tan, M.G.; Wang, Y.; Huang, H.Q. Numerical and experimental studies of hydraulic noise induced by surface dipole sources in a centrifugal pump. *J. Hydrodyn.* **2016**, *28*, 43–51. [CrossRef]
20. Langthjem, M.A.; Olhoff, N. A numerical study of flow-induced noise in a two-dimensional centrifugal pump. Part I. Hydrodynamics. *J. Fluid Struct.* **2004**, *19*, 349–368. [CrossRef]
21. Langthjem, M.A.; Olhoff, N. A numerical study of flow-induced noise in a two-dimensional centrifugal pump. Part II. Hydroacoustics. *J. Fluid Struct.* **2004**, *19*, 36–386. [CrossRef]
22. Huang, J.X.; Geng, S.J.; Wu, R.; Liu, K.; Nie, C.Q.; Zhang, H.W. Comparison of noise characteristics in centrifugal pumps with different types of impellers. *Acta Acoust.* **2010**, *35*, 113–118.
23. Ma, Z.L.; Chen, E.Y.; Guo, Y.L.; Yang, A.L. Numerical Simulation of the Influence of the Diameter at the Outlet of an Impeller on the Noise Level Induced by the Flow Inside a Centrifugal Pump. *J. Eng. Therm. Energy Power* **2016**, *31*, 93–98.
24. Liu, H.L.; Ding, J.; Tan, M.G.; Cui, J.B.; Wang, Y. Analysis and experimental of centrifugal pump noise based on outlet width of impeller. *Trans. Chin. Soc. Agric. Eng.* **2013**, *29*, 66–73.
25. Gao, M.; Dong, P.X.; Lei, S.H.; Turan, A. Computational Study of the Noise Radiation in a Centrifugal Pump When Flow Rate Changes. *Energies* **2017**, *10*, 221. [CrossRef]

26. Dong, P.X.; Gao, M.; Guan, H.J.; Lu, D.Y.; Song, K.Q.; Sun, F.Z. Numerical simulation for variation law of volute radiated noise in centrifugal pumps under variable rotating speed. *J. Vib. Shock* **2017**, *36*, 128–133.
27. Tao, J.Y.; Lu, X.N.; Wang, L.L. *Methods of Measuring and Evaluating Noise of Pumps*; Chinese Quality Supervision Bureau: Beijing, China, 2013.
28. Du, G.H.; Zhu, Z.M.; Gong, X.F. *Fundamentals of Acoustics*; Nanjing University Press: Nanjing, China, 2001.
29. Dong, P.X. Computational Study on the Noise Radiation of a Centrifugal Pump Based on 3-D Flow field with Varying Working Condition. Master's Thesis, Shandong University, Jinan, China, 2016.

energies

MDPI

Article

Numerical Study on the Acoustic Characteristics of a Axial Fan under Rotating Stall Condition

Lei Zhang *, Chuang Yan, Ruiyang He, Kang Li and Qian Zhang

School of Energy, Power and Mechanical Engineering, North China Electric Power University, Baoding 071003, China; yanchuang518@163.com (C.Y.); heruiyangdarry@163.com (R.H.); likang_ncepu@163.com (K.L.); zq_8299@163.com (Q.Z.)
* Correspondence: ncepu_zhanglei@163.com; Tel.: +86-138-3127-2833

Received: 23 October 2017; Accepted: 20 November 2017; Published: 23 November 2017

Abstract: Axial fan is an important piece of equipment in the thermal power plant that provides enough air for combustion of coal. This paper focuses on the aerodynamic noise characteristics of an axial fan in the development from stall inception to stall cells. The aerodynamic noise characteristic of monitoring region in time and frequency domains was simulated employing the large-eddy simulation (LES), with the addition of throttle setting and the Ffowcs Williams-Hawkings (FW-H) noise model. The numerical results show that, under the design condition, the acoustic pressure presents regular periodicity along with the time. The noise energy is concentrated with high energy of the fundamental frequency and high order harmonics. During the stall inception stage, the acoustic pressure amplitude starts fluctuating and discrete frequencies are increased significantly in the low frequency; among them, there are three obvious discrete frequencies: 27.66 Hz, 46.10 Hz and 64.55 Hz. On the rotating stall condition, the fluctuation of the acoustic pressure level and amplitude are more serious than that mentioned above. During the whole evolution process, the acoustic pressure peak is difficult to keep stable all the time, and a sudden increase of the peak value at the 34.5th revolution corresponds to the relative velocity's first sudden increase at the time when the valve coefficient is 0.780.

Keywords: axial fan; rotating stall; aerodynamic noise; numerical simulation; noise spectrum

1. Introduction

Axial fan is an important piece equipment in the thermal power plant that provides enough air for combustion of coal. There are two unstable flow phenomena for axial flow fan: rotating stall and surge. If the flow rate decreases below the critical value, the inner flow-field of impeller will be deteriorated and then the rotating stall occurs, which will lead to a phenomenon of surge. The surge will cause the variation of flow rate in a big scope and generate the impact of airflow. Once surge frequency agrees with natural frequency of fan, it can cause a big vibration of the fan and connecting pipes, and then the blade fracture will appear. Thus, for the safe operation of an axial fan, it is necessary to find a method to predict and delay the occurrence of stall inception [1].

Rotating stall is caused by the collapse of the steady flow field within the fan. Therefore, finding the stall inception accurately and rapidly, and learning its mechanism are the keys to the active control of rotating stall during process from stabilization to collapse of the flow field [2–4]. Mcdougall et al. [5] considered that, when the rate of flow within impeller decreases, part of the blade passages' flow will decrease and the rest will keep the large flow, which leads to the rotating stall. Through the three-dimensional numerical simulation about the transonic fan rotor considering the external disturbances. Niazi [6] found that the reflux appeared at the blade tip clearance firstly. The interplay of the reflux and shock wave leads to the rotating stall. Moore [7] and Greitzer [8] developed linear and nonlinear models of instability of the compression system, and obtained correlation between the compressor aerodynamic stability and the stall inception, which develops and spreads along the circumferential. Kosuge et al. [9] conducted the experiment of stall and surge phenomena on two different airfoil centrifugal impellers without diffuser under the small flow, and studied the influence of structure on stall. Yamada et al. [10] conducted a numerical analysis of the rotating stall inception on a multi-stage axial flow compressor for an actual gas turbine by large eddy simulation. The results showed that the first observed flow phenomenon in the stalling process is the hub corner separation when approaching the stall point. This hub corner separation expanded with time, and eventually leaded to the leading-edge separation on the hub side for the stator. Salunkhe and Pradeep [11] analyzed the stall inception mechanism in a single stage axial fan under undistorted and distorted inflow condition by the Morlet wavelet transform. Toge and Pradeep [12] attempted to study the stall inception mechanism of the low speed, contra rotating axial fan stage by wavelet transforming the unsteady pressure date from the casing wall sensor. The result clearly showed that stall inception occurs mainly through the long-length scale perturbations of both rotors. Li et al. [13] used the characteristic of vortex noise, the bar theory and nonlinear aerodynamics theory to analyze force acting on the blade in theoretical method. It analyzed the affecting factors of the vortex noise in the model qualitatively. There have been many studies on the induced mechanism of rotating stall in turbomachinery. However, it is worthy of further study on the new method to early detect stall for control.

The internal flow field of the fan is related to the aerodynamic noise. The noises of the fan mainly consist of aerodynamic noise, machinery noise and radiation noise [14]. Among them, aerodynamic noise occupies the main position, which includes the rotational noise and vortex noise. Liu et al. [15] obtained the spectrum characteristic of the double rotor by simulating different axial gaps, to determine the optimal range of the rotor spacing. Diaz [16] proposed a noise prediction method, and a single stage axial flow fan in the far field noise was predicted, and he proved the reliability of the method. Fukano and Jang [17] conducted contrast experiments in different flow coefficient under different tip clearance in a low-pressure axial flow fan. The results show that the tip leakage vortex mutual interference leads to the increase of fan noise. Wang et al. [18] used the technique of synchronous averaging with time-base stretching to decompose the original sound signal into various noise sources and analyze the noise characteristics of two identical small axial cooling fans in series with distorted inlet flow. It is found that the inlet flow distortion increases the total sound pressure level and rotational noise of the upstream fan, but has little effect on the rotation noise of the downstream fan. Feng et al. [19] predicted noise at different points by the point force model, and then used the hybrid method based on computational fluid dynamics (CFD) simulation and FW-H equation to predict the noise of the fan model. The results show that geometric asymmetry may significantly increase fan noise at some frequencies compared to symmetrical rotation. Mao et al. [20] studied the effect of the end- plate on the static and noise characteristics of the small axial fan by numerical simulation and experiment. He showed that the mechanism of noise reduction is due to the decrease of the vorticity of the blade surface caused by the end plate. At present, the research on the axial fan noise is mainly focused on steady operating conditions. When the unsteady phenomenon of rotating stall occurs, the flow field in axial fan will be significantly different with no stall conditions. Thus, the acoustic characteristic in the whole unsteady process from near stall to stall condition may be different.

This paper, through the large eddy simulation coupled with the throttle valve model, simulates the rotating stall of an axial fan and calculate the unsteady flow field. Then, to predict the noise of the fan, the FW-H equation based on Lighthill acoustic analogy theory is used. The time domain characteristic of acoustic pressure and acoustic pressure peak in the rotor blade domain are researched under four different conditions: the design condition, the near stall condition, the development of stall inception condition and the rotating stall condition. Finally, after obtaining the source data in time domain integration and then applying fast Fourier transform, the noise spectrums of the acoustic pressure level and amplitude at the rotor blade domain are investigated under the four conditions mentioned above.

2. Numerical Simulation

2.1. Subsection with Geometry

Figure 1 shows the single-stage variable pitch axial fan, which is taken as the research object in this paper. The facility consists of an inlet collector, a rotor, a stator and an outlet diffuser. The three-dimensional geometric model of the axial fan is established by ICEM CFD software (ANSYS 15.0, ANSYS Inc., Pittsburgh, PA, USA). The main parameters are shown in Table 1.

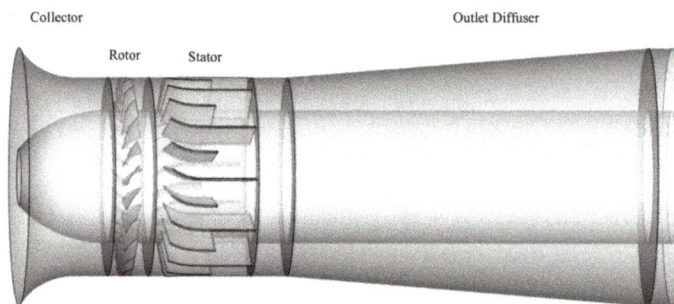

Collector   Outlet Diffuser

Rotor   Stator

Figure 1. Geometrical model of the axial fan.

Table 1. Parameters of the axial fan.

Parameter	Value
Rated speed n (r/min)	1490
Number of moving blade N_r	24
Number of stationary blade N_s	23
Inlet diameter D_1 (m)	2.312
Outlet diameter D_2 (m)	2.305
Rotor diameter D (m)	1.778
Hub ratio	0.668

2.2. Figures, Tables and Schemes

ICEM CFD (The Integrated Computer Engineering and Manufacturing code for Computational Fluid Dynamics) software is used to generate the multi-block structure grids of the whole axial fan. Near the leading-edge and trailing-edge of each blade, the separation of boundary layer will form some separation vortexes. The grids of those parts are encrypted as the tip clearance grids to improve the precision of calculation. Figure 2 shows the grids around the leading-edge and trailing-edge. In addition, the sliding mesh model is applied at the interface between the rotor and the stator. Moreover, to ensure that the average wall distance y+ is less than 5, boundary layer grids are used on the surface of the blade. Figure 3 shows local grid of the impeller blade and the enlarged view around the impeller. Grid-independence verification was conducted to ensure a high accuracy of the simulation results [21].

2.3. Governing Equations

Governing equations include the fluid continuity equation, Navier-Stokes equations and turbulence model. Because the Mach number of the air flow is below 0.3, the density is regarded as constant. In steady calculation, SST turbulence model is selected as turbulence model. In unsteady simulation, to solve the unsteady phenomenon of rotating stall, the large eddy simulation (LES) is conducted. Equation of momentum is calculated by the two order upstream, and the physical time step of the unstable computation is 1.677×10^{-4} s, which means 240 time steps per revolution. To predict the noise of the fan, FW-H equation based on Lighthill acoustic analogy theory is adopted. Noise spectrum distribution from the axial fan is acquired after the source data have been obtained in time domain integration, and then fast Fourier transform processing are conducted.

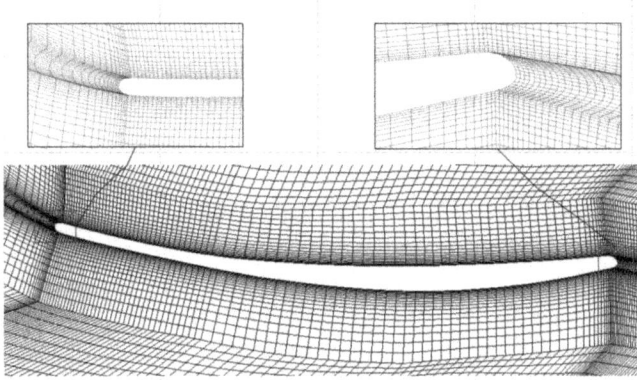

Figure 2. Computational grid around main blade edge.

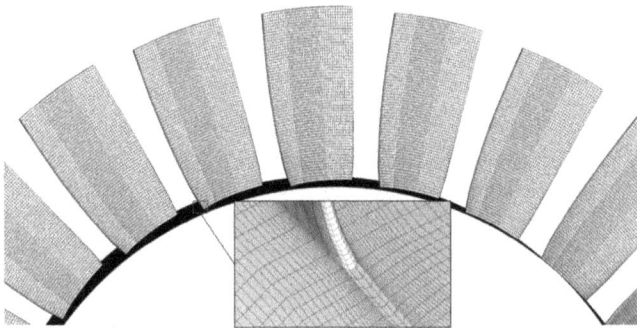

Figure 3. Local grid of the impeller blade and enlarged view around the impeller.

2.4. Boundary Conditions

The inlet cross section of the collector and the outlet section of the diffuser are the inlet and outlet of the calculation domain. The inlet boundary condition of the fan is the pressure inlet, given the total inlet pressure; airflow direction angle is axial admission; and the outlet boundary condition is pressure outlet. For steady calculation, the MRF model is applied to the static and dynamic region coupling, and the throttle valve model is adopted to control the outlet pressure during the unsteady calculation. The function of the throttle model is as follows:

$$Ps_{out}(t) = Pi_{in} + \frac{1}{2}\frac{k_0}{k_1}\rho U^2, \tag{1}$$

where k_0 is a constant, k_1 is throttle coefficient, ρ is density of the air, and U is velocity component of the z axial. Details of boundary conditions and throttle models are presented in [14].

3. Results and Discussion

3.1. Simulation Process of Rotating Stall

The variation of outlet static pressure is shown in Figure 4 when the valve coefficient k_1 is 0.781 and 0.780, respectively. When the valve coefficient k_1 is 0.781, the outlet pressure achieves a stable value quickly and has little change. However, when k_1 is 0.780, outlet pressure tends to a stable value after five rotation cycles and then the outlet pressure declines sharply after the T = 34.5th revolution. Then, the values of outlet pressure fluctuate after about 10 rotation cycles. The throttle value k_1 is 0.780, corresponding to the process of the occurrence and development of rotating stall.

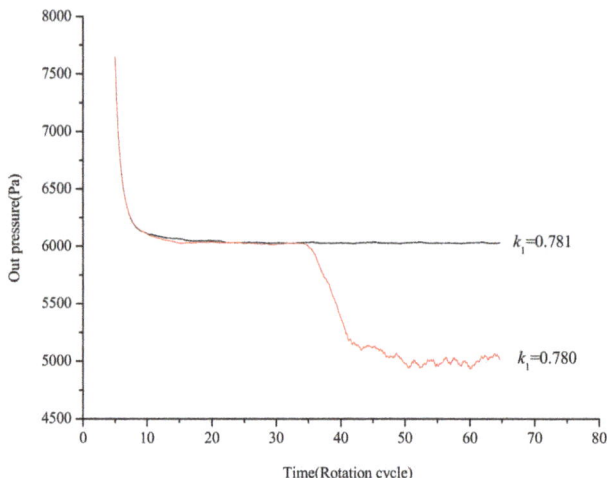

Figure 4. The history of outlet pressure on two valve coefficients.

To monitor the relative velocity inside the fan before and after rotating stall and acquire reliable evidence to analyze stall inception, three monitoring points, which are distributed at 50% of blade height and the circumferential interval is three impeller passages, are set up and marked as m1, m2 and m3, as shown in Figure 5.

Figure 5. The distribution of monitoring points.

Figure 6 shows the relative velocity of the three monitoring points in impeller when k_1 is 0.780. The original relative velocity value of monitoring point m2 and m3 are moved up 60 units and 120 units, respectively, for comparison. As shown in Figure 6, a stall does not occur before the 34th revolution, while, after the 39.5th revolution, stall inception transforms to mature stall cell. In the growth process from the stall inception to the stall cell, peaks appear in the relative speeds of monitoring points. As m1, m2 and m3 are distributed evenly in the circumferential direction, there exists time difference when one blade passes through the three monitoring points. There are 0.375 revolutions between monitoring points m1 and m3, and they have the same waveform. Rotation speed of stall inception is obtained from Equation (2), and the rotation speed of stall inception is 0.67 times the rotating speed of the rotor:

$$w_s = \frac{2 \times 45}{0.375 \times 360} w_r = 0.667 w_r, \tag{2}$$

where w_s is the rotational speed of stall inception, and w_r is the rotational speed of the rotor.

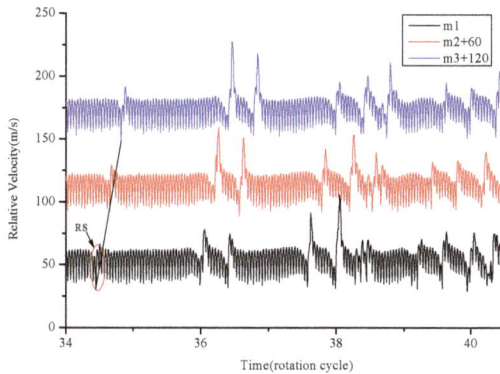

Figure 6. Relative velocities of the monitoring points.

The streamline of relative velocity is analyzed for the section (Figure 7) where Z = 0 cm at four different conditions. The distribution of relative velocity streamline at Z = 0 section is shown in Figure 8 when the valve coefficient is 0.780. Figure 8a shows the formation of the first vortex on the 34.5th revolution and it occupies 1/3 passage at the location of V_1. The velocity streamline of V_1 is uneven, while the distribution in the other passages are uniform, indicating that stall inception occurs in the location V_1 and does not spread along the circumferential direction.

When the second condition is reached, as show in Figure 8b, there are two locations, V_1 and V_2, where the velocity streamline is uneven. The interval between V_1 and V_2 is four passages on counterclockwise direction. The area of disturbance at V_2 ranges from 90% blade height to blade tip, which means the appearance of the second vortex. This phenomenon corresponds to the obvious increase of relative velocity at 36.05th revolution, as shown in Figure 6.

Figure 8c shows the occurrence of the third vortex at the location V_3, and there appear uneven distribution on four passages. At the same time, a corresponding peak also appears in the relative velocity at 37.6th revolution, as shown in Figure 6. After the process of two revolutions, there distribute four stall cells within the impeller and they are uniform on circumferentially direction, as shown in Figure 8d. The evolution from stall inception to the mature stall cell lasts seven revolutions.

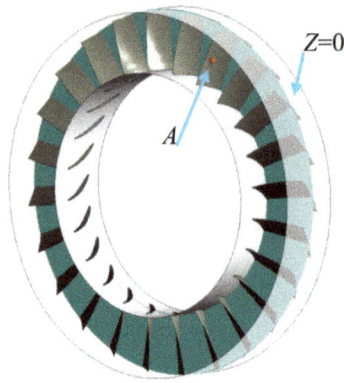

Figure 7. The location of Z = 0 cm section and monitoring point A.

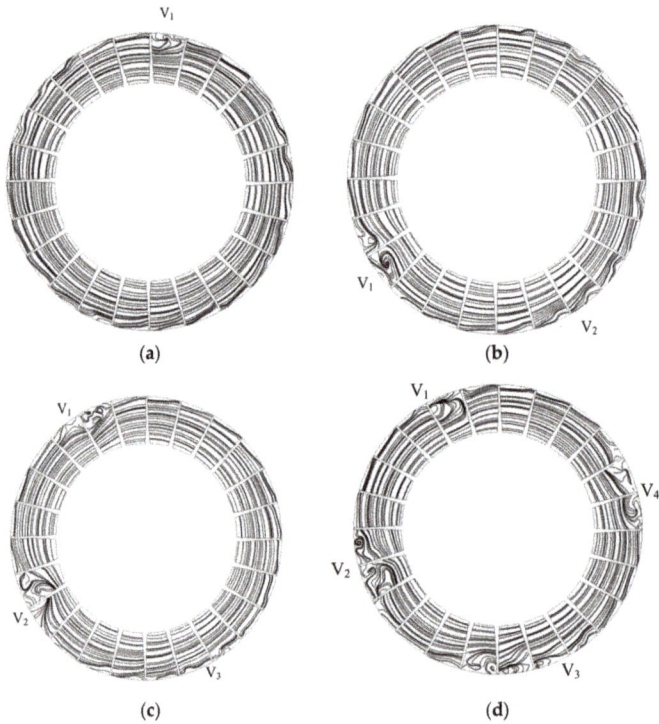

(a)

(b)

(c)

(d)

Figure 8. The distribution of relative velocity streamline at Z=0 Section: (**a**) T = 34.5th; (**b**) T = 36th; (**c**) T = 37.5th; and (**d**) T = 39.5th.

3.2. The Time Domain Characteristic of Acoustic Pressure

Figure 9 shows the time domain characteristic of acoustic pressure at monitoring point A, shown in Figure 7, under the four different conditions mention above. The time domain characteristic of acoustic pressure under the design condition are illustrated in Figure 9a; the two revolutions in this figure are selected randomly. Figure 9a shows that the periodicity of acoustic pressure is regular and its fluctuation amplitude is basically equal. The uniform distribution of the 48 equal pulse peaks corresponds to the number of passing blades in two cycles. This phenomenon indicates that the internal flow field of the fan is stable, with fluid flowing into each channel uniformly. The incentive effect between ambient air and each leaf blade is the same, so the production of acoustic pressure is equal.

Figure 9b shows the time domain characteristic of acoustic pressure when the value coefficient k_1 is 0.780; the two revolutions are selected randomly before the 34.5th revolution. It is the condition near stall stage. The acoustic pressure in the monitoring point A is almost regular, and the waveform of sound pressure shows little fluctuation during the two revolutions. Compared with Figure 9a, the acoustic pressure amplitude becomes large and has few fluctuations due to the decreasing of throttle coefficient leading to the decrease of fan flow rate.

As shown in Figure 9c, the stall inception condition is achieved; two revolutions are selected when stall inception occurs. The acoustic pressure amplitude shows obvious increase on the 34.5th revolution. Then, the two other acoustic pressure amplitudes increase obviously in the 34.6th and the 34.9th revolution, respectively. The acoustic pressure amplitude values of the three obvious increases have little difference. In addition, compared with Figure 9b, the fluctuation of acoustic pressure amplitude increases. This phenomenon indicates that, as the flow rate decreases, the uneven distribution of the fluid flow in the impeller leads to the damage of flow field, the acoustic pressure fluctuation becomes larger and the acoustic pressure amplitude increase sharply.

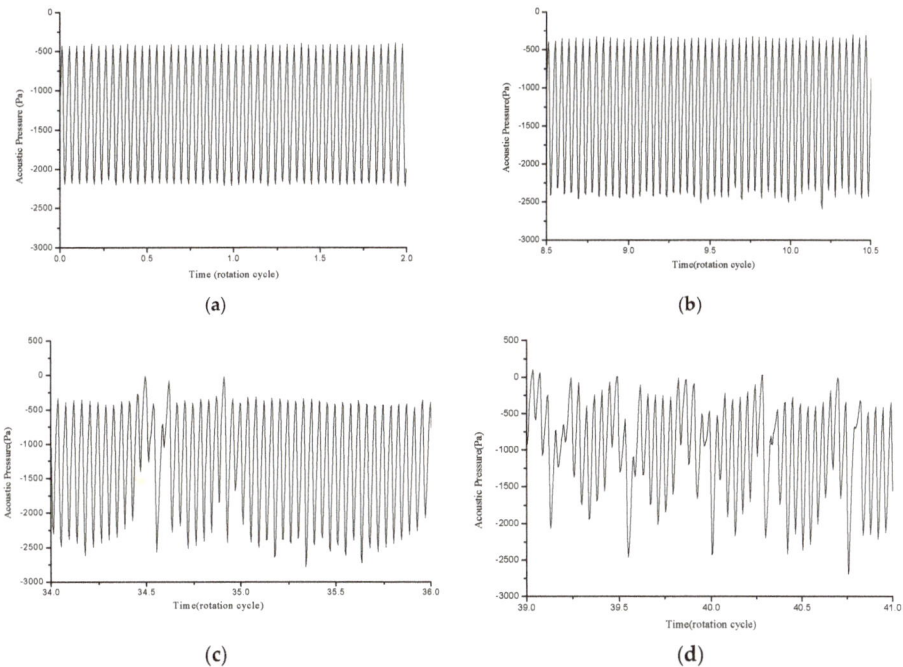

Figure 9. The time domain characteristic of acoustic pressure at monitoring points A: (**a**) the design condition; (**b**) near the stall inception condition; (**c**) stall inception condition; and (**d**) the mature stall cell condition.

The time domain characteristic of acoustic pressure on the mature stall cell condition is shown in Figure 9d. As can be seen, the fluctuation of acoustic pressure amplitude becomes more serious compared to the three other conditions. The acoustic pressure waveform has a periodic fluctuation.

However, the fluctuation cycle of acoustic pressure is not a specific value. This phenomenon shows that the stall cells within the impeller are not stable.

3.3. The Time Domain Analysis of the Acoustic Pressure Peak

The acoustic pressure peak pulsation at monitoring point A under three valve coefficients are shown in Figure 10. Figure 10a shows the acoustic pressure peak pulsation under design condition at monitoring point A. As can be seen from the graph, the acoustic pressure peak has a little fluctuation near the value −410 Pa and basically remains a stable value. Figure 9b shows the acoustic pressure peak pulsation when k_1 is 0.780 and 0.660, respectively. We can see that, when the valve coefficient is 0.780, the fluctuation range of acoustic pressure peak during the first 30 rotation cycles is small and it is similar to the distribution under design condition (as shown in Figure 9a). However, the stable value of acoustic pressure peak is −360 Pa, owing to the decrease of flow rate. The peak of acoustic pressure is difficult to keep stable after a period of time. An obvious increase of the peak value appears at the 34.5th revolution, at the moment stall inception occurs and then enters the rotating stall stage, as shown in the red circle marked RS, and the variation of peak value is 341 Pa. Afterwards, the fluctuation of acoustic pressure peak becomes more serious, and the maximum variation of peak value is 533 Pa. When the valve coefficient k_1 is 0.660, at the beginning of the rotation cycles, the fluctuation of peak value is also in a small range, and the stable value is about −330 Pa. however, after fewer than two rotation cycles, the value of peak increases obviously and the variation of peak value is 169 Pa, as shown in the red circle marked RS. Compared with when the valve coefficient k_1 is 0.780, the oscillation of acoustic pressure peak becomes larger, and the maximum variation of peak value is 873 Pa. To sum up, when the valve coefficient is 0.66, the peak of acoustic pressure presents an obvious increase after fewer than two revolutions. However, the appearance of this phenomenon needs quite a long time for the value of peak k_1 to be 0.780. The peak of acoustic pressure presents little fluctuation and remains at a certain value under design condition.

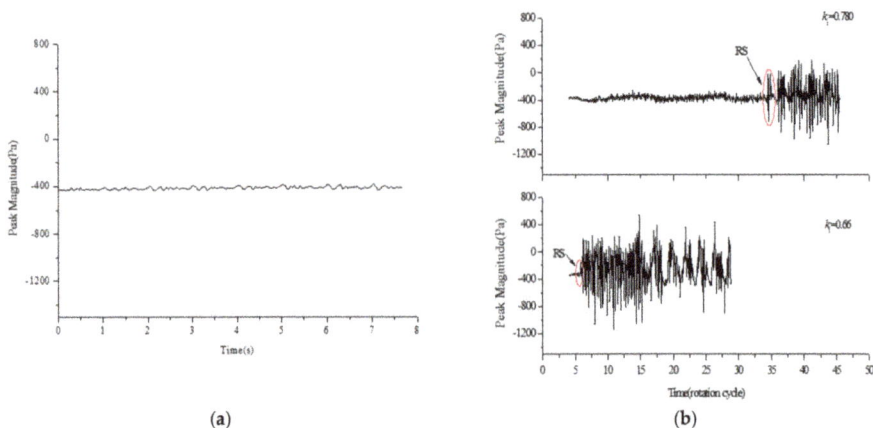

Figure 10. Acoustic pressure peak in time domain under three conditions: (**a**) acoustic pressure peak under design condition; and (**b**) acoustic pressure peak for when *k* is 0.780 and 0.660.

3.4. Spectrum Analysis of the Noise

Fast Fourier transform (FFT) is used on the acoustic pressure distribution in time domain to obtain the noise spectrum of monitoring points (as shown in Figure 11). Then, the noise spectrum of different monitoring points is analyzed to obtain the distribution characteristics of noise in each frequency band. The noise spectrums of the acoustic pressure level and amplitude under the design condition are shown in Figure 11a. The blade passing frequency (BPF) of axial flow fan is f = 1490 × 24/60 = 596 Hz. As can be seen from the graph, the location of the blade passing frequency (BPF) and fundamental frequency is consistent. The harmonic distribution is consistent in the simulation and the theory. When the frequency is more than 2500 Hz, the change of the acoustic pressure level is very small, and the acoustic pressure level tends to be stable. This phenomenon shows that the fan aerodynamic noise has discrete and broadband characteristics at the same time.

When k is 0.78, the noise spectrum of the acoustic pressure level and amplitude is as shown in Figure 11b, and, in this period, the fan is at near stall condition. The acoustic pressure level presents a W shape in the low frequency and has a maximum value at f = 248.8 Hz. As can be seen from the graph, near the maximum value, the acoustic pressure level gradually decreases at both sides. Apart from frequency multiplication with the increase of frequency, the acoustic pressure level presents fluctuations, but the range of fluctuations in small, as displayed in the frequency domain of amplitude, and the very small amplitude tends to zero. Compared with Figure 11a, the fluctuation of acoustic pressure level enhances, as the decreasing of the flow rate increases the internal disturbance of axial flow fan.

The noise spectrum of the acoustic pressure level and amplitude, when stall inception occurs, is shown in Figure 11c. The discrete frequency increases significantly in the low frequency, as shown in Figure 11c, and there are three obvious discrete frequency. Figure 11c is partially enlarged, showing that the peaks of amplitude occurred at 27.66 Hz, 46.10 Hz and 64.55 Hz. Combined with Figure 7, during the evolution process of the rotation stall, the stall cells increase gradually, so these three frequencies are superposition results between frequency of stall inception and rotating stall. Compared with Figure 11b, the fluctuation of amplitude increases in the low frequency. This is because the fan has difficulty maintaining stability after a period, enters the rotating stall, and the damage of fan inner flow field leads to the enhancement of fan noise.

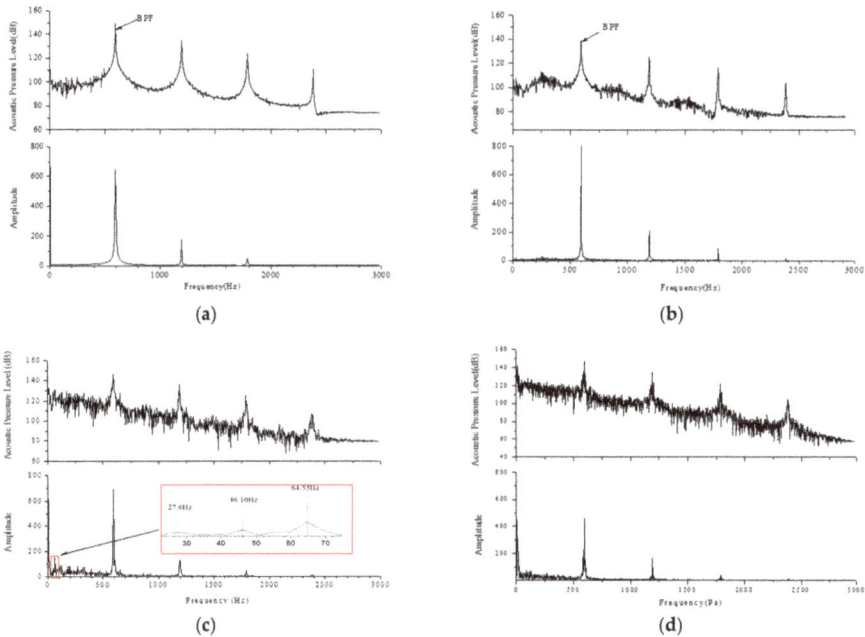

Figure 11. Noise spectrum of the acoustic pressure level and amplitude at four conditions: (**a**) the design condition; (**b**) near the stall inception condition; (**c**) the development of stall inception condition; and (**d**) deep rotating stall condition.

As shown in Figure 11d, mature stall cell come into being. The oscillation of the acoustic pressure level and amplitude are serious. Otherwise, the fluctuations of acoustic pressure level are enhanced with the increase of frequency in the high frequency. The deterioration of the internal flow field is the reason for the further enhancement of noise.

Figure 11 shows different characteristics of noise frequency spectrum at the four conditions mentioned above. However, the acoustic pressure level of each condition reduces with the increase of frequency. Using the noise intensity to measure, the strongest is fundamental frequency and the high order harmonic gradually weakened with the increase of frequency. The spectrum characteristics of noise is generally divided into three frequency bands: low-frequency noise is low than 500 Hz, and from 500 Hz to 1000 Hz is the mid-frequency noise, and high-frequency noise is high than 1000 Hz. The fluctuation of acoustic pressure level and amplitude are serious when the frequency is low than 500 Hz in each condition. The blade passing frequency is in the mid-frequency, so the aerodynamic noise of the axial fan is mainly low-frequency noise and mid-frequency noise. The rotation noise with discrete peak features are main noise in this frequency band. The peak value of low-frequency noise is the result of the periodic vortex shedding.

The physiological characteristics of human are sensitive to high-frequency noise but insensitive to low-frequency noise. Several kinds of filters are designed based on the loudness curve, and A-weighted acoustic pressure level is one of them. A-weighted (A-weighted is a standard developed by the American National Standards Institute in 1936, which describes the sensitivity of human ears to sound changes in different frequency bands.) acoustic pressure level is closest to the sensory characteristics of voice to the human ear. Thus, the A-weighted acoustic pressure level, combined with the 1/3 octave spectrum under the four conditions mentioned above, is used to observe the changes of the fan noise spectrum. The A-Weighted 1/3 octave spectrum of the four conditions as shown in Figure 12. The aerodynamic noise of axial flow fan mainly contains the discrete noise, and is mixed with the wide-band noise, the main component of the discrete noise are fundamental frequency and high order harmonics. Compared with the design condition, the fluctuation of acoustic pressure level is more intense in the low frequency after the occurrence of stall inception.

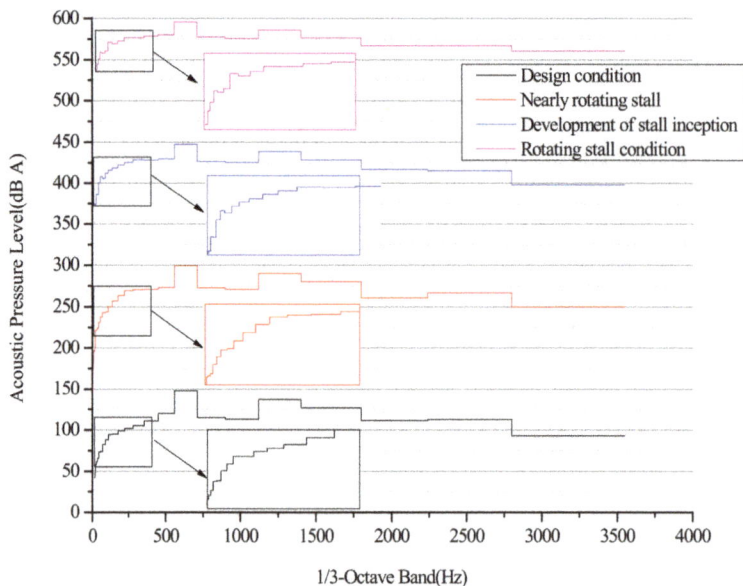

Figure 12. A-weighted 1/3 octave spectrum of the four conditions.

4. Conclusions

In this paper, the unsteady phenomenon of rotation stall in an axial fan is numerically studied through large eddy simulation with throttle valve model and FW-H noise model. In the case of axial fan rotating stall, the aerodynamic noise performance was analyzed. The main results can be listed as follows.

(1) Flow fields are analyzed in detail in the evolution process of rotating stall. The relative velocity first increases rapidly at the 34.5th rotation revolution, and the relative velocity of streamline forms a vortex at the location V1. In the whole evolution process, one stall cell gradually developed into four stall cells that last for seven revolutions.

(2) The time domain characteristics of acoustic pressure during the production and growth of rotation stall inception, at monitoring point A, are analyzed. Under the design condition, the acoustic pressure presents regular periodicity. During the growth procedure from the rotation stall inception to the stall cell, the acoustic pressure amplitude presents fluctuation, and the acoustic pressure peak has an obvious increase at the 34.5th revolution, which corresponds to the increase of relative velocity at this time. On the rotating stall condition, the fluctuation of acoustic pressure amplitude becomes more serious than the other conditions.

(3) The noise spectrum of the acoustic pressure level and amplitude is also analyzed in this paper. The noise energy is concentrated on the fundamental frequency with high energy and high order harmonics under design condition. When stall inception occurs, discrete frequency increased significantly in the low frequency, which has three obvious discrete frequencies: 27.66 Hz, 46.10 Hz and 64.55 Hz. The frequency domain of A-weighted sound pressure level provides further evidence: the fluctuation of acoustic pressure level is more intense in the low frequency after the occurrence of stall inception, compared with the design conditions.

Acknowledgments: This research was supported by National Natural Science Foundation of China (Grant No. 11602085), and Natural Science Foundation of Hebei Province, China (Grant No. E2016502098).

Author Contributions: Lei Zhang and Chuang Yan conceived and analyzed the data; Lei Zhang wrote the paper; Ruiyang He and Qian Zhang provided advice for this method and revised this paper; and Kang Li contributed simulation tools. All authors have read and approved the final manuscript.

Conflicts of Interest: The authors declare no conflict of interest.

References

1. Peng, C. Numerical Predictions of Rotating Stall in an Axial Multi-Stage-Compressor. In *ASME 2011 Turbo Expo, Proceedings of the Turbine Technical Conference and Exposition, Vancouver, BC, Canada, 6–10 June 2011*; Volume 6: Structures and Dynamics, Parts A and B; American Society of Mechanical Engineers: New York, NY, USA, 2011.

2. Lei, Z.; Rui, W.; Wei, Y.; Wang, S. Simulation of air jets for controlling stall in a centrifugal fan. *J. Mech. Eng. Sci.* **2015**, *229*, 2045–2055. [CrossRef]

3. Ohta, Y.; Fujita, Y.; Morita, D. Unsteady Behavior of Surge and Rotating Stall in an Axial Flow Compressor. *J. Therm. Sci.* **2012**, *21*, 302–310. [CrossRef]

4. Chen, J.P.; Hathaway, M.D.; Herrick, G.P. Prestall Behavior of a Transonic Axial Compressor Stage via Time-Accurate Numerical Simulation. *J. Turbomach.* **2007**, *130*, 353–368.

5. Mcdougall, N.M.; Cumpsty, N.A.; Hynes, T.P. Stall Inception in Axial Compressors. *J. Turbomach.* **1988**, *112*, 116–123. [CrossRef]

6. Niazi, S.; Stein, A.; Sankar, L. Numerical studies of stall and surge alleviation in a high-speed transonic fan rotor. In Proceedings of the 38th Aerospace Sciences Meeting and Exhibit, Reno, Nevada, 10–13 January 2000.

7. Moore, F.K. A Theory of Post-Stall Transients in Axial Compression Systems: Part I—Development of Equations. *J. Eng. Gas Turbines Power* **1985**, *108*, 68–76. [CrossRef]

8. Greitzer, E.M.; Moore, F.K. A Theory of Post-Stall Transients in Axial Compression Systems: Part II—Application. *J. Eng. Gas Turbines Power* **1986**, *108*, 231–239. [CrossRef]

9. Kosuge, H.; Ito, T.; Nakanishi, K. A Consideration Concerning Stall and Surge Limitations within Centrifugal Compressors. *J. Eng. Power* **1982**, *104*, 782–787. [CrossRef]

10. Yamada, K.; Furukawa, M.; Tamura, Y.; Saito, S.; Matsuoka, A.; Nakayama, K. Large-Scale DES Analysis of Stall Inception Process in a Multi-Stage Axial Flow Compressor. In *ASME Turbo Expo 2016, Proceedings of the Turbomachinery Technical Conference and Exposition, Seoul, Korea, 13–17 June 2016*; American Society of Mechanical Engineers: New York, NY, USA, 2016; p. V02DT44A021.

11. Salunkhe, P.B.; Pradeep, A.M. Stall Inception Mechanism in an Axial Flow Fan under Clean and Distorted Inflows. *J. Fluids Eng.* **2010**, *132*, 121102. [CrossRef]

12. Toge, T.D.; Pradeep, A.M. Experimental Investigation of Stall Inception Mechanisms of Low Speed Contra Rotating Axial Flow Fan Stage. *Int. J. Rotating Mach.* **2015**, *3*, 1–14. [CrossRef]

13. Li, L.; Huang, Q.; Qiao, Y. Research on model of the vortex noise of axial fan blade and its characteristics. *China Mech. Eng.* **2006**, *17*, 1056–1059.

14. Zhang, L.; Wang, R.; Wang, S. Simulation of Broadband Noise Sources of an Axial Fan under Rotating Stall Conditions. *Adv. Mech. Eng.* **2014**, 1–11. [CrossRef]

15. Liu, P.; Jin, Y.; Wang, Y. Effects of rotor structure on performance of small size axial flow fans. *J. Therm. Sci.* **2011**, *20*, 205–210. [CrossRef]

16. Diaz, A.K.M.; Fernandez, O.J.M.; Marigorta, E.B.; Morros, C.S. Numerical prediction of tonal noise generation in an inlet vaned low-speed axial fan using a hybrid aeroacoustic approach. *J. Mech. Eng. Sci.* **2009**, *223*, 2081–2098. [CrossRef]

17. Fukano, T.; Jang, C.M. Tip clearance noise of axial flow fans operating at design and off-design condition. *J. Sound Vib.* **2004**, *275*, 1027–1050. [CrossRef]

18. Wang, C.; Zhang, W.; Huang, L. Noise Source Analysis and Control for Two Axial-Flow Cooling Fans in Series. In *Fluid-Structure-Sound Interactions and Control*; Springer: Berlin/Heidelberg, Germany, 2016.

19. Feng, T.; Wang, J.; Liu, B.; Li, N.; Wu, X. Computational simulation of rotating noise of fan. In Proceedings of the 2011 IEEE 3rd International Conference on Communication Software and Networks, Xi'an, China, 27–29 May 2011; pp. 728–731.

20. Mao, H.; Wang, Y.; Lin, P.; Jin, Y.; Setoguchi, T.; Kim, H.D. Influence of Tip End-plate on Noise of Small Axial Fan. *J. Therm. Sci.* **2017**, *26*, 30–37. [CrossRef]

21. Zhang, L.; Lang, J.H.; Liang, S.F.; Wang, S.L. Dynamic characteristic study on variable pitch axial fan impeller of the power plant. *Proc. CSEE* **2014**, *34*, 4118–4128.

MDPI

St. Alban-Anlage 66

4052 Basel

Switzerland

Tel. +41 61 683 77 34

Fax +41 61 302 89 18

www.mdpi.com

Energies Editorial Office

E-mail: energies@mdpi.com

www.mdpi.com/journal/energies

www.ingramcontent.com/pod-product-compliance
Lightning Source LLC
Chambersburg PA
CBHW051726210326
41597CB00032B/5629